PROCEEDINGS OF THE 1ST INTERNATIONAL CONGRESS ON ENGINEERING TECHNOLOGIES

Mosharaka for Research and Studies International
Conference Proceedings

ISSN: 2768-5675
eISSN: 2768-5683

Series Editor

Mohammad M. Banat

ENGITEK 2020, 16–18 JUNE 2020, IRBID, JORDAN

Proceedings of the 1st International Congress on Engineering Technologies

Edited by

Suhil Kiwan
Department of Mechanical Engineering,
Jordan University of Science and Technology,
Irbid, Jordan

Mohammad M. Banat
Department of Electrical Engineering,
Jordan University of Science and Technology,
Irbid, Jordan

CRC Press is an imprint of the
Taylor & Francis Group, an **informa** business
A BALKEMA BOOK

CRC Press/Balkema is an imprint of the Taylor & Francis Group, an informa business

Typeset by MPS Limited, Chennai, India

Library of Congress Cataloging-in-Publication Data

Applied for

Published by: CRC Press/Balkema
Schipholweg 107C, 2316 XC Leiden, The Netherlands
e-mail: enquiries@taylorandfrancis.com
www.routledge.com – www.taylorandfrancis.com

ISBN: 978-0-367-77630-5 (hbk)
ISBN: 978-1-032-01339-8 (pbk)
ISBN: 978-1-003-17825-5 (ebk)
DOI: 10.1201/9781003178255

Proceedings of the 1st International Congress on Engineering Technologies – Kiwan & Banat (Eds)
© 2021 Taylor & Francis Group, London, ISBN 978-0-367-77630-5

Table of contents

Proceedings of the 1st International Congress on Engineering Technologies – Kiwan & Banat (Eds)
 ISBN 978-0-367-77630-5

Preface

Welcome to the Proceedings of the 1st International Congress on Engineering Technologies (EngiTek 2020), remotely held 16-18 June 2020, and organized by the Faculty of Engineering (FoE) of the Jordan University of Science and Technology (JUST).

I have had the pleasure of acting as the Organizing Committee Chairperson, which allowed me to be in frequent and close touch with all people in the congress community. This wonderful community has included authors, speakers, reviewers, members of congress committees, chairpersons at all levels, technical and administrative assistants, volunteers, and other fine colleagues. Before leaving this part of the preface, I am obliged to say that I owe each and every one of these great contributors a big thank you. Without them, we would not have a congress to start with, let alone having a remarkable event that has left everybody looking forward to being part of the next edition.

EngiTek 2020 has represented a major forum for professors, students, and professionals from all over the world to present their latest research results, and to exchange new ideas and practical experiences in the most cutting-edge areas of the field of engineering technologies. Topics covered include communication systems and networks, electronic circuits, microwaves and electromagnetics, signal processing, energy and power systems, biomedical engineering, nanotechnology, computer engineering, mechatronics, robotics and automation, control engineering, thermal power systems, fluid mechanics and heat transfer, mechanical system dynamics, industrial engineering, and materials engineering.

Initial plans for EngiTek 2020 were based on welcoming the participants to the JUST campus in the beautiful and historic north Jordan city of Irbid. We were hoping that we would demonstrate to participants that coming to Jordan represents a unique opportunity to witness the embrace of history and future. As small in area as it is, Jordan has a rich history spanning thousands of years, a history that a visitor can only admire. Places like Petra and the Dead Sea are only examples of the many one-of-a-kind places to visit in Jordan. On top of all that, Jordan, as a peaceful country in a turbulent area, enjoys extreme levels of safety and security. Jordanian people are internationally known for generosity, hospitality and cultural awareness and tolerance.

However, the outbreak of Covid-19 did not allow our initial plans to go ahead, and a remote congress was the only realistic option for us as organizers. Therefore, we used Microsoft Teams as a means to hold the congress and run the sessions. All keynote talks and all papers in the technical program were presented precisely on time. Speakers and authors presented results of some of the finest research on hot state-of-the-art engineering technologies. Each contributed paper has been rigorously peer-reviewed by 3 or more reviewers per paper to guarantee the highest possible quality of the accepted papers.

Having already thanked EngiTek 2020 contributors at all levels, I cannot close before thanking CRC Press/Balkema for their giving us the chance to work with them. It has been a memorable experience through which they have provided hard work, commitment, and real unlimited support. On behalf of the whole EngiTek 2020 community, I would like to express my utmost gratitude to you. I am certainly looking forward to working with you on other future events.

Last but not least, I am very hopeful of witnessing a future edition of EngiTek that is held in the loving land of Jordan. You are all invited to be with us then.

EngiTek 2020 Organizing Committee Chairperson
Mohammad M. Banat
Department of Electrical Engineering
Jordan University of Science and Technology
Irbid, Jordan

Proceedings of the 1st International Congress on Engineering Technologies – Kiwan & Banat (Eds)
© 2021 Taylor & Francis Group, London, ISBN 978-0-367-77630-5

1st International Congress on Engineering Technologies (EngiTek 2020) Chairpersons and Committees

General Chair
Prof. Suhil Kiwan, *Jordan University of Science and Technology, Jordan*

Steering Committee Chair
Prof. Eduard Babulak, *Sungkyunkwan University, Korea*

Organizing Committee Chair
Dr. Mohammad M. Banat, *Jordan University of Science and Technology, Jordan*

Patronage Chair
Prof. Ahmad Harb, *German-Jordanian University, Jordan*

Local Committee Chair
Mr. Khalil Al-Refaie, *Jordan University of Science and Technology, Jordan*

Conference Chairs
Dr. Khaled Al-Shboul, *Jordan University of Science and Technology, Jordan*
Dr. Areen Al-Bashir, *Jordan University of Science and Technology, Jordan*
Dr. Issam Smadi, *Jordan University of Science and Technology, Jordan*
Prof. Ahmad Abu El-Haija, *Jordan University of Science and Technology, Jordan*
Dr. Yazan Taamneh, *Jordan University of Science and Technology, Jordan*
Dr. Mohammed Almomani, *Jordan University of Science and Technology, Jordan*
Dr. Saud Khashan, *Jordan University of Science and Technology, Jordan*

Steering Committee
Prof. Eduard Babulak, *Sungkyunkwan University, Korea*
Prof. Nael Barakat, *University of Texas at Tyler, USA*
Dr. Domenico Ciuonzo, *Università degli Studi di Napoli Federico II, Italy*
Dr. Victor Monzon, *Universidad Carlos III de Madrid, Spain*
Prof. Qinmin Yang, *Zhejiang University, China*

Organizing Committee
Prof. Khairedin Abdalla, *Jordan University of Science and Technology, Jordan*
Dr. Areen Al-Bashir, *Jordan University of Science and Technology, Jordan*
Prof. Majdi Al-Mahasneh, *Jordan University of Science and Technology, Jordan*
Dr. Aslam Al-Omari, *Jordan University of Science and Technology, Jordan*
Prof. Bashar Al-Omari, *Jordan University of Science and Technology, Jordan*
Dr. Mohammed Al-Saleh, *Jordan University of Science and Technology, Jordan*
Dr. Khaled Al-Shboul, *Jordan University of Science and Technology, Jordan*
Dr. Mohammed Almomani, *Jordan University of Science and Technology, Jordan*
Dr. Mohammad M. Banat, *Jordan University of Science and Technology, Jordan*
Dr. Saud Khashan, *Jordan University of Science and Technology, Jordan*
Dr. Jaser K. Mahasneh, *Jordan University of Science and Technology, Jordan*
Dr. Issam Smadi, *Jordan University of Science and Technology, Jordan*

Dr. Yazan Taamneh, *Jordan University of Science and Technology, Jordan*
Prof. Ghassan Tashtoush, *Jordan University of Science and Technology, Jordan*

Local Arrangements Committee
Ms. Sanaa Abu Alasal, *Jordan University of Science and Technology, Jordan*
Mr. Ahmed Abuqaoud, *Jordan University of Science and Technology, Jordan*
Mr. Rowaid Al Hourani, *Jordan University of Science and Technology, Jordan*
Mr. Khalil Al-Refaie, *Jordan University of Science and Technology, Jordan*
Ms. Ilham Al-Saidi, *Jordan University of Science and Technology, Jordan*
Ms. Shatha Alghueiri, *Jordan University of Science and Technology, Jordan*
Mr. Mohammad Alothman, *Jordan University of Science and Technology, Jordan*
Mr. Abdallah Alsaleem, *Jordan University of Science and Technology, Jordan*
Mr. Emran Bashabsheh, *Jordan University of Science and Technology, Jordan*
Ms. Ayat Dagher, *Jordan University of Science and Technology, Jordan*
Ms. Mai Farjallah, *Jordan University of Science and Technology, Jordan*
Mr. Adnan Masadeh, *Jordan University of Science and Technology, Jordan*
Mrs. Marwa Tuffaha, *Jordan University of Science and Technology, Jordan*

Technical Program Committee
Dr. Abdallah Ababneh, *Yarmouk University, Jordan*
Prof. Ahmad Ababneh, *Jordan University of Science and Technology, Jordan*
Dr. Qammer Abbasi, *University of Glasgow, UK*
Dr. Nisrin Abdelal, *Jordan University of Science and Technology, Jordan*
Dr. Bilal H. Abed-Alguni, *Yarmouk University, Jordan*
Dr. Hatim G. Abood, *University of Diyala, Iraq*
Dr. Wisam Abu Jadayil, *American University of Ras Al Khaimah, UAE*
Dr. Ali Abu Odeh, *Khawarizmi International College, UAE*
Dr. Rabie Abu Saleem, *Jordan University of Science and Technology, Jordan*
Dr. Amjad Abu-Baker, *Yarmouk University, Jordan*
Dr. Ahmad Abuelrub, *Jordan University of Science and Technology, Jordan*
Dr. Said Abushamleh, *University of Arkansas at Little Rock, USA*
Dr. Anamika Ahirwar, *Jayoti Vidyapeeth Women's University, India*
Prof. Noura Aknin, *Université Abdelmalek Essaadi, Morocco*
Prof. Salih Akour, *University of Jordan, Jordan*
Prof. Hasan Al Dabbas, *Philadelphia University, Jordan*
Dr. Bilal Al Momani, *Mohawk College for Art and Technology, Canada*
Prof. Mohammed Al Salameh, *Jordan University of Science and Technology, Jordan*
Prof. Raid Al-Aomar, *German-Jordanian University, Jordan*
Prof. Omar Al-Araidah, *Jordan University of Science and Technology, Jordan*
Dr. Muntasir Al-Asfoor, *University of Al-Qadisiyah, Iraq*
Dr. Mohammed Al-Hayanni, *University of Technology, Iraq*
Dr. Mohammad Al-Khedher, *Abu Dhabi University, UAE*
Dr. Hussein Al-Masri, *Yarmouk University, Jordan*
Dr. Intisar Al-Mejibli, *University of Information Technology and Communications, Iraq*
Dr. Anwar Al-Mofleh, *Al-Balqa Applied University, Jordan*
Dr. Ahmad N. Al-Omari, *Yarmouk University, Jordan*
Dr. Nedhal Al-Saiyd, *Applied Science University, Jordan*
Dr. Khaled Al-Shboul, *Jordan University of Science and Technology, Jordan*
Prof. Yahia Al-Smadi, *Jordan University of Science and Technology, Jordan*
Dr. Hamid Alasadi, *University of Basrah, Iraq*
Prof. Abdulgani Albagul, *College of Electronic Technology Baniwalid, Libya*
Dr. Saher Albatran, *Jordan University of Science and Technology, Jordan*

Reviewers

Proceedings of the 1st International Congress on Engineering Technologies – Kiwan & Banat (Eds)

Detailed simplified implementation of filter bank multicarrier modulation using sub-channel prototype filters

Mohammad M. Banat & Ahmed A. Al-Shwmi
Department of Electrical Engineering, Jordan University of Science and Technology, Jordan

ABSTRACT: Filter bank multicarrier modulation is one of the most researched techniques in recent wireless communications literature. It is robust against channel frequency selectivity, but it does not require cyclic prefix to compensate for intersymbol interference. Therefore, filter bank multicarrier modulation is superior to orthogonal frequency division multiplexing in terms of spectral efficiency. However, this comes at the cost of higher implementation complexity.

Despite the vast published research on filter bank multicarrier modulation, little attention has been given to providing a detailed implementation supported by sufficient mathematical analysis. This is particularly true about the prototype filter section, which is the main source of implementation complexity. In this paper we try to fill this gap by presenting a detailed implementation based on the use of simple sub-channel prototype filters. Sub-channel prototype filters are shorter than a single prototype filter by a factor that is equal to the number of data symbols per multicarrier symbol, resulting in reduced complexity and added design flexibility. Detailed mathematical analysis is provided to support the validity of the proposed implementation.

Keywords: multicarrier modulation, filter banks orthogonal frequency division multiplexing

1 INTRODUCTION

Future wireless systems should support a large range of possible use cases, such as high data rate services, point to-point or point-to-multipoint communication and low-latency transmissions. This requires efficient utilization of the available time/frequency resources. In orthogonal frequency division multiplexing (OFDM), this is difficult to achieve due to its weak spectral efficiency, which is due to the inserted cyclic prefix.

A multicarrier modulation (MCM) signal consists of the superposition of a number of sinusoidal subcarriers [1] This is a process through which the total signal bandwidth is divided into many narrowband sub-channel signals to be transmitted simultaneously. In other words, an MCM signal is the result of splitting up a wideband signal with a high symbol rate into several lower rate signals, each one having a lower symbol rate and occupying a narrower bandwidth [2].

MCM signals are known to be robust to multipath fading [2]. For this reason, MCM signals are usually used when transmission is performed over frequency selective fading channels. The bandwidths of MCM sub-channels are chosen so that they are lower than the coherence bandwidth of the communication channel. Therefore, even though the overall communication channel may be frequencyselective, the sub-channels can be made frequencyflat.

Recently MCM has become very important in wireless communications, specifically for multimedia transmission. This is mainly due to the everincreasing human need to transmit signals at higher data rates. Applications involving multimedia signals are top candidates in this regard. Next generation (and beyond) wireless communication systems place great emphasis on applications involving high data rate multimedia signals [3]. There are various types of MCM, and some of them have already been used in some applications, including spatial modulation (SM) and orthogonal frequency division multiplexing with index modulation (IM) [4].

Filter bank multicarrier (FBMC) modulation is an alternative to OFDM that has better spectral properties [5] Contrary to OFDM, FBMC does not need the so-called cyclic prefix (CP) to avoid intersymbol interference (ISI). The new concepts of dynamic access spectrum management (DASM) and cognitive radio require high resolution spectral analysis, a functionality in which filter banks are superior to the discrete Fourier transform of OFDM [6]. FBMC requires higher implementation complexity than OFDM.

The main contribution of this paper is the development of a detailed simplified implementation of the FBMC modulator and demodulator with sub-channel prototype filters Sub-channel prototype filters are shorter than a single prototype filter by a factor that is equal to the number of data symbols per multicarrier symbol, resulting in reduced complexity and added design flexibility.

The rest of this paper is organized as follows: in section II, we describe the pre-processing stage of FBMC; in section III, we present the inverse discrete Fourier transform (IDFT) stage; in section IV, we present the filtering stage and derive the signal to be transmitted; in section V, we describe the reception stage; and in Section VI we present our conclusions.

2 FBMC PRE-PROCESSING

Consider a block of $M = 2Q$serial (generally complex-valued) quadrature amplitude modulation (QAM) symbols $x_{m,n}\big|_{m=0}^{M-1}$, where n represents discrete time index of the block of symbols. If each QAM symbol has a durationT_s, the duration of the whole symbol block is then equal to

$$T = MT_s \tag{1}$$

The QAM symbols can be decomposed into real and imaginary parts by using complex to real conversion as follows:

$$x_{m,n} = x_{m,n}^R + jx_{m,n}^I \tag{2}$$

where

$$\begin{aligned} x_{m,n}^R &= \mathrm{Re}\{x_{m,n}\} \\ x_{m,n}^I &= \mathrm{Im}\{x_{m,n}\} \end{aligned} \tag{3}$$

Let

$$\begin{aligned} \tau &= \tfrac{T}{2} \\ &= QT_s \end{aligned} \tag{4}$$

The continuous-time multicarrier signal at the output of the modulator is given by [2]

$$v(t) = \sum_{l=-\infty}^{\infty} \sum_{q=0}^{Q-1} \begin{pmatrix} v_{2q,l}(t)e^{j2\pi(2q)\Delta ft} \\ +v_{2q+1,l}(t)e^{j2\pi(2q+1)\Delta ft} \end{pmatrix} \tag{5}$$

where

$$v_{2q,l}(t) = \left. \begin{matrix} x_{2q,l}^R z(t-2l\tau) \\ +jx_{2q,l}^I z\left(t-(2l+1)\tau\right) \end{matrix} \right|_{q=0}^{Q-1} \tag{6}$$

$$v_{2q+1,l}(t) = \left. \begin{matrix} jx_{2q+1,l}^I z(t-2l\tau) \\ +x_{2q+1,l}^R z\left(t-(2l+1)\tau\right) \end{matrix} \right|_{q=0}^{Q-1} \tag{7}$$

where $z(t)$ is an even real-valued pulse shape that generally extends for $-\infty < t < \infty$, and Δf is the subcarrier spacing, given by

$$\Delta f = \frac{1}{T} \tag{8}$$

Note that Δf is equal to the reciprocal of the multicarrier symbol period T. Let

$$\begin{aligned} d_{2q,2l} &= x_{2q,l}^R \\ d_{2q,2l+1} &= jx_{2q,l}^I \end{aligned} \tag{9}$$

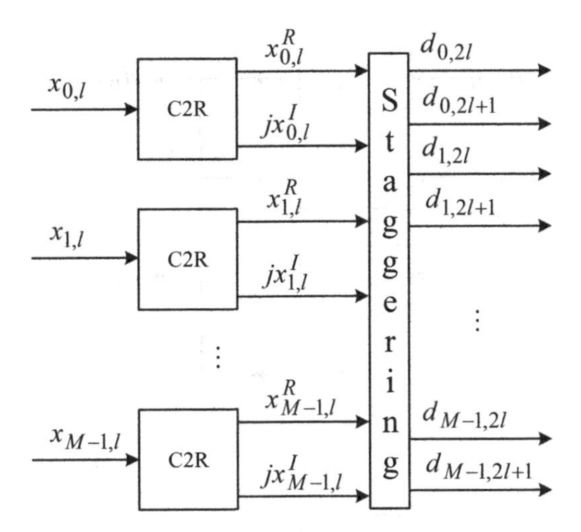

Figure 1. The first part of the pre-processing operation.

$$\begin{aligned} d_{2q+1,2l} &= jx^I_{2q+1,l} \\ d_{2q+1,2l+1} &= x^R_{2q+1,l} \end{aligned} \tag{10}$$

The operation performed in (9) is the first part of the pre-processing (staggering) stage, and is represented in Figure 1

Substituting (9) into (6) and (7) yields

$$\begin{aligned} v_{2q,l}(t) = d_{2q,2l}z(t - 2l\tau) \\ +d_{2q,2l+1}z\,(t - (2l+1)\tau) \end{aligned} \tag{11}$$

$$\begin{aligned} v_{2q+1,l}(t) = d_{2q+1,2l}z(t - 2l\tau) \\ +d_{2q+1,2l+1}z\,(t - (2l+1)\tau) \end{aligned} \tag{12}$$

Let $m = 2q$ in (11) and $m = 2q + 1$ in (12) to get

$$v_{m,l}(t) = \left. \begin{aligned} d_{m,2l}z(t - 2l\tau) \\ +d_{m,2l}z\,(t - (2l+1)\tau) \end{aligned} \right|_{m=0}^{M-1} \tag{13}$$

The last result allows (5) to be rewritten in the form

$$v(t) = \sum_{l=-\infty}^{\infty} \sum_{m=0}^{M-1} v_{m,l}(t) e^{j2\pi m \Delta f t} \tag{14}$$

Similar steps to the above can be performed on the index l in (13), so that (5) simplifies to

$$v(t) = \sum_{n=-\infty}^{\infty} \sum_{m=0}^{M-1} d_{m,n} z(t - n\tau) e^{j2\pi m \Delta f t} \tag{15}$$

The sequence $d_{m,n}$ in (15) is obtained from the outputs of Figure 1 through the second part of the pre-processing stage, as shown in Figure 2.

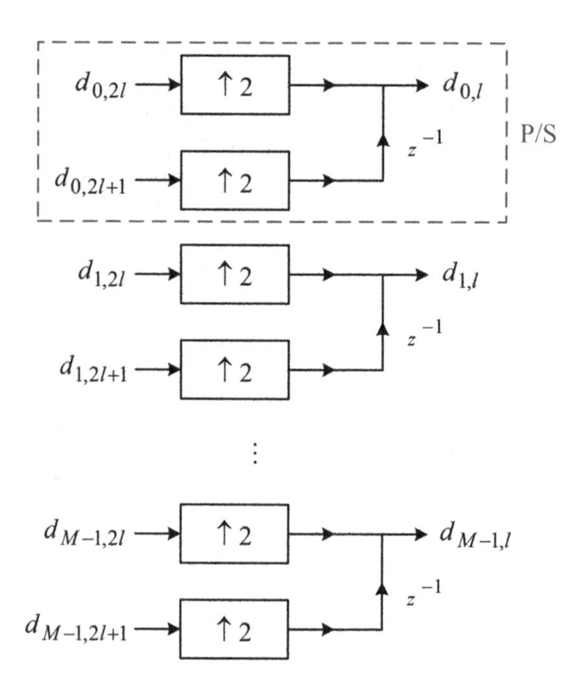

Figure 2. The second part of the pre-processing operation.

3 IDFT STAGE

The pulse shaping waveform $z(t)$ is obviously noncausal. Let $z(t)$ be negligibly small for $|t| > (L-1)T_s/2$, for some odd integer L, that is chosen such that

$$L = 2\Gamma + 1 \tag{16}$$

where $\Gamma = FM$ and F is an integer number. Note that with the last assumption on L, $z(t)$ is negligibly small for $|t| > \Gamma T_s$ Let $z_{tr}(t)$ be a truncated version of $z(t)$, such that

$$z_{tr}(t) = \begin{cases} z(t), & |t| \leq \Gamma T_s \\ 0, & \text{otherwise} \end{cases} \tag{17}$$

Let's define the causal waveform $z_c(t)$, extending for $0 \leq t \leq 2\Gamma T_s$, as follows

$$z_c(t) = z_{tr}(t - \Gamma T_s) \tag{18}$$

Without loss of generality, and to simply signal expressions that will be obtained later, we let

$$z_c(2\Gamma T_s) = 0 \tag{19}$$

Let $z_c(t)$ be sampled every T_s, for all $0 \leq k \leq 2\Gamma$, to produce the sequence $p[k]$, according to

$$p[k] = z_c(kT_s) \tag{20}$$

Then we have

$$p[k] = z_{tr}((k - \Gamma)T_s) \tag{21}$$

From (19),

$$p[2\Gamma] = 0 \tag{22}$$

To create a causal discrete-time transmitted sequence, let

$$s[k] = v((k - \Gamma)T_s) \tag{23}$$

Using $z_{tr}(t)$ in place of $z(t)$ and substituting (21), (1) and (8) in (23) yields

$$s[k] = \sum_{n=-\infty}^{\infty} \sum_{m=0}^{M-1} d_{m,n} p[k - nQ] e^{j\frac{2\pi}{M}mk} \tag{24}$$

Note that $p[k - nQ]$ in (24) is non-zero only when

$$0 \leq k - nQ < 2\Gamma \tag{25}$$

This can be rearranged into the form

$$\frac{k}{Q} - 4F < n \leq \frac{k}{Q} \tag{26}$$

Let the high rate time index k be decomposed in terms of a lower rate time index l as follows

$$k = lQ + q \tag{27}$$

where $q = 0, \cdots, Q - 1$. Then,

$$l - 4F + \frac{q}{Q} < n \leq l + \frac{q}{Q} \tag{28}$$

Since n is an integer and $0 \leq q \leq Q - 1$, the lowest possible value of n is $l - 4F + 1$ and the largest value of n is l Hence, we can rewrite (24) in the form

$$s_{q,l} = \sum_{n=l-4F+1}^{l} p_{q,l-n} \sum_{m=0}^{M-1} d_{m,n} e^{j\frac{2\pi}{M}m(lQ+q)} \tag{29}$$

where

$$s_{q,l} = s[lQ + q] \tag{30}$$

$$p_{q,n} = p[nQ + q] \tag{31}$$

Changing variables in the outer summation in (29), we get

$$s_{q,l} = \sum_{n=0}^{4F-1} p_{q,n} \sum_{m=0}^{M-1} d_{m,l-n} e^{j\frac{2\pi}{M}m(lQ+q)} \tag{32}$$

Simple manipulation of (32) yields

$$\begin{aligned} s_{q,l} = \sum_{n=0}^{4F-1} p^{e}_{q,n} \sum_{m=0}^{Q-1} d_{2m,l-n} e^{j\frac{2\pi}{Q}mq} \\ + \sum_{n=0}^{4F-1} p^{o}_{q,n} \sum_{m=0}^{Q-1} d_{2m+1,l-n} e^{j\pi(l-n)} e^{j\frac{2\pi}{Q}mq} \end{aligned} \tag{33}$$

where

$$p^{e}_{q,n} = p_{q,n} \tag{34}$$

$$p^{o}_{q,n} = p_{q,n} e^{j\frac{2\pi}{M}(nQ+q)} \tag{35}$$

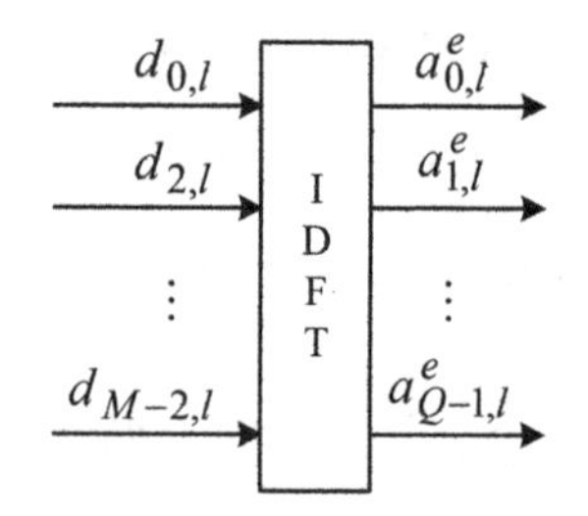

Figure 3. Even IDFT.

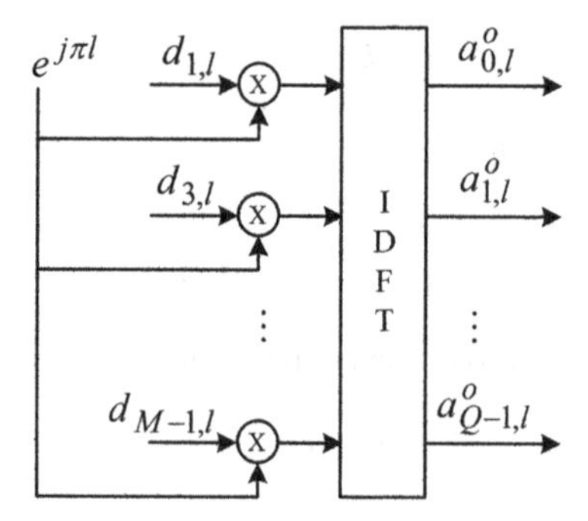

Figure 4. Odd IDFT.

Let

$$\begin{aligned} a^e_{q,l} &= \sum_{m=0}^{Q-1} d_{2m,l} e^{j\frac{2\pi}{Q}mq} \\ &= IFFT\left\{ d_{2m,l}\Big|_{m=0}^{Q-1} \right\} \end{aligned} \tag{36}$$

The IDFT operation in (36) is illustrated in Figure 3

$$\begin{aligned} a^o_{q,l} &= \sum_{m=0}^{Q-1} d_{2m+1,l} e^{j\pi l} e^{j\frac{2\pi}{Q}mq} \\ &= IFFT\left\{ d_{2m+1,l} e^{j\pi l}\Big|_{m=0}^{Q-1} \right\} \end{aligned} \tag{37}$$

The IDFT operation in (37) is illustrated in Figure 4.

4 FILTERING STAGE AND TRANSMITTED SIGNAL

Equation (33) can be rewritten in the form

$$s_{q,l} = \sum_{n=0}^{4F-1} p^e_{q,n} a^e_{q,l-n} + \sum_{n=0}^{4F-1} p^o_{q,n} a^o_{q,l-n} \tag{38}$$

Equivalently, we have

$$s_{q,l} = p^e_{q,l} * a^e_{q,l} + p^o_{q,l} * a^o_{q,l} \tag{39}$$

The filtering of the even and odd IDFT sequences, that is performed in (39) is represented in Figure 5.

A parallel-to-serial operation is used to sequentially arrange the filtered IDFT outputs, as shown in Figure 6.

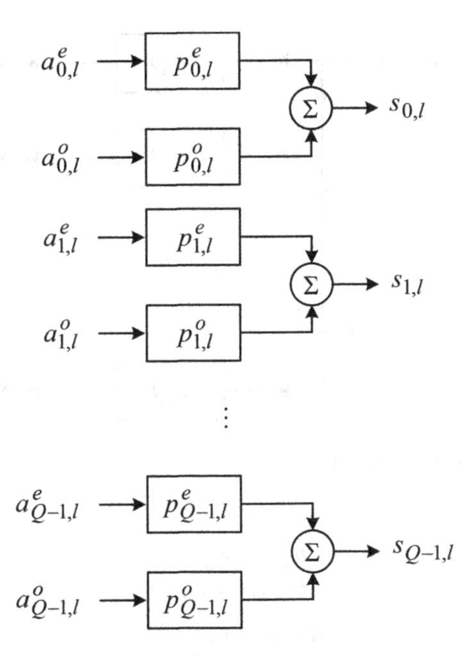

Figure 5. Filtering of IDFT outputs.

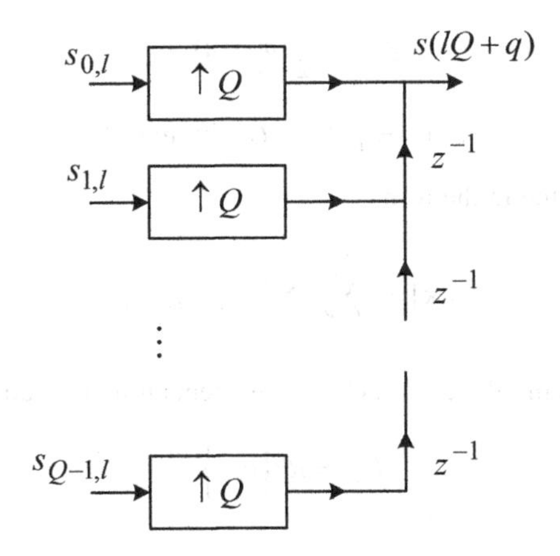

Figure 6. Parallel-to-serial conversion.

As can be seen from (39), and compared to OFDM, the size-M IDFT operation is replaced by two size-Q IDFT operations, followed by convolutions with sequences of length $4F$ at each IDFT output.

5 RECEPTION

The first step in the reception process it to convert the sequence $s(lQ + q)$from serial to parallel form. This is illustrated in Figure 7

Note that $d_{m,n}$ can be rewritten in the form

$$d_{m,n} = b_{m,n} e^{j\varphi_{m,n}} \tag{40}$$

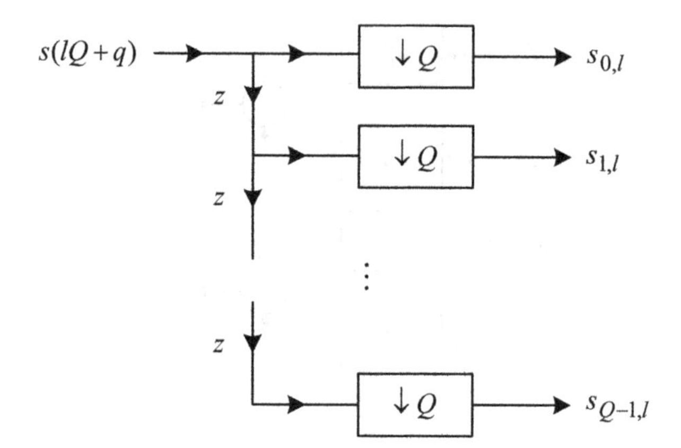

Figure 7. The serial-to-parallel conversion.

where

$$\begin{aligned} b_{2q,2l} &= x^R_{2q,l} \\ b_{2q,2l+1} &= x^I_{2q,l} \end{aligned} \tag{41}$$

$$\begin{aligned} b_{2q+1,2l} &= x^I_{2q+1,l} \\ b_{2q+1,2l+1} &= x^R_{2q+1,l} \end{aligned} \tag{42}$$

and

$$\varphi_{m,n} = \frac{\pi}{2}(m+n)_2 \tag{43}$$

Letting

$$\gamma_{m,n}[k] = p[k - nQ]e^{j\varphi_{m,n}}e^{j\frac{2\pi}{M}mk} \tag{44}$$

equation (24) can be rewritten in the form

$$s[k] = \sum_{n=-\infty}^{\infty}\sum_{m=0}^{M-1} b_{m,n}\gamma_{m,n}[k] \tag{45}$$

Reception of transmitted symbols can be achieved by generating the sequence

$$\hat{b}_{r,u} = Re\left\{\tilde{b}_{r,u}\right\} \tag{46}$$

where

$$\tilde{b}_{r,u} = \sum_{k=-\infty}^{\infty} s[k]\gamma^*_{r,u}[k] \tag{47}$$

Substituting (44) into (47) yields

$$\begin{aligned} \tilde{b}_{r,u} &= \sum_{k=-\infty}^{\infty} s[k]p[k-uQ]e^{-j\varphi_{r,u}}e^{-j\frac{2\pi}{M}rk} \\ &= \sum_{l=-\infty}^{\infty} e^{-j\varphi_{r,u}}e^{-j\pi rl} \\ &\times \sum_{q=0}^{Q-1} s_{q,l}p_{q,l-u}e^{-j\frac{2\pi}{M}rq} \end{aligned} \tag{48}$$

Note that $p_{q,l-u}$ is non-zero in the range

$$u \le l < 4F + u \tag{49}$$

This allows (48) to be written as

$$\begin{aligned}\tilde{b}_{r,u} &= \sum_{l=u}^{4F-1+u} e^{-j\varphi_{r,u}} e^{-j\pi rl} \\ &\times \sum_{q=0}^{Q-1} s_{q,l} p_{q,l-u} e^{-j\frac{2\pi}{M}rq}\end{aligned} \tag{50}$$

By a change of variables, (50) becomes

$$\begin{aligned}\tilde{b}_{r,u} &= \sum_{l=0}^{4F-1} e^{-j\varphi_{r,u}} e^{-j\pi rl} \\ &\times \sum_{q=0}^{Q-1} s_{q,l+u} p_{q,l} e^{-j\frac{2\pi}{M}rq}\end{aligned} \tag{51}$$

For even $r = 2i$

$$\tilde{b}_{2i,u} = e^{-j\varphi_{2i,u}} \sum_{q=0}^{Q-1} g^{e}_{q,u} e^{-j\frac{2\pi}{Q}iq} \tag{52}$$

where

$$\begin{aligned}g^{e}_{q,u} &= \sum_{l=0}^{4F-1} s_{q,l+u} p^{e}_{q,l} \\ &= s_{q,l} * p^{e}_{q,-l}\end{aligned} \tag{53}$$

Based on the last result,

$$\tilde{b}_{2i,u} = e^{-j\varphi_{2i,u}} DFT\left\{ g^{e}_{q,u}\Big|_{q=0}^{Q-1} \right\} \tag{54}$$

For odd $r = 2i + 1$

$$\tilde{b}_{2i+1,u} = e^{-j\varphi_{2i+1,u}} \sum_{q=0}^{Q-1} g^{o}_{q,u} e^{-j\frac{2\pi}{M}iq} \tag{55}$$

$$\begin{aligned}g^{o}_{q,u} &= \sum_{l=0}^{4F-1} s_{q,l+u} p^{o*}_{q,l} \\ &= s_{q,l} * p^{o*}_{q,-l}\end{aligned} \tag{56}$$

Based on the last result,

$$\tilde{b}_{2i+1,u} = e^{-j\varphi_{2i+1,u}} DFT\left\{ g^{o}_{q,u}\Big|_{q=0}^{Q-1} \right\} \tag{57}$$

The first step in the receiver is to generate the sequences $g^{e}_{q,u}$ and $g^{o}_{q,u}$. This is done using (53) and (56), and is illustrated in Figure 8.

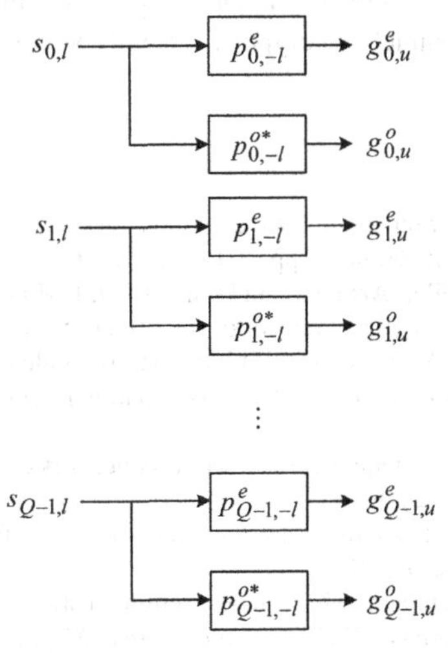

Figure 8. Filtering of DFT inputs.

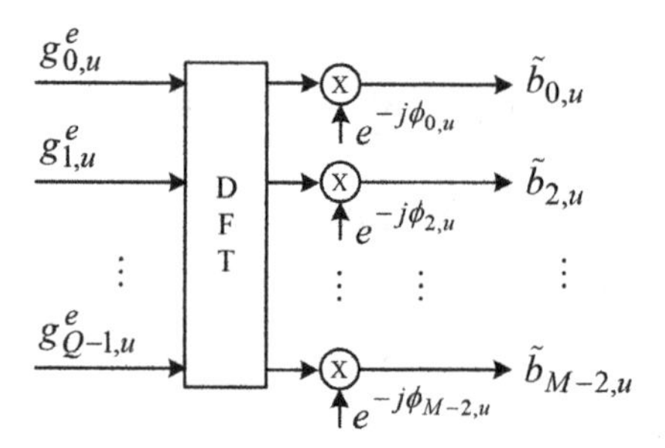

Figure 9. Even DFT.

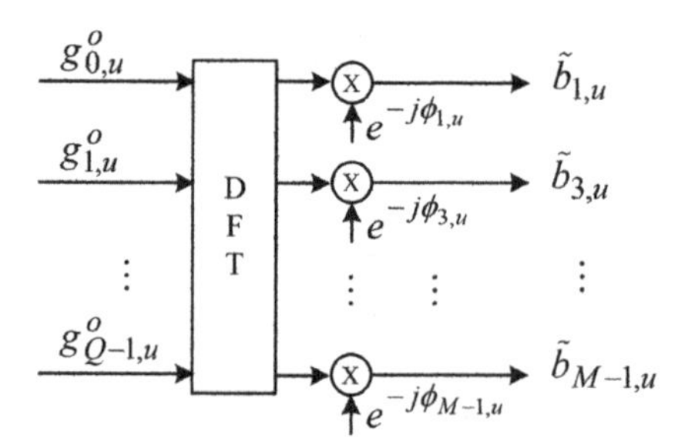

Figure 10. Odd DFT.

Generation of even- and odd-numbered samples of the sequence $\tilde{b}_{r,u}$ is illustrated in Figure 9 and Figure 10.

6 CONCLUSION

This paper has presented a detailed simplified implementation of a filter bank multicarrier transmitter and receiver with sub-channel prototype filters. Sub-channel prototype filters are shorter than a single prototype filter by a factor that is equal to the number of data symbols per multicarrier symbol, resulting in reduced complexity and added design flexibility. Exploitation of the flexibility resulting from the use of sub-channel prototype filters will be considered in future research.

REFERENCES

[1] J. A. C. Bingham, "Multicarrier modulation for data transmission: An idea whose time has come," *IEEE Communications Magazine,* vol. 28, no. 5, pp. 5–14, May 1990.
[2] P. Siohan, C. Siclet & N. Lacaille, "Analysis and Design of OFDM/OQAM Systems Based on Filterbank Theory," *IEEE Transactions on Signal Processing,* vol. 50, no. 5, pp. 1170–1183, May 2002.
[3] D. Bethanabhotla, G. Caire & M. J. Neely, "WiFlix: Adaptive Video Streaming in Massive MU-MIMO Wireless Networks," *IEEE Transactions on Wireless Communications,* vol. 15, no. 6, pp. 4088–4103, Jun. 2016.
[4] E. Basar, "Index modulation techniques for 5G wireless networks," *IEEE Communications Magazine,* vol. 54, no. 7, pp. 168–175, Jul 2016.
[5] B. Farhang-Boroujeny, "Filter bank multicarrier modulation: A waveform candidate for 5G and beyond," *Advances in Electrical Engineering,* 2014.
[6] B. Farhang-Boroujeny & R. Kempter, "Multicarrier communication techniques for spectrum sensing and communication in cognitive radios," *IEEE Communications Magazine,* vol. 46, no. 4, pp. 80–85, Apr. 2008.

Proceedings of the 1st International Congress on Engineering Technologies – Kiwan & Banat (Eds)
 ISBN 978-0-367-77630-5

Academic Rewards Coin (ARC) concept: Mining academic contribution into bitcoin

Mohammed M. Banat
Department of Electrical Engineering, Jordan University of Science and Technology, Irbid, Jordan

Haifaa Omar Elayyan
ITC Middle East University, Amman, Jordan

ABSTRACT: Intellectualizing internet data is easy when anyone can find basic information about their identity upon conducting a straightforward search. This paper assesses value for the information related to academic contributions in cyberspace and tries to construct an Online Academic Identity (OAI) structure. The OAI represents the academic persona and the academic contributions that have been added to cyberspace. The paper also evaluates the influential impact of an OAI and encourages the digital compensation of academic contributions using the Bitcoin concept. Academic Rewards Coin (ARC) could quantify and reward academic contributions using a customized mining process based on the bitcoin concept. This paper proposes a system of classifying academic contributions and creating missions to be accomplished, and then rewardeding success by amounts of ARC. Mission examples include reviewing a book or submitting a paper to a scientific conference. The ARCs earned upon completing a mission can be used as partial payments for conference registrations or other similar paid services.

Keywords: Bitcoin, Online academic identity, Cyberspace, Mining processes, Academic rewards coin.

1 INTRODUCTION

The internet is a technological revolution on which most of us depend in many ways because it is a robust resource, efficient, and easy to use, Social networks have created stable' virtual environments where individuals can customize entity relationships, easily communicate, and keep up with family and contacts [1]. Many people claim that they cannot live, work, and achieve without the internet digital world through which they form their online (digital) presence, or their "online identity."

Primarily, online identity is a digital representation that depends on and interacts with socio-cultural components framing the digital appearance of the self [2]. Since the advent of the internet, the "contribution" to the digital world has been scattered among various categories. Academics and researchers have promoted academic online identities and proved the theory of its positive impact on learning and self-development experiences. It is normal to recognize academic achievements for quality assurance and control purposes. In this regard, we argue that the bitcoin concept is the best means to recognize academic contributions.

In June 2019, Facebook, the famous social network, presented the "Libra" white paper that includes a brief, potentially seismic nod of digital standards and almost presented a jump-start of a globally accepted digital ID instead of linking bitcoins to the exchangeable products. Facebook decided later to postpone "Libra" and save the global financial system pending another bold step. Facebook described the consortium that will govern the Libra coin; the white paper states at the top of page nine: "An additional goal of the association is to develop and promote an open identity

DOI 10.1201/9781003178255-2

standard. We believe that decentralized and portable digital identity is a prerequisite to financial inclusion and competition".

The foremost reason behind the Libra being postponed is the major deceptiveness of online identities in the Facebook social network. The internet identity is poorly framed and the questions will be always there: "who are you, online?" and "are you really who you claim to be?" The natural question here is how much social networks contribute to your digital presence and your perception of your reality. The answer worries most of us and justifies the need for more studies about the impact of emerging technology in our lives, that is, how emerging technology has a disruptive impact on businesses now and over time. Facebook as a business entity needs to guard against fraud and frames the copious amounts of personal data consumers must share as necessary to prove that they are who they claim to be.

In our research concept, we cooperate with some publishers to frame and present the online academic identity as an acknowledged structure rooted with recognized online academic contributions. This involves personal academic achievements, academic contributions, and academic influential impacts. This structure can actively perform significant interactions in the "internet of people" social network to circulate professional academic high standards throughout the digital world.

Digital academic recognition exalts gaining knowledge process, learning, and developing experience while being awarded digitally. The research asserts the potential of accumulating and aggregating academic contributions all over the digital world into a digital wallet with an academic reward value that can enrich performance aiming for superior academic exploits.

The paper introduces the online academic identity (OAI) and proposes a digital wallet system for getting rewarded and/or earning distinguished opportunities for exchanging these services. Once the academic identity is structured, it has all the privileges to contribute and raise the digital wallet value. This can also include digital exchanges like payments for purchases and accumulating academic rewards coins (ARCs). The paper presents a brief methodology of the digital wallet and proposes algorithms for mining the academic contributions into the information-driven system to decide the ARCs that are to be awarded and saved in the blockchain at the back system.

Constructing an online academic identity is crucial for any academic nowadays. The internet provides various academic platforms to publicize the relevant contributions, academic achievements, and academic forecasting as well. Being an academic was not a choice for some people as they work in the education sector for many reasons. However, being an influential academic persona is a passion for most motivated academics. Therefore, they always try to excel to meet their dreams of being pioneers in their fields.

Most agree that academia is a crucial pillar for any community; as it plays crucial roles in technological development, and in improving people's lives in general. Research, studies, and on-going investigations are non-stoppable processes that we consider, evaluate, and accumulate our knowledge and experiences based on. Structuring an Online Academic Identity is a procedure that needs to be embedded in our technology revolution. Affording more technology facilitates the process of constructing online academic identities, presents their performance, and securely evaluates their impacts within a group of academic entities. The "internet of people" platforms rely on people entities and identities for communication purposes. Each is expected to contribute and touch solid ground craving more success.

The rest of this paper is organized as follows: Section 2 presents some basic relevant definitions. The ARC concept is explained in section 3. The main conclusions of the paper are drawn in section 4.

2 SOME DEFINITIONS

- **The online academic identity (OAI)** is a structured computer file that organizes relative values to represent the online academic identity's interactions and contributions in the digital world.
- **The online academic contribution (OAC)** is the conceptualizing of the online academic identity which is expected to raise the potentiality of characterizing and standardizing the academic achievements and the contributions this identity adds to the academic sector. The academic

online contribution helps to trace, rank, and measure the influential side of the relevant AOI on the internet. It is a linked structure that aggregates more features of the corresponding OAI.

- **The online academic reward** (OAR) is compensation of the online academic contribution that is also expected to abet the AOI for more recognition and cognition.

3 ACADEMIC REWARDS COIN (ARC) CONCEPT

3.1 *Bitcoin concept*

Bitcoin is a digital cryptocurrency using which people can trade and buy products with in the virtual world. People can send Bitcoins to their digital wallets to the digital wallets of others. Every single transaction is recorded in a public list called the blockchain which is publicly listed for real-time exchanges [3].

Bitcoin was developed by Satoshi Nakamoto in 2008 and became popular as the first digital currency that can solve the problems associated with the decentralization using the mining knowledge. Bitcoin's mining solution is trending, as it simply makes the transaction history public so each user can store a local copy of the transaction history and can check this ledger before accepting payment.

3.2 *Bitcoin mining process*

Opposite to all other tradable products, bitcoins are not stored centrally or locally, and therefore, no one entity is their custodian. Like computer files, bitcoins are represented as records on a distributed ledger called the blockchain, copies of which are shared by a volunteer network of connected computers. Anyone can enjoy the bitcoin's ability to transfer control of it to someone else by creating a record of the transfer in the blockchain. This can be done only by accessing an Elliptic Curve Digital Signature Algorithm (ECDSA) private and public key pair. The algorithm uses an elliptic curve and a finite field to sign data in such a way that third parties can verify the authentication of the signature while the miner retains the exclusive ability to create the signature.

The Bitcoin concept declares that the data is signed if the transaction involves transfers of ownership. The mathematical procedures create the one-way trap door functions necessary to preserve the information asymmetry which defines ownership of a bitcoin and preserves the robustness of the system, provided safeguarding the private keys [4], [5].

Bitcoin is not regulated by a central authority, it is backed by millions of computers across the world called "Miners" [6], [7]. These "Miners" interact to create a computer platform operating similarly to Federal Reserve, MasterCard, and Visa. The platform mines record transactions and checks accuracy in recording data in a public list that can be accessed by anyone. Despite the fact of miners being spread out across the world, digital information can be reproduced relatively easily every time the mining process takes a place. In addition to adding blocks of transactions to the blockchain, miners make sure that bitcoin is not being duplicated using a unique quirk of digital currencies to avoid the risk of a digital currency being spent twice.

The mining process is composed of the following steps [8], [9]:

1. To be able to update the ledger, the miner guesses a random number that solves an equation generated by the system, the more powerful the computer miner is, the more numbers can be guessed per second, increasing chances of winning.
2. Once the computer miner makes a correct guess, the mining process determines which of the current pending transactions will be grouped in the next block of transactions.
3. Compiling the block grants the computer miner a temporary banker option of bitcoin, which privileges the miner to update the Bitcoin transaction ledger (blockchain).
4. The created block is associated with the solution and sent to the whole network so other miners can validate it.
5. Each miner validates the solution, updates its stored copy with the transactions the winner miner chose to include in the block.

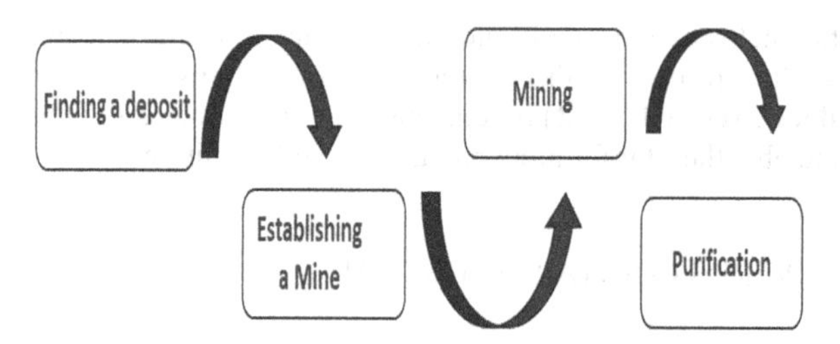

Figure 1. Major steps in the mining process.

6. The system generates a fixed number of bitcoins (currently 12.5) and rewards them to the winner miner as compensation.
7. The updates made to the block are confirmed by the bitcoin network and are virtually irreversible.

Figure 1 (Source: MIT website: https://web.mit.edu/12.000/www/m2016/finalwebsite/solutions/mining.html) breaks the mining process into four major steps: finding a deposit, establishing a mine, mining, and purification.

The miners get paid any transaction fees that were attached to the transactions the winner miner inserted into the following block. The mining process has inspired the idea of academic rewards coin that will be presented below.

3.3 *Academic Rewards Coin (ARC)*

The idea of rewarding academic contributions is not new at all. Many institutions and publishers have long been after recognizing and rewarding academic achievements. Both real-life rewards and digital world rewards have been used. A leading publisher that has standardized the benefits of being an author is the Institute of Electrical and Electronics Engineers (IEEE – http://ieee.org), founded in 1884. Some platforms, like Orvium - https://orvium.io, and PLUTO - https://pluto.network/, provide economic rewards for reviewing scientific work in the form of tokens that are staked by the authors.

At the time of writing, no professional associations accredit the academic contribution using bitcoin.

The concept of rewarding academic contribution using ARC, as a customized version of bitcoin, is introduced in the below. ARCs can be earned and kept in the digital wallet of the academic contributor. Mosharaka for research and studies is taken as an example professional association that is working towards remunerating academic contributions with ARCs within its portal.

The rewarding process requires creating an online academic identity that can make academic contributions under the umbrella of Mosharaka scientific activities. ARCs are rewarded every time a contribution is mined in the Mosharaka blockchain. This blockchain is visible and accessible by members of Mosharaka.

3.4 *ARC mining process*

Data mining is very important in this research. In light of the recent advances in data collection and data processing techniques, associations like Mosharaka store large amounts of data. Data mining has been recognized as a technology that can discover previously unknown and potentially useful information from databases [10]. The academic contribution presents the dataset for the associative classifier we are investigating to generate a mission imitating the problem in the bitcoin mining process.

The process classifies the user's preferences to generate a mission that can be accomplished by the online academic contributor. The mission is categorized by the general services featured by the

Mosharaka platform. Current mission categories include reviewing and donating. More categories like authoring and joining conference committees will be added in the future.

The ARC mining process consists of the following steps:

1. Generating a mission, where the contributor is the miner to accomplish the mission.
2. Once the mission is accomplished, the mining process determines which of the current pending transactions will be grouped in the next block of transactions.
3. Compiling the block grants the contributor a temporary banker option of bitcoin which privileges the miner to update the Bitcoin transaction ledger (blockchain).
4. The created block is associated with the mission metadata and sent to the whole network so other contributors can validate it.
5. Each contributor validates the mission and updates the stored copy with the transactions the accomplisher chose to in include in the block.
6. The system generates a fixed number of ARCs (currently 10) and rewards them to the miner as compensation.
7. The updates contributor has made to the block are confirmed by the Mosharaka system and are virtually irreversible.

3.5 *Creating the digital wallet*

The digital wallet is a designed system that securely stores a person's payment information. The digital wallet consists of software and data [11], [12]. The software can be an online service or a device that works to keep the entire information stored in this digital wallet and secure the data and information by encryption algorithms. Transaction processes may differ according to the software or online service but in general, such a system is basic in any conference management system like Mosharaka. Authors argue that system promotion by having a cryptocurrency wallet is the best theme for the presented ARC system.

A cryptocurrency wallet is a software program that stores private and public keys and enables users to send and receive digital currency and to monitor their balance. ARC cryptocurrency wallet should be able to store the public key associated with the mission once it is generated. It should also be able to store the private key that is mapped from the contributor side before interacting with the blockchain. ARC cryptocurrency wallet stores all relevant information from the contributor side, and links it to the private key that is attached to the transaction once the mission is completed. The system defines a personal wallet accumulating the ARC values, that are accessible to and viewable by only the contributor with updating permissions without affecting the blockchain. At the Mosharaka platform, the digital wallet is not currently active, but work is ongoing to activate it and enable it to interact directly with the blockchain.

The compensation requires a procedure of constructing an Online academic identity that can interact and cooperate academically according to Mosharaka membership conditions and affiliation. The OAI is rewarded with ARC every time a contribution is mined in the Mosharaka blockchain. This blockchain is visible and accessible by members of the Mosharaka platform. Mosharaka platform generates a public key and allows membership to map it to a private key kept for verification purposes.

3.6 *The Academic Reward Coin mining process*

The data mining concept is important to our research. With the advent of technology in data collection and data processing, enterprises such as the Mosharaka platform store a large amount of data. Data mining has been recognized as a technology that can discover previously unknown and potentially useful information from databases (Witten et al. 2011). The academic contribution presents the dataset for the associative classifier we are investigating to generate a mission imitating the problem in the original bitcoin mining process. The process classifies the user's preferences to generate a mission that can be accomplished by the online academic contributor. The mission is categorized by the general services featured by the Mosharaka platform that is limited for now to: reviewing and donating missions given that Mosharaka is a professional association for research and studies affords sponsorships of conferences, educational activities and peer-reviewing activities and authorships as well.

3.7 *The suggested ARC mining process*

1. Generating a mission will be automatically done by the system design and the assigned contributor will be the miner to accomplish the mission.
2. Once the mission is accomplished, the mining process determines which of the current pending transaction will be grouped in the next block of transactions.
3. Compiling the block grants the contributor a temporary banker option of bitcoin which privileges the miner to update the Bitcoin transaction ledger (blockchain).
4. The created block is associated with the mission metadata and sent to the whole network so other contributors can validate it.
5. Each contributor validates the mission and updates the stored copy with the transactions the accomplisher chose to in include in the block.
6. The system generates a fixed amount of ARC (currently 10) and rewards them to the winner miner as a compensation.
7. The updates contributor has made to the block are confirmed by the Mosharaka network and are virtually irreversible.

To Generate the public and the private keys, every member registered in the Mosharaka.net gets a membership number. This number consists of a serial number (8 digits). Once the mission is created by the admin, it holds a serial number that is auto-generated.

1. **Public key**: This is a serial/sequential number generated by the system for this mission. The public key should consist of:
   a. **A category of service:** For example, "A" stands for Reviewing a paper, "B" stands for reviewing a book, and so on.
   b. **A category of the field**: to identify the field of this book or paper was submitted under. For example, "1" for data mining, "2" for communication, and so on.
   c. **A paper Id:** which is automatically generated once the submission is completed.
   Example of what was mentioned above: The auto-generated mission number is **345** and the is paper in the data mining field. The paper Id is 56. So, the mission ID is 345A1-56.
2. **A private key** is a random number of 3 digits out of the membership digits.
3. The public key is published, and the private key is kept for the miner (linked to his IP for security reasons).
4. A mission linked to a miner represents the problem for this miner to solve (accomplish).
5. The same mission can be generated for reviewing the same paper, but the private key would then be different.
6. The private key is changeable every time a mission is assigned to the miner.
7. Transferring to the blockchain then:
   a. Creates a new record.
   b. The record will be added as:
   c. A Serial number of the mission / private key (3 digits) / status (one digit specifying that this mission is accomplished) can be (0/1) / ARC amount.

For donating, the mission is created by specifying the donatorID, receiver, and ARC amount that should be fixed for each transaction. The algorithm checks the donator's digital wallet and creates an if statement:

If the digital wallet.total > Donation
   Then

```
{Donator.digitalwallate = donator.digitalwallet-donation;
Reciever.digitalwallet=receiver.digitalwallet
  +donation;
}
  Else
Return 0;
```

Donation is fixed which should be equal to 10 for example. If the person wants to donate 20, then the transaction happens twice. Donating is also a mission that should have a public and private key. The private key is a combination of random digits generated out of donator and receiver membership numbers.

3.8 *Nanotechnology emerging*

Nanotechnology as a concept refers to the manipulation of matter on an atomic, molecular, and supramolecular scale [13]. A Nano concept can be deployed with bitcoin as a digital currency designed for speed and efficiency. It is widely considered to have solved some of the problems that have plagued the bitcoin blockchain because it does not contain the information from the entire chain every time it processes a transaction.

Nano cryptocurrency promises instant transactions with zero fees. This means it is considered a strong candidate for micropayment solutions that are expected to encourage people to consider bitcoins in general. Nano is fee-less and uses much less electricity to run its blockchain network. Unlike Bitcoin, Nano spreads its blockchain over each account, building a peer-to-peer trust network of account chains that reduces clogging and bottlenecks [14]. Nano cryptocurrency can be customized such that a customer can pay for one service without the need for a full subscription. For example, a user is not registered in a specific Mosharaka event but can be assigned for reviewing a book chapter and get paid with academic bitcoins.

ARC is a computer file that can be used as a whole or in parts to fit a nanoscale service. Like in the case of trading, an ARC value can be traded with a similar ARC value in a different type of academic contribution. ARCs can be used to promote a paper reviewer to a book reviewer in a field in which they have earned sufficient ARCs for paper reviewing. This is a major advancement of the ARC concept, compared to the simple process of being rewarded with ARCs for doing paper reviews.

4 CONCLUSION

ARC is a nano-based concept that encrypts the recognized academic contribution into information to build the academic bitcoin. It can be traded for paying full or part fees required by other services within the Mosharaka conference management system.

ARC allows people to aggregate individual contributions using the concept of the "internet of people". This requires academic connections and academic interactions/interconnections for specific services and allows participants to trade the ARC into services the platform affords.

REFERENCES

[1] J. Caron & J. Light, "Social media have opened a world of 'open communication:' Experiences of adults with cerebral palsy who use augmentative and alternative communication and social media.," *Augmentative and Alternative Communication,* vol. 32, no. 1, pp. 25–40, 2016.

[2] M. N. O. Sadiku, A. E. Shadare & S. M. Musa, "Digital Identity," *International Journal of Innovative Science, Engineering & Technology,* vol. 3, no. 12, pp. 192–193, 2016.

[3] J. Herrera-Joancomartí, "Research & challenges on bitcoin anonymity," in *Data Privacy Management, Autonomous Spontaneous Security, and Security Assurance*, 2014.

[4] "Federal Information Processing Standards (FIPS) 186-4, Digital Signature Standard (DSS)," National Institute of Standards and Technology, Gaithersburg, MD, 2013.
[5] T. Pornin, "RFC 6979 – Deterministic usage of the digital signature algorithm (DSA) and elliptic curve digital signature algorithm (ECDSA)," 2013.
[6] A. H. Fay, A glossary of the mining and mineral industry, Washington, Govt. Print. Off., 1920.
[7] R. W. Raymond, Glossary of mining and metallurgical terms, HANSEBOOKS, 2016.
[8] W. M. P. van der Aalst & A. J. M. M. Weijters, "Process mining: a research agenda," *Computers in Industry,* vol. 53, no. 3, pp. 231–244, April 2004.
[9] F. A. Yasmin, F. A. Bukhsh, & P. A. Silva, "Process Enhancement in Process Mining: A Literature Review," in *Proceedings of the 8th International Symposium on Data-driven Process Discovery and Analysis (SIMPDA 2018)*, Seville, Spain, 2018.
[10] I. H. Witten, E. Frank, M. A. Hall, & C. J. Pal, Data Mining: Practical Machine Learning Tools and Techniques, Morgan Kaufmann, 2016.
[11] K. Tiwari, *Secure Digital Wallet Authentication Protocol (Master Thesis),* Halifax, Nova Scotia, Canada: Dalhousie University, 2017.
[12] H. Rezaeighaleh & C. Zou, "Deterministic Sub-Wallet for Cryptocurrencies," in *IEEE International Conference on Blockchain*, Atlanta, GA, USA, 2019.
[13] A. V. Deshpande & R. Meghe, "Nanomaterials and nanotechnology: Future emerging technology," *International Journal of Advanced Research in Engineering and Technology,* vol. 5, no. 12, pp. 41–47, December 2014.
[14] E. Budish, "The economic limits of bitcoin and the blockchain," *National Bureau of Economic Research,* no. w24717, 2018.

Proceedings of the 1st International Congress on Engineering Technologies – Kiwan & Banat (Eds)
 ISBN 978-0-367-77630-5

Analytical modeling of dispersion penalty for NRZ transmission

Chris Matrakidis & Dimitris Uzunidis
OpenLightComm Ltd., London, UK

Alexandros Stavdas
University of Peloponnese, Tripolis, Greece

Gerasimos Pagiatakis
School of Pedagogical and Technological Education, Athens, Greece

ABSTRACT: In this work we present an analytical model that can be used to evaluate the dispersion effects and subsequent penalty for NRZ pulse transmission through a given length of fiber taking into account the finite bandwidth of a direct detection receiver. This is done by using a pulse shape that is similar to the shape of NRZ pulses but that its propagation through single mode fibers can be solved analytically. The model is evaluated by using numerical simulations to obtain values of the dispersion penalty for NRZ pulses, showing very good agreement.

Keywords: Optical fiber communication, Optical fiber dispersion.

1 INTRODUCTION

The use of analytical methods has several applications in the study of optical communication systems. Such methods are fast and can be used to assess the performance of links at both system and network levels. Any of the physical parameters can be used as an optimization variable in order to ascertain its impact (and potential penalty) in the performance of the link. Due to their flexibility, analytical methods can also be applied in a scalable and dynamic network, especially as part of routing algorithms that take into account physical layer impairments [1]. Typically, this is done by obtaining the Q-factor of the signal at the end of the link.

Several models are available for various effects that degrade the performance of a link, such as ASE accumulation [2], cross-phase modulation (XPM) [3] and four-wave mixing (FWM) [4]. However, no satisfactory model exists to estimate the performance impact of the link's dispersion. The common technique is to obtain the pulse broadening and use this to estimate the power penalty. This method is described in detail in [5]. While this method can give good results for RZ signals it is not suitable for NRZ ones.

The objective of this work is to propose an analytical model that can be used to evaluate the dispersion effects and subsequent penalty for NRZ pulse transmission through a given length of fiber. This is done by using the finite bandwidth of a direct detection receiver in a way that can be combined with models for other effects in order to improve estimates for the Q-factor of the link. In currently deployed networks, legacy direct detection channels are used to convey an important portion of network traffic. This is especially true in the Metro domain, where sub-1G/1G/10G rates are dominant [6], [7]. Thus, an analytical model which can be used to optimize their performance is an important asset for network engineers.

DOI 10.1201/9781003178255-3

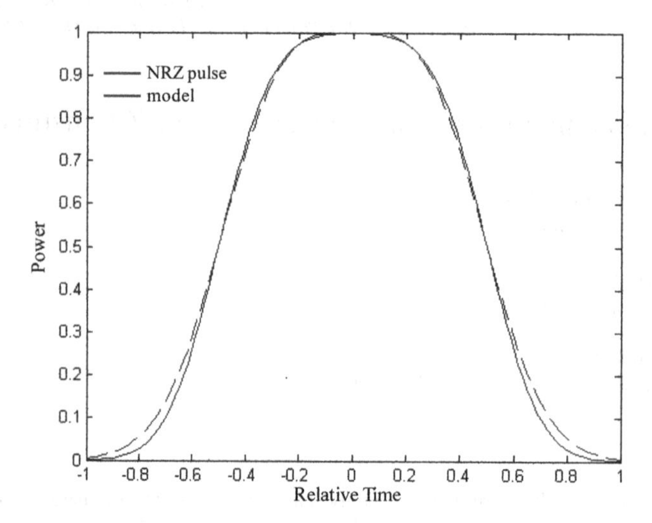

Figure 1. Comparison of pulse shape used (dashed) and NRZ 'one' pulse with rise time $0.4/B_R$ (solid).

2 METHODOLOGY

We assume that the worst penalty at the 'one' level occurs when a single 'one' bit is transmitted in the middle of a large sequence of 'zeros', and that similarly, the worst penalty at the 'zero' level occurs when a single 'zero' bit is transmitted in the middle of a large sequence of 'ones'. We use a pulse shape similar to a rise-time limited square pulse (whose propagation can be calculated analytically) thereby obtaining the pulse shape at the end of the link. This can then be used to calculate the pulse energy in a specific time interval, approximating the receiver output after filtering.

The pulse shape used for a single 'one' pulse is:

$$A(t) = e^{-\frac{t^2}{2T_0^2}} \cosh\left(\frac{t}{T_0}\right) \tag{1}$$

The width of the pulse depends on the value of parameter T_0. For a given full-width half-maximum duration T_{FWHM}, we obtain $T_0 = 0.3087 \bullet T_{FWHM}$ by numerically solving $A(T_{FWHM}/2) = 1/2$. The resulting pulse (Figure 1) is a very good fit to a single NRZ bit of the same T_{FWHM} with rise time 40% of the bit duration (or $0.4/B_R$ with B_R the bit rate of the NRZ signal). Obviously, the pulse shape used has a fixed rise time. How this affects the modeling of the propagation of NRZ pulses with different rise time is discussed later.

Similarly, the pulse shape used for a single 'zero' pulse is:

$$A(t) = 1 - e^{-\frac{t^2}{2T_0^2}} \cosh\left(\frac{t}{T_0}\right) \tag{2}$$

In this case $T_0 = 0.2058 \bullet T_{FWHM}$. The resulting pulse is a very good fit to a single 'zero' between consecutive NRZ 'one' bits with rise time 40% of the bit duration (Figure 2).

The propagation of a signal through a single mode fiber of length z, ignoring non-linear effects and losses, is given in the frequency domain by (3) where β_2 is the second-order dispersion coefficient.

$$A(\omega, z) = A(\omega, 0)e^{\frac{i\beta_2\omega^2 z}{2}} \tag{3}$$

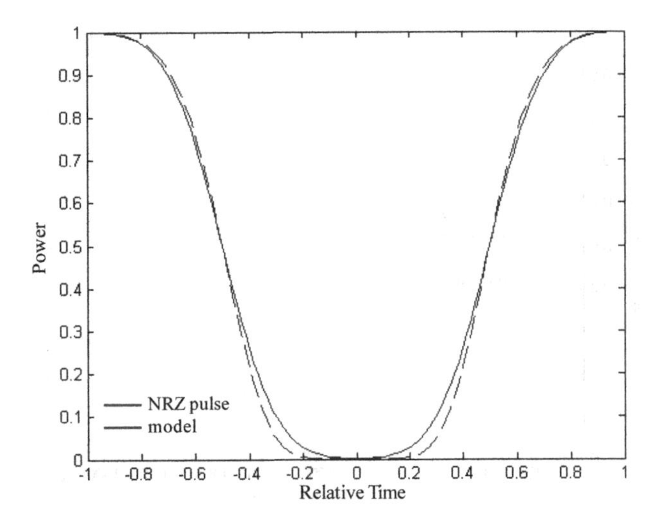

Figure 2. Comparison of the pulse shape used (dashed) and the NRZ 'zero' pulse with rise time $0.4/B_R$ (solid).

Using (3), it is easy to obtain the pulse shape after propagation of a 'one' pulse through a fiber span. The result is:

$$A(t,z) = \frac{T_0}{\sqrt{T_0^2 - i\beta_2 z}} e^{-\frac{t^2 + i\beta_2 z}{2(T_0^2 - i\beta_2 z)}} \cosh\left(\frac{tT_0}{T_0^2 - i\beta_2 z}\right) \tag{4}$$

$$\begin{aligned} E_1(z) = \frac{\sqrt{\pi}}{4} T_0 \Bigg(& e \cdot erf\left(\frac{T_0(2T_0 + T_f)}{2\sqrt{T_0^4 + \beta_2^2 z^2}}\right) - e \cdot erf\left(\frac{T_0(2T_0 - T_f)}{2\sqrt{T_0^4 + \beta_2^2 z^2}}\right) \\ & + erf\left(\frac{2i\beta_2 z + T_0 T_f}{2\sqrt{T_0^4 + \beta_2^2 z^2}}\right) - erf\left(\frac{2i\beta_2 z - T_0 T_f}{2\sqrt{T_0^4 + \beta_2^2 z^2}}\right)\Bigg) \end{aligned} \tag{5}$$

$$\begin{aligned} E_0(z) = {} & T_f + \frac{e\sqrt{\pi}}{4} T_0 \left(erf\left(\frac{T_0(2T_0 + T_f)}{2\sqrt{T_0^4 + \beta_2^2 z^2}}\right) - e \cdot erf\left(\frac{T_0(2T_0 - T_f)}{2\sqrt{T_0^4 + \beta_2^2 z^2}}\right)\right) \\ & - \frac{\sqrt{\pi}}{2} T_0 Re\left(erf\left(\frac{2i\beta_2 z - T_0 T_f}{2\sqrt{T_0^4 + \beta_2^2 z^2}}\right)\right) \\ & - \sqrt{2e\pi}\, T_0 Re\left(erf\left(\frac{2T_0 + T_f}{2\sqrt{2}\sqrt{T_0^2 - i\beta_2 z}}\right) - erf\left(\frac{2T_0 - T_f}{2\sqrt{2}\sqrt{T_0^2 - i\beta_2 z}}\right)\right) \end{aligned} \tag{6}$$

3 RESULTS

To estimate the output of a direct detection receiver with finite bandwidth, we calculate the pulse energy by integrating the pulse power over a time interval that is symmetrical around the pulse centre ($t = 0$). An integration interval equal to the bit duration gives the performance of a matched filter, while different filtering can be estimated using other integration intervals.

Following this, the pulse energy for a single "one" bit is given by (5) where T_f is the integration interval. Similarly, for a single "zero" bit the pulse energy is given by (6). Comparing this model

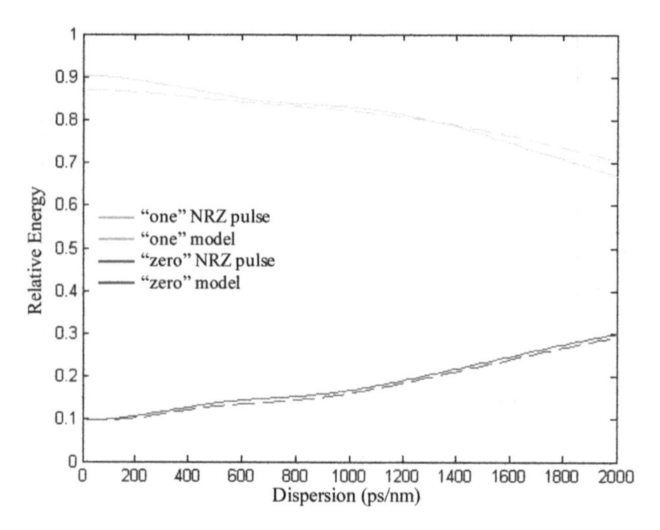

Figure 3. Pulse energy for "one" (top) and "zero" (bottom) bits for model (dashed) and 10 Gb/s NRZ pulse (solid) with rise time 30 ps as a function of the total link dispersion for matched receiver filter.

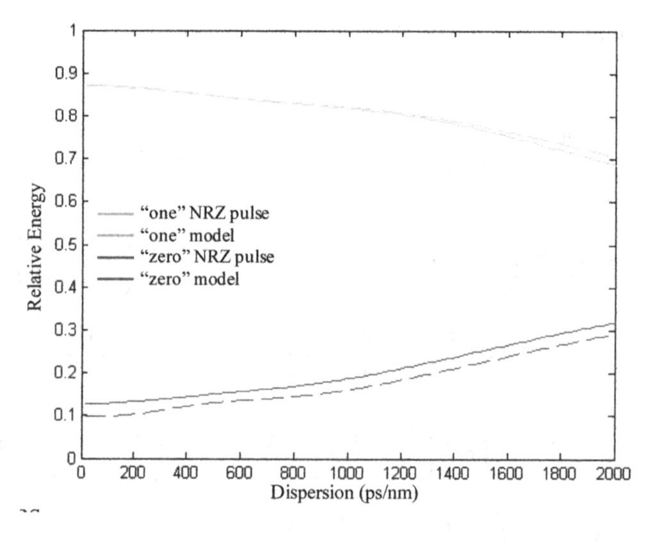

Figure 4. Pulse energy for "one" (top) and "zero" (bottom) bits for model (dashed) and 10 Gb/s NRZ pulse (solid) with rise time 40 ps as a function of the total link dispersion for matched receiver filter.

with energy values obtained through direct numerical calculation of the NRZ pulse transmission, we see from Figures 3 and 4 that the results are very close for a large range of dispersion values and for different values of the NRZ pulse rise time. The numerical results were obtained by calculating, using (3), the result of transmitting a pulse sequence through a lossless fiber with the given dispersion, ignoring non-linear effects. An ideal source was assumed at a wavelength of 1550nm and the sequence was detected using an ideal receiver with matched filter. The bit rate used was 10 Gb/s. The results are shown in Figure 3 and 4 as a function of the total dispersion (in ps/nm) for a pulse rise time of 30 ps and 40 ps respectively.

To calculate the dispersion penalty (dp) that can be directly used in the Q-factor calculation, we find the eye opening by subtracting the energy of the 'zero' and 'one' bits normalized to its value at z =,0.

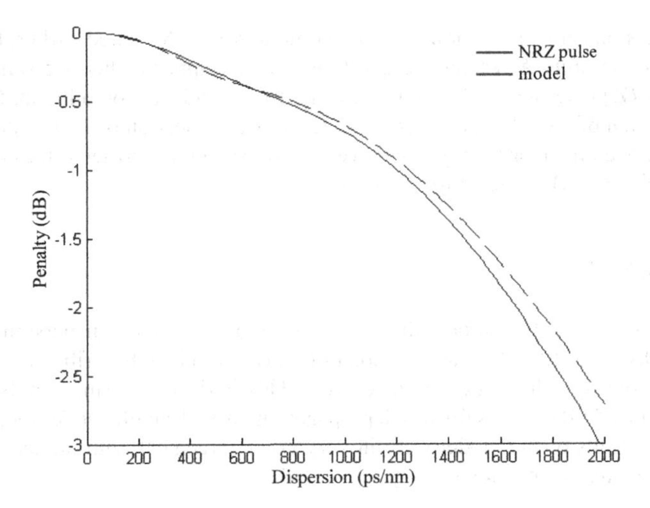

Figure 5. Dispersion penalty for model (dashed) and 10 Gb/s NRZ pulse (solid) with rise time 40ps as a function of the total link dispersion and with a matched receiver filter.

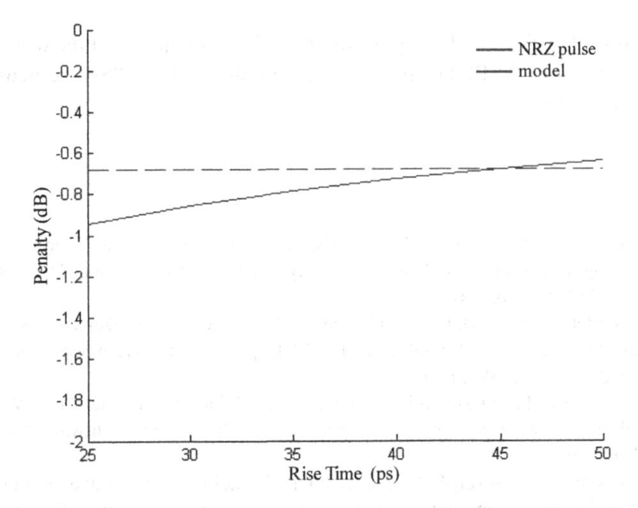

Figure 6. Dispersion penalty for model (dashed) and 10 Gb/s NRZ pulse (solid) with at the output of a link with D=1000ps/nm total dispersion and matched receiver filter as a function of the NRZ pulse rise time.

$$dp(z) = \frac{E_1(z) - E_0(z)}{E_1(0) - E_0(0)} \tag{7}$$

Then the Q factor calculation becomes

$$Q = dp(z)\frac{\mu_1 - \mu_0}{\sigma_1 + \sigma_0} \tag{8}$$

In (8), μ and σ respectively are the current level and standard deviation at the decision point (after the receiver) for 'zero' and 'one' bits. For $z = 0$, where $dp(z)=1$, the Q factor calculation is obviously equal to the one given by its traditional definition.

Again, comparing the model with energy values obtained through numerical simulation of an NRZ pulse sequence we see very good agreement. The results are plotted in Figure 5 as a function of the total dispersion (in ps/nm) for a pulse rise time of 40 ps. As stated earlier, the pulse shape used has a fixed rise time. A reasonable question is what happens when we want to model the propagation of NRZ pulses with a different rise time. To this end we plot in Figure 6 the dispersion penalty as a function of the NRZ pulse rise time for a total dispersion of 1000 ps/nm. In this case we see that the modeling error is very small even for considerably shorter values of the rise time, those not exceeding 0.3dB in the 25 to 50 ps range.

4 CONCLUSION

We have presented an analytical model that can be used to evaluate the dispersion effects and the subsequent penalty for NRZ pulse transmission through a given length of fiber taking into account the finite bandwidth of a direct detection receiver. This is done by using a pulse shape that is similar to the shape of NRZ pulses for which propagation through single mode fibers can be solved analytically. The results of the model were in very good agreement with numerically calculated values of the NRZ pulse dispersion penalty.

ACKNOWLEDGMENT

The authors acknowledge financial support for the dissemination of this work from the Special Account of Research of ASPETE through the funding program "Strengthening Research of ASPETE Faculty Members".

REFERENCES

[1] V. Anagnostopoulos, C. (T.) Politi, C. Matrakidis, & A. Stavdas, "Physical Layer Impairment Aware Wavelength Routing Algorithms based on Analytically Calculated Constraints," Optics Communications, Vol. 270, pp. 247–254, 15 Feb. 2007.

[2] A. Stavdas, S. Sygletos, M. O'Mahoney, H. Lee, C. Matrakidis, & A. Dupas: "IST-DAVID: Concept Presentation and Physical Layer Modeling of the Metropolitan Area Network", Journal of Lightwave Technology, Vol. 21, pp. 372–383, Feb. 2003.

[3] A. V. T. Cartaxo, "Cross-Phase modulation in intensity modulation-direct detection WDM systems with multiple optical amplifiers and dispersion compensators," Journal of Lightwave Technology, vol. 17, pp. 178–190, Feb. 1999.

[4] W. Zeiler, F. D. Pasquale, P. Bayvel, & J. E. Midwinter, "Modeling of four-wave mixing and gain peaking in amplified WDM optical communication systems and networks," Journal of Lightwave Technology, vol. 14, pp. 1933–1942, Sept. 1996.

[5] G. P. Agrawal, "Nonlinear Fiber Optics", Academic Press, 2001

[6] G. Shen et al., "Ultra-Dense Wavelength Switched Network: A Special EON Paradigm for Metro Optical Networks," in IEEE Communications Magazine, vol. 56, no. 2, pp. 189–195, Feb. 2018.

[7] D. Uzunidis, E. Kosmatos, C. Matrakidis, A. Stavdas & A. Lord, "DuFiNet: Architectural Considerations and Physical Layer Studies of an Agile and Cost-Effective Metropolitan Area Network," in Journal of Lightwave Technology, vol. 37, no. 3, pp. 808–814, 1 Feb.1, 2019.

Proceedings of the 1st International Congress on Engineering Technologies – Kiwan & Banat (Eds)
 ISBN 978-0-367-77630-5

On the attainable transparent length of multi-band optical systems employing rare-earth doped fiber amplifiers

Dimitris Uzunidis & Chris Matrakidis
OpenLightComm Ltd., London, UK

Alexandros Stavdas
University of Peloponnese, Tripolis, Greece

Gerasimos Pagiatakis
School of Pedagogical and Technological Education, Athens, Greece

ABSTRACT: We investigate the linear and nonlinear transmission performance of a multi-band system employing commercially available rare-earth doped fiber optical amplifiers. First, the available spectrum is partitioned into amplification zones with an ideal, flat gain spectrum that is attainable with existing rare-earth amplifier technology. A universal multi-band physical layer model is then exploited that allows us to quantify the impact of amplified spontaneous emission (ASE) noise and four wave mixing (FWM) crosstalk on signal quality. Finally, the maximum transparent length, under these physical layer constraints, is estimated for each transmission band.

Keywords: four wave mixing (FWM), nonlinear transmission performance, multi-band systems, rare-earth doped fiber amplifiers

1 INTRODUCTION

The proliferation of the Internet of Things and other Machine-to-Machine communication systems as well as the advances in access technologies and the upcoming 5G deployment will put optical transport networks under significant strain. To overcome the upcoming capacity crunch and system gridlock [1], the ultimate bandwidth potential of the single mode fiber (SMF) has to be unlocked [2]. Multi-Band (MB) systems, which can fully exploit the second and third low attenuation windows of SMF, may increase the available capacity by an order of magnitude.

Research on MB systems is still in its infancy and the related experimental work primarily addresses C and L bands [3]–[7]. In this work we first propose an amplification scheme based on commercially available rare-earth doped fiber amplifiers for O, S, C, and L bands. Secondly, using a currently introduced closed-form expression, we assess the attainable transparent length of each band of an MB system considering both the impact of ASE accumulation and FWM crosstalk. This closed-form expression was previously applied to systems employing C-band only whilst here we show that it is equally applicable to all transmission bands showing very good accuracy against numerical results. In our study, we ignore the impact of other important nonlinearities like Stimulated Raman Scattering (SRS), which is the subject of ongoing work.

2 DOPED FIBER AMPLIFICATION

To date, transmission systems are exploiting primarily the C band and to a lesser degree the L band, both employing Erbium Doped Fiber Amplifiers (EDFA). However, the other currently neglected bands shown in Figure 1 can also be employed if there are optical amplifiers with sufficiently flat gain over a substantial spectral range along with a relatively low noise Figure Existing rare-earth

DOI 10.1201/9781003178255-4

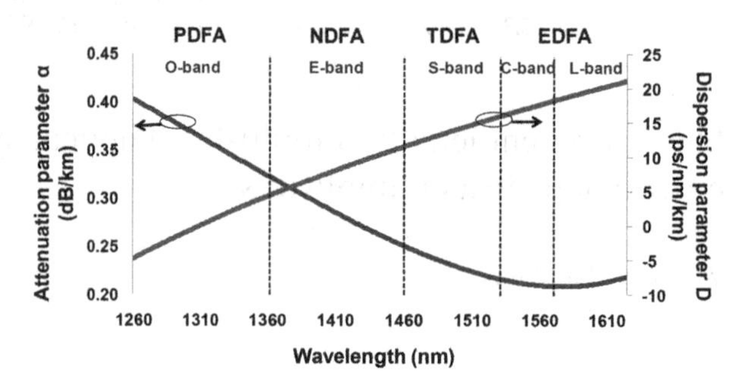

Figure 1. Attenuation and dispersion parameter of G.652D fiber.

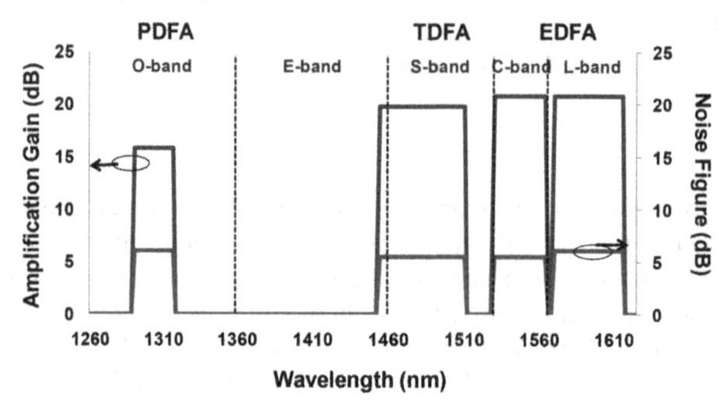

Figure 2. Amplification bands of an MB system using commercially available rare-earth amplifiers for a total input power in each amplifier of 0 dBm.

doped fiber amplifiers (DFAs) in O and S bands provide these characteristics at different degrees while for the E band a neodymium-doped fiber amplifier (NDFA) has been proposed in [8].

Figure 1 illustrates the dispersion (D) and attenuation parameters (α) in the wavelength range 1260–1615 nm of a G.652D fiber. As it is obvious, the wavelengths in the second transmission window suffer from a higher value of fiber attenuation whilst for the wavelengths in the third transmission window, the value of local dispersion is higher.

Figure 2 depicts the gain and noise figure of commercially available rare-earth doped fiber amplifiers, for 0 dBm input power, for praseodymium-DFA (PDFA) in O band, thulium-DFA (TDFA) in S band and EDFA in the C and L bands.

A key consideration when designing an optically amplified transmission system is the amplifier placement. Two configurations are shown: parallel and serial as shown in Figure 3. Both configurations have their pros and cons.

In the parallel configuration of Figure 3a, band- multiplexers/demultiplexers split the amplification bands while all amplifiers must be collocated. As two band splitters placed in tandem are needed in every amplification stage, the loss figure of C-band systems must be re-iterated and the EDFAs have also to be re-regulated to compensate the extra loss. Therefore, the optical signal to noise ratio (OSNR) performance of existing systems would be somewhat degraded and the operational complexity would increase. The advantage of this scheme is that the bands are totally decoupled and that it makes the most from the existing ducts for EDFAs: the optical amplifiers for the new transmission bands are housed in these ducts, preserving the huge investment made on the existing transportation infrastructure.

The placement of optical amplifiers serially, in a cascade as in Figure 3b, is appealing as it may minimize any disruption of C-band links. Moreover, it may also allow for shorter spans between amplification stages benefiting the bands employing lower performance amplifiers. On the other hand, the main question for this configuration is that the impact of Gain Flattening Filters (GFF) in the propagation characteristics of other bands has not yet been fully analyzed. In prior art [9]–[13],

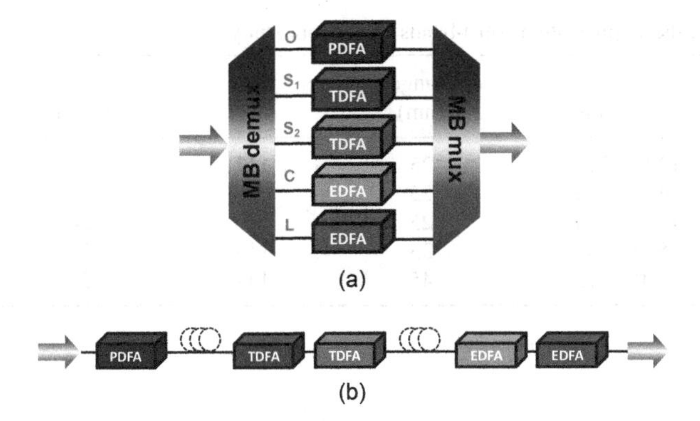

Figure 3. Amplifier connectivity schemes: a) Parallel and b) serial.

a number of serial configurations have been proposed. However, no more than two bands (C&L or S&C bands) were cascaded. Assuming no impact from GFFs, in the serial configuration the OSNR performance of C-band channels is marginally affected when additional bands are utilized. This is because no band filters are necessary to separate the bands and as such no additional gain is needed to compensate for the band filter loss. However, the configuration of Figure 3b may require new amplifier ducts at different locations from those currently employed. Finally, a drawback of the serial configuration is that the channels that correspond to the optical pumping wavelengths for the different doped-fiber amplifiers of Figure 2 must remain unused for data transportation.

In this work we adopt the parallel configuration of Figure 3a.

3 UNIFIED PHYSICAL LAYER MODELING FOR ALL BANDS

The maximum transparent length of a system that is limited by ASE accumulation and FWM crosstalk in an MB system is different between each band. This is because the strength of these degradation effects depends on the local values of fiber attenuation and dispersion as well as the noise figure of the corresponding DFA employed in each band. The impact of these effects is estimated by means of the optical signal to noise plus interference ratio (OSNIR):

$$OSNIR = \frac{P_{ch}}{P_{ASE} + P_{FWM}} \tag{1}$$

where P_{ch} is the power of the investigated channel and P_{FWM} is the power of FWM. A universal formalism for all bands is needed so that the comparative studies are made under the same set of assumptions and approximations. Regarding the calculation of P_{FWM}, we employ a generalized expression for the FWM which was derived in [14]. Accordingly, the power of FWM is given by

$$\begin{aligned}
P_{FWM} = {} & \frac{32}{27}\frac{\gamma^2 L_{eff}^2 P_{ch} N_s^2 c}{\lambda^2 B^2 D\sqrt{z_1}}\left(1+\frac{4e^{-aL}}{\left(1-e^{-aL}\right)^2}\right)\left(P_{ch}^2 asinh\left(\frac{\pi\lambda^2 DB^2}{8c}\sqrt{z_2}\right)\right. \\
& \left. + \sum_{n=-\frac{N_{ch}-1}{2},\, n\neq 0}^{\frac{N_{ch}-1}{2}} P_n^2\left(1-\frac{5}{6}\Phi_n\right)\left|Log\left(\frac{n+1/2}{n-1/2}\right)\right|\right) \\
& -\frac{32}{27}\frac{\gamma^2 L_{eff}^2 P_{ch} N_s^2 c}{\lambda^2 B^2 D\sqrt{z_1+12L^2}}\frac{4e^{-aL}}{\left(1-e^{-aL}\right)^2}\left(P_{ch}^2 asinh\left(\frac{\pi\lambda^2 DB^2}{8c}\sqrt{z_2+12L^2}\right)\right. \\
& \left. + \sum_{n=-\frac{N_{ch}-1}{2},\, n\neq 0}^{\frac{N_{ch}-1}{2}} P_n^2\left(1-\tfrac{5}{6}\Phi_n\right)\left|Log\left(\frac{n+1/2}{n-1/2}\right)\right|\right)
\end{aligned} \tag{2}$$

Table 1. Details of the amplification (sub)-bands used in our study.

	Used Range (nm)	Range (nm)	N_{ch}	$P_{out,max}$ (dBm)	$P_{ch,max}$ (dBm)
O band	1290–1315	25	117	20	-0.68
S_1 band	1455–1480	25	92	20	0.36
S_2 band	1485–1510	25	89	20	0.51
C band	1530–1565	35	116	22	1.36
L band	1570–1615	45	141	22	0.51

where $z_1 = \left(\frac{2}{a}\right)^2 + 2L^2\left(N_s^2 - 1\right) / \left(\sum_{k=x_1}^{x_2} \frac{1}{1+(2k\pi/(aL))^2}\right)^2$, $z_2 = \left(\frac{2}{a}\right)^2 + 2L^2\left(N_s^2 - 1\right)$, $x_1 = -\frac{\lambda^2 B^2 DLN_{ch}^2}{16c}$, $x_2 = \frac{\lambda^2 B^2 DLN_{ch}^2}{2c}$, with x_1 and x_2 rounded to the nearest integer less than or equal to their values. The index $n$is the channel index of all active channels within the band and the corresponding values are ranging as$-\left(N_{ch} - 1\right)/2 \leq n \leq \left(N_{ch} - 1\right)/2$. P_n denotes the power of the nth interfering channel, B is the optical bandwidth of the channel, γ is the nonlinear fiber coefficient, L is the span length and N_s is the number of fiber spans. Φ_n depends on the modulation format of the nth channel taking the value of 1 for BPSK and QPSK, 17/25 for 16-QAM and 13/21 for 64-QAM. A limitation of (2) is that it cannot be applied to a range of $\sim$ 2 nm around the zero dispersion wavelength since as D in (2) approaches zero, (2) goes to infinity.

Finally, the ASE power for a PM system is given by $P_{ASE} = N_s hf\,(NF \cdot G - 1)\,B_0$ with G and NF being the amplifier gain and noise figure respectively, equally applicable to all bands.

4 RESULTS AND DISCUSSION

4.1 *System parameters*

Table I shows the (sub)-band partitioning used in our study. The S-band is split into two sub-bands since the maximum output power of a single TDFA cannot ensure sufficient power per channel at its output as this would be limited to < -2 dBm per channel. The maximum attainable power per channel for each band is shown in the last column of Table 1.

We assume a polarization multiplexed- quadrature phase shift keying (PM-QPSK) system with channel spacing of 37.5 GHz and transmitter symbol-rate equal to 32 GBaud. The distance between amplification stages is L=40 km, and the gain of each amplifier compensates both the fiber and band filter losses (with the later taken to total 2 dB).

4.2 *Benchmarking against numerical solutions*

The accuracy of the analytical OSNIR calculation is benchmarked against numerical results where the signal transmission in the fiber was estimated using the Split Step Fourier Method (SSFM). SSFM solves the nonlinear Schrödinger equation (NLSE) numerically with very good accuracy [1]. For this purpose, a system with 9 PM-QPSK channels is considered for the bands of Table I. The power of FWM crosstalk was estimated at the wavelengths of 1302.5 nm, 1467.5 nm and 1547.5 nm for the O, S_1 and C bands, respectively. Figure 4 is illustrating the OSNIR vs $P_{ch}$for these channels. The considerable OSNIR performance degradation observed for the O-band is mainly due to a) the higher fiber loss, leading to a faster ASE noise accumulation, and b) the lower value of local dispersion that leads to higher FWM crosstalk.

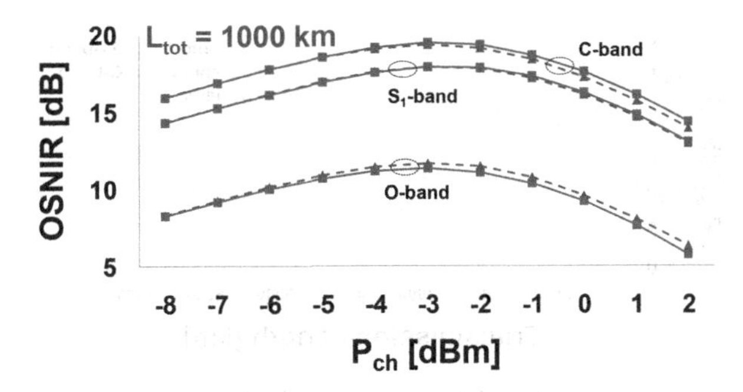

Figure 4. Comparison of the analytical OSNIR calculation (red, dashed lines) against the numerical results (blue, solid lines) for the O, S_1 and C bands.

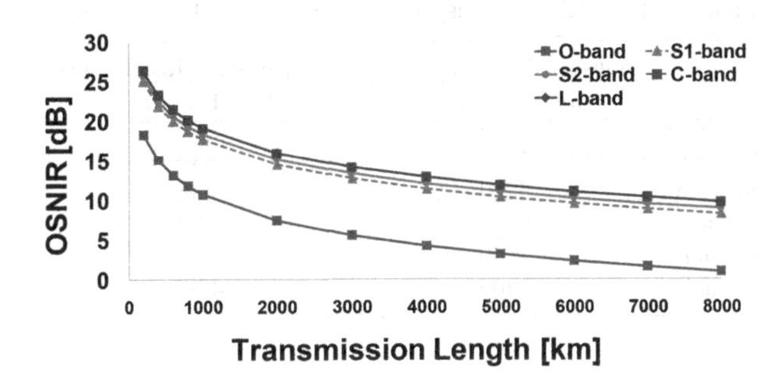

Figure 5. OSNIR evolution for different amplification bands over distance. The combined impact of FWM crosstalk and ASE noise is considered.

4.3 *The combined impact of ASE noise and FWM crosstalk*

The ASE noise and FWM crosstalk are two antagonistic effects that need to be balanced since from (1) as P_{ch} increases the P_{ch}/P_{ASE} ratio increases whilst the P_{ch}/P_{FWM} ratio decreases since $P_{FWM} \propto P_{ch}^3$. The optimum power P_{opt},as calculated in [15], is the one that counterbalances these two constraints. In this way, the OSNIR is maximized and the resulting Bit-Error-Rate (BER) is minimized. RegardingP_{opt}, for $P_{ch}<P_{opt}$ the link is OSNR limited while for $P_{ch}>P_{opt}$ it is nonlinearity limited, as shown in Figure 4. In our approach, we first calculate P_{opt} for a channel in each sub-band and then use this value to estimate the attainable transparent length of this band. This is illustrated in Figure 5 whilst the evolution of optimum power per channel for different transparent lengths is shown in Figure 6. P_{opt} differs between bands, as expected, since both the rate of ASE noise accumulation and the strength of FWM power is different in each band. Another observation that can be drawn from Figure 6 is that P_{opt} decreases with distance. Although the ASE noise generated in different spans adds up linearly, that is $P_{ASE} \propto N_s$, the FWM frequencies generated between different spans add up non-linearly (coherently) following the $P_{FWM} \propto N_s^{1+\varepsilon}$ scaling law [16], [17]. Consequently, FWM crosstalk accumulates faster than ASE noise over distance and thus the P_{opt} has to be re-iterated each time the transmission length differs in order to ensure optimum performance.

As deduced from Figure 5, the performance of the channels in the C and L bands is similar whilst those in the S_1 and S_2 sub-bands demonstrate ~1.4 and ~0.8 dB lower OSNIR, respectively. The two reasons for this are: a) $P_{ASE}$is higher in the S band due to the higher fiber losses ($a \approx 0.25$ dB/km in the S_1 band) and the subsequent higher amplifier gain, resulting to faster noise accumulation; b) $P_{FWM}$is also higher due to the relatively lower fiber dispersion in these two sub-bands. A low local

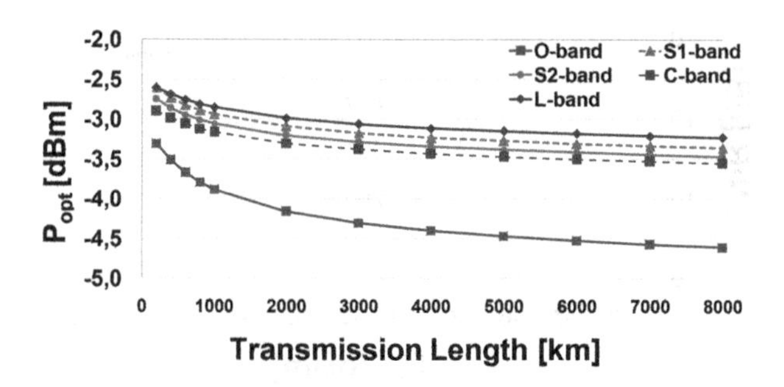

Figure 6. Optimum power per channel for different amplification bands and transmission distance.

Table 2. The attainable transparent length for two BER thresholds.

	Transmission Reach (km) for BER<10^{-3}	Transmission Reach (km) for BER<10^{-5}
O band	1160	640
S_1 band	5400	3000
S_2 band	6200	3400
C band	7400	4200
L band	7400	4200

dispersion value increases the strength of the FWM interference generated within a single span while the contributions between different spans do add-up coherently [16], [18], [19]. Specifically, when the dispersion parameter D approaches zero (around the nominal 1310 nm, see Figure 1) there is rapid signal deterioration due to perfect phase matching conditions that allow FWM to scale as N_s^2 [20].

As in [15], for selected values of the BER, dictated by the performance of the Forward-Error-Correction (FEC) employed, the corresponding OSNIR values are derived and from them the maximum transparent length is deduced. For example, with BER values of 10^{-3} and 10^{-5}, the transparent length is as listed in Table 2: for S, C and L bands it is more than 5000 km for a system with an amplifier spacing of 40 km. Similarly, most channels in O-band may reach over 1100 km in transmission distance. It is worth mentioning that a pre-FEC BER of 10^{-3} can be improved to a post-FEC value less than 10^{-12} depending on the FEC overhead and algorithmic complexity [21].

5 CONCLUSION

We have explored the impact of ASE accumulation and FWM in an MB system employing commercially available optical amplifiers. The assessment was made with the aid of a universal analytical FWM formula applicable to all bands that allows to deduce comparative conclusions. The obtained formalism allows to estimate the impact of FWM over a diverse set of operational conditions. Based on the performance of commercially available doped fiber amplifiers, the second and third transmission windows were segmented in sub-bands. The channels in the O-band are the worst affected ones with respect to both FWM and ASE noise accumulation. C and L bands are the premium bands while the OSNIR performance of two S sub-bands may differ by more than 1 dB among them, due to their dissimilar dispersion and fiber loss characteristics.

ACKNOWLEDGMENTS

This work was partially funded by the 5G-PPP METRO-HAUL project, EU grant agreement number: 761727. The authors also acknowledge financial support for the dissemination of this work from the Special Account of Research of ASPETE through the funding program "Strengthening Research of ASPETE Faculty Members".

REFERENCES

[1] P. Bayvel *et al.*, "Maximizing the optical network capacity," *Philos. Trans. R. Soc. A Math. Phys. Eng. Sci.*, vol. 374, no. 2062, Mar. 2016.

[2] A. Stavdas, "Architectural Solutions Towards a 1,000 Channel Ultra - Wideband WDM Network," vol. 2, no. 1, pp. 51–60, 2001.

[3] S. Okamoto *et al.*, "5-band (O, E, S, C, and L) WDM transmission with wavelength adaptive modulation format allocation," in *Europ. Conf. on Opt. Commun. (ECOC)*, 2016.

[4] J. Renaudier *et al.*, "First 100-nm Continuous-Band WDM Transmission System with 115Tb/s Transport over 100km Using Novel Ultra-Wideband Semiconductor Optical Amplifiers," in *Europ. Conf. on Opt. Com. (ECOC)*, 2017.

[5] G. Saavedra *et al.*, "Experimental Analysis of Nonlinear Impairments in Fibre Optic Transmission Systems up to 7.3 THz," *J. Light. Technol.*, vol.35, no. 21, pp. 4809–4816, 2017.

[6] A. Ghazisaeidi *et al.*, "Advanced C+L-Band Transoceanic Transmission Systems Based on Probabilistically Shaped PDM-64QAM," *J. Light. Technol.*, vol. 35, no. 7, pp. 1291–1299, Apr. 2017.

[7] J.-X. Cai *et al.*, "70.46 Tb/s Over 7,600 km and 71.65 Tb/s Over 6,970 km Transmission in C+L Band Using Coded Modulation With Hybrid Constellation Shaping and Nonlinearity Compensation," *J. Light. Technol.*, vol. 36, no. 1, pp. 114–121, Jan. 2018.

[8] J. W. Dawson *et al.*, "E-band Nd^3+ amplifier based on wavelength selection in an all-solid micro-structured fiber," *Opt. Express*, vol. 25, no. 6, p. 6524, Mar. 2017.

[9] M. Karasek & M. Menif, "Serial topology of wide-band erbium-doped fiber amplifier for WDM applications," *IEEE Photonics Technol. Lett.*, vol. 13, no. 9, pp. 939–941, Sep. 2001.

[10] Q. Jiang *et al.*, "Dynamical gain control in the serial structure C+L wide-band EDFA," *IEEE Photonics Technol. Lett.*, vol. 16, no. 1, pp. 87–89, Jan. 2004.

[11] B. A. Hamida *et al.*, "Wideband and flat-gain amplifier using high concentration Erbium doped fibers in series double-pass configuration," in *Int. Conf. on Comp. and Commun. Eng. (ICCCE)*, 2012.

[12] T. Sakamoto *et al.*, "Hybrid fiber amplifiers consisting of cascaded TDFA and EDFA for WDM signals," *J. Light. Technol.*, vol. 24, no. 6, pp. 2287–2295, Jun. 2006.

[13] T. Sakamoto *et al.*, "Rare-earth-doped fiber amplifier for eight-channel CWDM transmission systems," in *Opt. Fib. Com. Conf.*, 2004.

[14] D. Uzunidis *et al.*, "Closed-form FWM expressions accounting for the impact of modulation format," *Opt. Commun.*, vol. 440, 2019.

[15] D. Uzunidis *et al.*, "Optimizing power, capacity, transmission reach and amplifier placement in coherent optical systems using a physical layer model," in *Pan-Hellenic Conf. on Inf.*, 2016.

[16] D. Uzunidis *et al.*, "An improved model for estimating the impact of FWM in coherent optical systems," *Opt. Commun.*, vol. 378, pp. 22–27, Nov. 2016.

[17] A. Carena *et al.*, "EGN model of non-linear fiber propagation," *Opt. Express*, vol. 22, no. 13, pp. 16335–62, Jun. 2014.

[18] D. Uzunidis *et al.*, "Simplified FWM model for flexible-grid optical systems," in *24th Telecom. Forum (TELFOR)*, 2016.

[19] D. Uzunidis *et al.*, "Analytical FWM Expressions for Coherent Optical Transmission Systems," *J. Light. Technol.*, vol. 35, no. 13, pp. 2734–2740, 2017.

[20] D. Uzunidis *et al.*, "Simplified model for nonlinear noise calculation in coherent optical OFDM systems," *Opt. Express*, vol. 22, no. 23, 2014.

[21] F. Chang *et al.*, "Forward error correction for 100 G transport networks," *IEEE Commun. Mag.*, vol. 48, no. 3, pp. S48–S55, Mar. 2010.

Proceedings of the 1st International Congress on Engineering Technologies – Kiwan & Banat (Eds)
 ISBN 978-0-367-77630-5

Complex magnetic and electric dipolar resonances of subwavelength GaAs prolate spheroids

Georgios D. Kolezas and Grigorios P. Zouros
School of Electrical and Computer Engineering, National Technical University of Athens (NTUA), Athens, Greece

Gerasimos K. Pagiatakis
Department of Electrical and Electronic Engineering Education, School of Pedagogical and Technological Education (ASPETE), Athens, Greece

ABSTRACT: In this work we investigate the complex magnetic and electric dipolar resonances of a subwavelength gallium arsenide (GaAs) prolate spheroid. We employ a formulation based on spheroidal vector wave functions (VWFs) for the expansion of the electromagnetic fields in the interior and exterior of the spheroid. Satisfying the boundary conditions at the spheroidal surface, we finally obtain a linear system of equations. The resonance wavelengths are obtained by setting the system's determinant equal to zero and utilizing an efficient root-finding algorithm. Comparisons with HFSS commercial software are performed, and numerical results are given for different aspect ratio nanospheroids.

1 INTRODUCTION

Subwavelength dielectric particles with high refractive index can support strong electric and magnetic resonances in the optical region. This feature, along with their very low losses at optical frequencies, has made high-permittivity dielectric nanoparticles very important for a number of advanced photonic applications, including nanoantennas, metasurfaces, solar cells, biosensors and novel nanophotonic devices (1–3).

Among other particle geometries, the spheroidal shape has received increased attention. For instance, nanospheres may be deformed to nanospheroids during fabrication, leading to shifts in the spectral positions of the resonances (4). Furthermore, varying the aspect ratio of high-index nanospheroids allows for the tuning of their optical response. In this context, optimized forward light scattering with minimal backscattering can be achieved (5), as well as observation of directional Fano resonances in individual nanospheroids (6). Thus, subwavelength spheroidal particles are promising candidates for the implementation of functional photonic nanoresonators.

The standard method for obtaining the resonance wavelengths of dielectric nanospheroids is through calculation or measurement of their scattering spectra (5, 6). However, in the present study we propose another approach that is based on a root-finding algorithm and allows for more accurate calculation of the electric and magnetic dipolar eigenmodes. In particular, we consider a prolate nanospheroid composed of GaAs, which is a high-permittivity material, and investigate its complex resonances. The geometry of the configuration is shown in Figure 1. The spheroid has a major axis of length $2a$, a minor axis of length $2b$, an interfocal distance $2d$, and a boundary surface S. The resonator occupies region I which is characterized by a permittivity $\epsilon_1 = \epsilon_{1r}\epsilon_0$, with ϵ_{1r} the relative value and ϵ_0 the free space permittivity, while its permeability is equal to the permeability of free space μ_0. Region II surrounding the spheroid is free space.

 DOI 10.1201/9781003178255-5

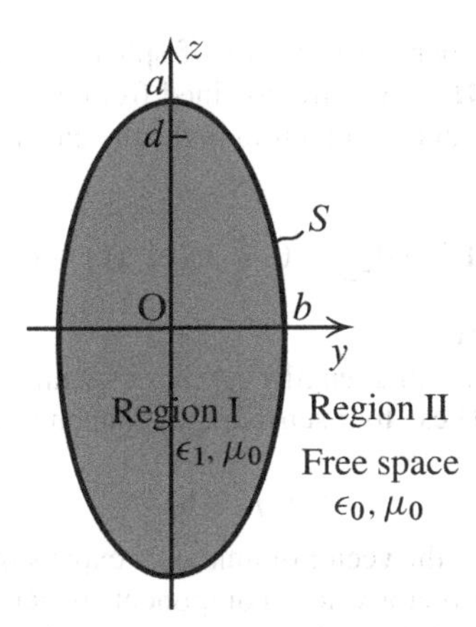

Figure 1. Geometry of the configuration.

In order to solve the problem, we employ the scattering formulation where the incident, scattered and interior fields are expanded in terms of spheroidal VWFs. Next, the boundary conditions are imposed at the spheroid's surface and a system of linear equations is obtained. The resonant wavenumbers can then be computed by setting the system's determinant equal to zero.

2 SOLUTION OF THE PROBLEM

We assume that a TE plane electromagnetic wave impinges on the spheroid. The incident electric field $\mathbf{E}^{\text{inc}}$ is polarized along the positive y-axis, the plane of incidence is xz-plane, and the direction of propagation forms an angle θ_0 with the positive z-axis. Then, the incident field has the form (7)

$$\mathbf{E}^{\text{inc}}(\mathbf{r}) = \sum_{m=0}^{\infty}\sum_{n=m}^{\infty}\left[c_{mn}(c_0,\theta_0)\mathbf{Me}_{mn}^{r(1)}(c_0,\mathbf{r}) + d_{mn}(c_0,\theta_0)\mathbf{No}_{mn}^{r(1)}(c_0,\mathbf{r})\right]. \tag{1}$$

In (1), $\mathbf{r} = (\xi, \eta, \varphi)$ is the position vector in the spheroidal coordinates, $\mathbf{Me}_{mn}^{r(1)}$ and $\mathbf{No}_{mn}^{r(1)}$ are the spheroidal eigenvectors of the first kind (7), $c_0 = k_0 d$ with k_0 the free space wavenumber, and c_{mn}, d_{mn} are the known expansion coefficients of the incident plane wave (7). The time dependence $e^{-i\omega t}$ is assumed and suppressed throughout.

The scattered field in region II is expressed in terms of spheroidal VWFs as (8)

$$\mathbf{E}^{\text{sc}}(\mathbf{r}) = \sum_{m=0}^{\infty}\sum_{n=m}^{\infty}\left[C_{mn}\mathbf{Me}_{mn}^{r(3)}(c_0,\mathbf{r}) + D_{mn}\mathbf{No}_{mn}^{r(3)}(c_0,\mathbf{r})\right], \tag{2}$$

where $\mathbf{Me}_{mn}^{r(3)}$ and $\mathbf{No}_{mn}^{r(3)}$ are the spheroidal eigenvectors of the third kind (7), and C_{mn}, D_{mn} are unknown expansion coefficients.

The field in the spheroid's interior is also written as an expansion of spheroidal VWFs. Thus, in region I we have (8)

$$\mathbf{E}^{\text{int}}(\mathbf{r}) = \sum_{m=0}^{\infty}\sum_{n=m}^{\infty}\left[A_{mn}\mathbf{Me}_{mn}^{r(1)}(c_1,\mathbf{r}) + B_{mn}\mathbf{No}_{mn}^{r(1)}(c_1,\mathbf{r})\right], \tag{3}$$

where $c_1 = k_1 d, k_1 = k_0\sqrt{\epsilon_{1r}}$, and A_{mn}, B_{mn} are unknown expansion coefficients.

After the electric fields are expanded in terms of spheroidal VWFs as described above, the respective magnetic fields $\mathbf{H}^{\text{inc}}$, $\mathbf{H}^{\text{int}}$, $\mathbf{H}^{\text{sc}}$, are obtained from Faraday's law $\mathbf{H} = -i/(\omega\mu_0)\nabla \times \mathbf{E}$. We then proceed to satisfy the boundary conditions for the electric and the magnetic field at the spheroid's surface S, i.e.,

$$\hat{n} \times \left[\mathbf{E}^{\text{sc}}(\mathbf{r}) + \mathbf{E}^{\text{inc}}(\mathbf{r}) - \mathbf{E}^{\text{int}}(\mathbf{r})\right]_{\mathbf{r}\in S} = 0,\ \hat{n} \times \left[\mathbf{H}^{\text{sc}}(\mathbf{r}) + \mathbf{H}^{\text{inc}}(\mathbf{r}) - \mathbf{H}^{\text{int}}(\mathbf{r})\right]_{\mathbf{r}\in S} = 0, \quad (4)$$

where $\hat{n}$ is the unit normal vector on S.

By imposing (4) and after performing lengthy algebraic manipulations, we finally arrive at four infinite sets of linear equations. These four sets comprise, upon truncation, a linear system having the form

$$\mathbb{A}(x_0)\mathbf{v} = \mathbf{b}. \quad (5)$$

In (5), $\mathbf{v} = [A_{mn}, B_{mn}, C_{mn}, D_{mn}]^T$ is the vector of unknown expansion coefficients (with T denoting transposition), $\mathbf{b}$ is the excitation vector whose components depend on the expansion coefficients of the incident wave c_{mn}, d_{mn}, and $\mathbb{A}(x_0)$ is the system matrix whose elements depend on the normalized wavenumber $x_0 = k_0 a$.

The solution to the resonance problem is obtained by considering zero excitation, i.e., vector $\mathbf{b}$ is set equal to zero. Then, (5) is transformed into the homogeneous system

$$\mathbb{A}(x_0)\mathbf{v} = 0. \quad (6)$$

In order for system (6) to have a non-trivial solution, its determinant must be equal to zero. The values $x_0 := u - iv \in \mathbb{C}$, $u \in (0, \infty)$, $v \in [0, \infty)$, for which $\det \mathbb{A}(x_0) = 0$, are the complex normalized resonant wavenumbers. The complex x_0 are computed by means of an efficient root-finding algorithm (9). Once x_0 are obtained, the respective complex resonance frequencies are given by $f = x_0/(2\pi a\sqrt{\mu_0\epsilon_0})$ and the resonance wavelengths by $\lambda = 2\pi a/x_0$.

3 NUMERICAL RESULTS

Herein, we calculate the magnetic and electric dipolar resonances for prolate subwavelength spheroids of various aspect ratios a/b. The spheroids are composed of GaAs whose dispersive properties are reported in (10). GaAs exhibits high permittivity values in the optical range, especially for wavelengths above 440 nm, thus making it a promising material for the realization of nanoantennas. The first two dipolar magnetic and electric resonances that appear in subwavelength structures are of interest since, their combination may act as a Huygens source and result in the synthesization of a nanoantenna with specified radiation characteristics. In particular, the knowledge of the location of the first two dipolar resonances is of importance since their simultaneous excitation leads to forward or backward unidirectional scattering, with the former being a desirable property in nanophotonics.

Table 1 tabulates the resonant wavelengths in nm, for GaAs spheroids having three different aspect ratio values. The major semi-axis of all spheroids is kept at 100 nm, thus, by increasing each time the aspect ratio, the semi-minor axis becomes smaller and the particle more elongated. The accuracy of the extracted wavelengths is given in five significant figures. This is achieved by employing the CCOMP algorithm (9). Although this procedure is time consuming due to the need of composition of fully populated matrices via the spheroidal formulation (see Section II), the search of the zero eigenvalues through (6) is still faster than the detection of the resonances through computation of the scattering spectra, because the latter will not yield a desirable accuracy unless a fine sweep versus frequency is performed, resulting in a huge CPU time. To validate the results obtained by CCOMP, we employ HFSS' eigenmode solver. As it can be seen from Table 1, the agreement is very good in both real and imaginary parts, up to four significant digits. It should be emphasized that although HFSS has been initialized with a coarse discretization, still it requires more than 20 GB of RAM to compute one eigenfrequency. The accuracy of HFSS can be increased

Table 1. Resonant magnetic and electric dipolar wavelengths λ in nanometers, for prolate GaAs spheroids. Values of parameters used: $a = 100$ nm, dispersive relative permittivity $\epsilon_{1r}(\lambda)$ as in (10), and $\mu = \mu_0$.

	Magnetic dipolar	Electric dipolar
Aspect ratio: 1.6667		
This work	$579.81 + 21.023i$	$497.83 + 10.006i$
HFSS	$579.75 + 21.042i$	$497.85 + 9.9797i$
Aspect ratio: 2		
This work	$524.33 + 18.364i$	$470.20 + 6.9537i$
HFSS	$524.23 + 18.349i$	$470.45 + 6.9069i$
Aspect ratio: 2.2942		
This work	$489.58 + 16.303i$	$451.35 + 5.3040i$
HFSS	$489.45 + 16.326i$	$452.00 + 5.2350i$

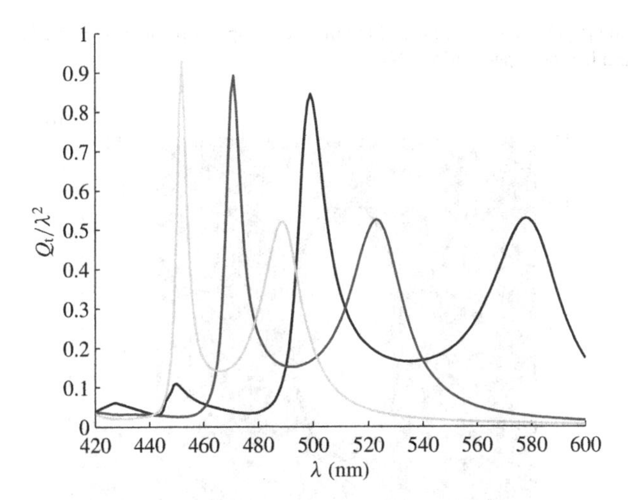

Figure 2. Normalized total scattering cross section Q_t/λ^2 versus wavelength λ for a GaAs prolate spheroid illuminated axially by a TE polarized plane wave. The semi-major axis is $a = 100$ nm. Blue: $a/b = 1.6667$; red: $a/b = 2$; green: $a/b = 2.2942$.

if a finer grid is employed, yet this scheme will require huge computer resources. To complete the results given in Table 1, we also report the normalized wavenumbers x_0, as computed with CCOMP. For $a/b = 1.6667, x_0 = 1.0822 - 0.039241i$ and $x_0 = 1.2616 - 0.025357i$ for magnetic and electric dipolar resonance, respectively. For $a/b = 2$ the respective values are $x_0 = 1.1968 - 0.041917i$ and $x_0 = 1.336 - 0.019758i$, and finally, for $a/b = 2.2942$, they are $x_0 = 1.282 - 0.042690i$ and $x_0 = 1.3919 - 0.016357i$. It is also evident from Table 1 that, as the aspect ratio is increasing, both magnetic and electric resonances experience a blue shift.

Response to plane wave stimulation of the spheroids considered in Table 1, is depicted in the scattering spectra of Figure 2. Here we plot the normalized total scattering cross section Q_t/λ^2 (8) versus the incident wavelength. The resonant wavelengths reported in Table 1 are now indicated by the peaks in the total scattering cross section. In terms of frequency, the magnetic dipolar resonance frequency for, e.g., the case of $a/b = 2.2942$, is $\mathrm{Re}(f) = 611.67$ THz and the linewidth, given by $-2\mathrm{Im}(f)$, is 40.738 THz. Obviously, this linewidth is larger than the 15.609 THz linewidth of the respective electric dipolar mode (again for the $a/b = 2.2942$ case), as can be also deduced from Figure 2 by observing the full-width at half-maximum values of the green curve.

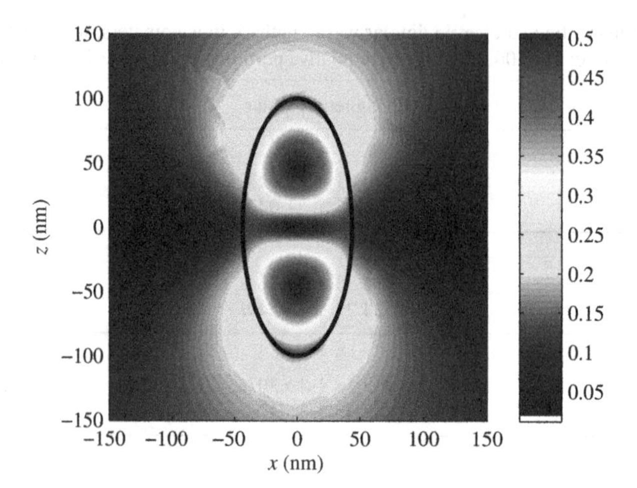

Figure 3. Eigenmode |**E**| (V/m) on xz-plane of the magnetic dipolar resonance excited at 489.58 nm, for the prolate GaAs spheroid having aspect ratio 2.2942.

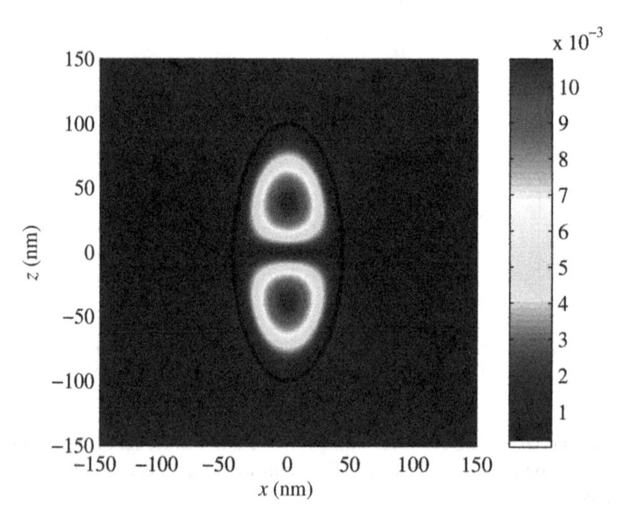

Figure 4. Eigenmode |**H**| (A/m) on xz-plane of the electric dipolar resonance excited at 451.35 nm, for the prolate GaAs spheroid having aspect ratio 2.2942.

In Figures 3 and 4 we plot two eigenmodes for the $a/b = 2.2942$ case. In particular, Figures 3 depicts the magnitude of the **E** field on xz-plane for the magnetic dipolar resonance, excited exactly at 489.58 nm. Figures 4 shows the magnitude of the **H** field, again on xz-plane, for the electric dipolar resonance at 451.35 nm. The dipolar pattern of the field distributions is evident. Moreover, the fields occupy almost the entire space within the resonator since the quality factor is not high. For instance, in this particular example, the quality factor is $\mathrm{Re}(\lambda)/[2\mathrm{Im}(\lambda)] = 15.01$ for the magnetic dipolar resonance and 42.55 for the electric dipolar resonance.

4 CONCLUSION

The magnetic and electric dipolar resonances of a subwavelength GaAS prolate spheroid are studied. The problem is formulated in terms of spheroidal VWFs and a root-finding algorithm is employed for the calculation of the complex resonance wavelengths. Numerical results obtained via our method are in good agreement with HFSS. Our approach provides an efficient alternative to the scattering problem solution, for revealing the resonance wavelengths with increased accuracy.

ACKNOWLEDGMENT

The authors acknowledge financial support for the dissemination of this work from the Special Account of Research of ASPETE through the funding program "Strengthening Research of ASPETE Faculty Members".

REFERENCES

[1] M. Decker and I. Staude, "Resonant dielectric nanostructures: a low-loss platform for functional nanophotonics," *J. Opt.*, vol. 18, p. 103001, 2016.
[2] A. I. Kuznetsov, A. E. Miroshnichenko, M. L. Brongersma, Y. S. Kivshar, and B. Luk'yanchuk, "Optically resonant dielectric nanostructures," *Science*, vol. 354, p. aag2472, 2016.
[3] Z.-J. Yang, R. Jiang, X. Zhuo, Y.-M. Xie, J. Wang, and H.-Q. Lin, "Dielectric nanoresonators for light manipulation," *Phys. Rep.*, vol. 701, pp. 1–50, 2017.
[4] Y. H. Fu, A. I. Kuznetsov, A. E. Miroshnichenko, Y. F. Yu, and B. Luk'yanchuk, "Directional visible light scattering by silicon nanoparticles," *Nature Commun.*, vol. 4, p. 2538, 2013.
[5] B. S. Luk'yanchuk, N. V. Voshchinnikov, R. Paniagua-Domínguez, and A. I. Kuznetsov, "Optimum forward light scattering by spherical and spheroidal dielectric nanoparticles with high refractive index," *ACS Photonics*, vol. 2, pp. 993–999, 2015.
[6] C. Ma, J. Yan, Y. Huang, and G. Yang, "Directional Fano resonance in an individual GaAs nanospheroid," *Small*, vol. 15, p. 1900546, 2019.
[7] C. Flammer, *Spheroidal wave functions*. Stanford, CA: Stanford University Press, 1957.
[8] G. P. Zouros, A. D. Kotsis, and J. A. Roumeliotis, "Efficient calculation of the electromagnetic scattering by lossless or lossy, prolate or oblate dielectric spheroids," *IEEE Trans. Microw. Theory Techn.*, vol. 63, pp. 864–876, 2015.
[9] G. P. Zouros, "CCOMP: An efficient algorithm for complex roots computation of determinantal equations," *Comput. Phys. Comm.*, vol. 222, pp. 339–350, 2018.
[10] D. E. Aspnes, S. M. Kelso, R. A. Logan, and R. Bhat, "Optical properties of $Al_xGa_{1-x}As$," *J. Appl. Phys.*, vol. 60, pp. 754–767, 1986.

Proceedings of the 1st International Congress on Engineering Technologies – Kiwan & Banat (Eds)

The partial disclosure of the gpsOne file format for assisted GPS service

Vladimir Vinnikov
Departmentt of Computer Sciences, Higher School of Economics, Moscow, Russian Federation

Ekaterina Pshehotskaya
Department of Information Security, Moscow Polytechnic University, Moscow, Russian Federation

ABSTRACT: This paper is concerned with decoding the data format of a gpsOneXTRA binary file for A-GPS web-service. We consider the mandatory data content of the file and reveal the changes of this content at different moments in time. The frequency of the changes hints at the location of records for current GPS date and satellite orbit information. Comparing the repeating data patterns against reference orbit information, we obtain the meaning of data fields of the orbit record for each operational satellite. The deciphered file header and GPS almanac data layout are provided as tables within the paper. The partiality of the disclosure due to ephemeris complexity is discussed in the respective section.

Keywords: assisted GPS, almanac, gpsOneXTRA service, decoding, data structure, binary format

1 INTRODUCTION

Recent advances in microelectronics and space technologies has enabled geo-positioning functionality to be implemented in various devices like smartphones, digital cameras and fitness trackers. The geo-positioning relies on global navigation satellite systems (GNSS) used to determine range (as time delay of travelling signal) from spaceborne emitting antenna to receiving antenna.

The performance of GNSS-positioning depends on the number and mutual disposition of satellites on the sky hemisphere. However, the coordinates of the satellites are a-priori unknown. By default, the receiver has to lock-on to satellite signal and download navigational messages, containing data on satellite orbits. The orbits are represented as rough almanacs and more accurate ephemeris parts. However, the locking-on and downloading takes significant time (up to 12.5 minutes for GPS in the worstcase scenario) due to low the throughput of the satellite signal broadcast with a data rate of only 50 bit/s. This circumstance impacts usability of the freshly started device. To overcome this inconvenience, it was proposed to deliver the almanac and the ephemerides onto receiver software via faster technical means. Since most of the consumer devices are within the reach of the internet, the downloading of respective data files from web-services was considered the best solution.

The above approach to improving performance of GNSS-positioning is known as "Assisted GPS" or A-GPS (e.g. see [1]). The A-GPS technology is implemented as a web-service, the frontend of which consists of a binary datafile and respective URL to retrieve it from a server.

Almost all smartphone manufacturers provide their own A-GPS services. The most widely used services with A-GPS content are provided by Google (for the GlobalLocate chipset), Qualcomm (for the gpsOne chipset), and Mediatek (for the SiRFStarIII chipset). The respective URLs are given in table 1. In smartphones with Android OS, these URLs are placed into ini-file as values for the keys "XTRA SERVER 1", "XTRA SERVER 2", "XTRA SERVER 3",

DOI 10.1201/9781003178255-6

"XTRA SERVER S" The example is provided in the string key-value expression (1).

$$\text{XTRA SERVER 1} = \text{http : //xtra1gpsonextranet/xtrabin} \tag{1}$$

However, to this day there is no standard format for A-GPS files. Moreover, binary files have proprietary formats with a-priori unknown mapping to data structures.

The exposure of A-GPS file data layouts is a significant factor for improving information security and reducing risks of various exploits designed to compromise end-point user devices. For example, the gpsOne service was considered vulnerable in at least two instances (see [2,3]). One instance was concerned with an MitM-attack through unsecured HTTP able to substitute the correct binary file with a fake. The other instance was upon ingestion of a fake binary file of large size Android OS would crash. These vulnerabilities allowed cumulative exploits undetectable by any antiviral scans due to the unknown structure of the binary files.

To resolve the MitM-issue the provider implemented a secured HTTPS access and introduced digital signatures for A-GPS files. However, the signature only indicates that the initial content is unchanged, since the integrity and validity of the underlying data can be only assessed through parsing, Moreover, the nature of A-GPS service with regularly provided files permits not only deciphering/decoding of the stored orbits data, but also deciphering the signature algorithm as well. Once the signature algorithm is revealed, one can alter or regenerate the content of the A-GPS file and resign it. The obvious solution to encrypt the A-GPS file completely seems feasible only at first glance. An encryption would require key-handling procedures within the decryption parts of the client-side decoding program installed on every chip of the respective A-GPS provider. Since the A-GPS file comes in essentially one instance for all respective devices (e.g. smartphones), there can only be a single easily extracted key to decrypt the contents, which profanes the whole idea.

Thus, the knowledge of the file structure permits one to safely parse the data fields and check for any inconsistencies thus facilitating protection against potential exploits.

Usually, there are various approaches to determine the layout of the A-GPS data format Namely data analysis, software analysis and the reverse engineering of the decoding software. The complexity of both techniques depends on many factors such as the available software and hardware resources as well as the level of complement for technical documentation. According to various open source gitrepositories with Android utilities for GNSS navigation, the applications only retrieve the A-GPS file from the respective URL and proceed with the injection of the file content into the proprietary provider library. So, the software analysis yields no relevant information on the layout of the considered binary file format. The library itself usually acts as an interface to the chip firmware. This circumstance significantly complicates the latter approach, since it requires specialized software and hardware tools to obtain and reverse engineer the decoding firmware from the chip for the following analysis. As officially stated, the details on the file format and how the digital signature is verified are only available to OEMs directly from Qualcomm. Therefore, only the former approach remains. Evidently, the data analysis does not require intrusion into proprietary Android applications or tampering with chip firmware. The requirements are limited only to large bulks of A-GPS binary files in public domain that are easy to obtain,

Our aim is to outline a decoding technique for A-GPS files in a tutorial manner using classical cryptography attacks based on data redundancy and repetition. This technique is applicable not only for the Qualcomm gpsOneXtra format but for all A-GPS formats of all other providers.

2 RELATED WORKS

To the best of our knowledge, there are no publications concerned with the describing of A-GPS data formats. An exhaustive bibliographical search yielded no relevant results. Most publications consider general uses of assisted GPS technologies and especially its extended ephemeris (ee) part (see e.g. [4]). Such scarcity of information can be explained through practical limitations, since the generation of a prognostic extended ephemeris is an expensive computational task. The extended

Table 1. Providers of free A-GPS services.

Provider	Ver	URL
Google	–	http://gllto.glpals.com/4day/glo/v2/latest/lto2.dat
	1	http://xtra1.gpsonextra.net/xtra.bin
		http://xtra2.gpsonextra.net/xtra.bin
		http://xtra3.gpsonextra.net/xtra.bin
		https://xtrapath1.izatcloud.net/xtra.bin
		https://xtrapath2.izatcloud.net/xtra.bin
		https://xtrapath3.izatcloud.net/xtra.bin
	1	SSL https://ssl.gpsonextra.net/xtra.bin
Qualcomm		http://xtrapath1.izatcloud.net/xtra2.bin
	2	http://xtrapath2.izatcloud.net/xtra2.bin
		http://xtrapath3.izatcloud.net/xtra2.bin
	2	SSL https://ssl.gpsonextra.net/xtra2.bin
		https://xtrapath1.izatcloud.net/xtra3grc.bin
	3	https://xtrapath2.izatcloud.net/xtra3grc.bin
		https://xtrapath3.izatcloud.net/xtra3grc.bin
	3	SSL https://ssl.gpsonextra.net/xtra3grc.bin
Mediatek	-	http://nsdu.atwebpages.com/packedephemeris.ee
	=	http://epodownload.mediatek.com/EPO.DAT
Sony	–	http://control.d-imaging.sony.co.jp/GPS/assistme.dat
Nikon	–	https://downloadcenter.nikonimglib.com/en/download/fw/111.html
Olympus	–	http://sdl.olympus-imaging.com/agps/index.en.html

ephemeris, contained within every A-GPS file, is a valuable asset, used in various commercial sectors, in particular in the IoT sector. A large portion of the IoT sector is critically dependent on cold start GNSS acquisition and positioning time intervals. Thus, the integrity and validity of relied upon A-GPS services is of paramount importance. For example, the recent GPS-week rollover issue caused A-GPS service inconsistency leading to severe IoT-problems and required firmware updates to more than 100 000 devices (see [5]). Given the nature of the GPS-week rollover (GPS-week number presentation as in modulo 1024)we regard this as a minor issue for modern devices, since it can be fixed while knowing the current date.

The lack of similar publications makes us believe that the present paper has a high level of originality. We are unable to point out other independent works on this topic.

3 GENERAL CONSIDERATIONS ON THE DATAFILE CONTENT

The mandatory contents of A-GPS files include the almanac and ephemerides of the considered GNSS for the actual timeframe. To this day there are four global satellite systems, namely: GPS, GLONASS (GLN), BEIDOU (BDS), GALILEO (GAL). The almanac represents the long-term valid orbit approximations for the respective GNSS. The ephemeris, as term suggests, is short-lived data. For the GPS it expires about every two hours and is updated about every hour. Other systems have similar expiration times for corresponding ephemerides. Due to short expiration times, the A-GPS file also contains various timestamps denoting data validity as well as ephemerides prediction for 7–28 days in order to reduce device dependence on web-connectivity.

The initial GNSS-broadcasted navigation messages have a rigid data format, so it is safe to assume that an A-GPS file is also similarly structured into fixed data-fields. Such kinds of data structures manifest themselves in binary files as periodic patterns with easily deducible sizes. We also assume that data is stored in efficient ways, so only numeric types are used, and the minimum required number of bytes is allocated for them.

We should emphasize the possibility of two different binary bitwise representations, namely "Little Endian" and "Big Endian" The former is usually implemented in CPUs of x86 architecture,

Table 2. Binary content of the file.

Offset	00	01	02	03	04	05	06	07	08	09	0A	0B	0C	0D	0E	0F	10	11	12	13	14	15	16	17	18	19	1A	1B	1C	1D	1E	1F
0x0000	01	1B	08	01	02	15	01	24	05	BB	13	00	00	96	DE	08	24	16	01	DF	FD	08	24	15	E3	07	00	06	1C	01	00	25
0x0020	14	07	10	0F	0E	0D	0C	0B	0A	0C	37	08	10	10	0F	0E	0D	0C	0A	09	0E	53	08	11	0F	0E	0E	0C	0C	09	08	0D
0x0040	96	02	0B	05	02	03	03	C2	1F	01	00	4B	6A	90	17	81	FD	62	00	A1	0C	CA	FF	F2	84	DF	00	1E	F2	F4	00	5B
0x0060	E1	04	FF	07	FF	FD	08	24	02	00	A0	8E	90	09	D6	FD	55	00	A1	0C	6D	FF	EF	8B	58	FF	BB	3F	64	00	67	D6
0x0080	9B	FE	7B	FF	FE	08	24	03	00	15	65	90	0E	51	FD	43	00	A1	0C	B8	00	1C	EE	72	00	1F	D7	39	00	2C	17	D1
0x00A0	FF	C4	FF	FE	08	24	04	FF	03	5D	7B	0B	15	FD	55	00	A1	0D	7A	00	48	E5	9A	FF	93	67	AF	FF	8F	0A	EC	FF
0x00C0	E8	FF	FF	08	24	05	00	2F	87	90	05	C9	FD	36	00	A1	0C	0A	00	1B	B8	75	00	20	CB	86	FF	D0	8D	52	FF	FB
0x00E0	00	00	08	24	06	00	0D	F9	90	17	40	FD	65	00	A1	0D	6B	FF	F2	2E	BB	FF	CF	03	85	00	6A	75	E5	FF	58	FF
0x0100	FD	08	24	07	00	6C	2A	90	07	B9	FD	50	00	A1	0D	52	00	72	C3	B3	FF	9D	60	35	00	3F	3A	B1	FF	49	FF	FE
0x0120	08	24	08	00	28	08	90	11	E2	FD	45	00	A1	0C	70	FF	C6	FB	81	FF	F6	DF	3E	FF	BD	BB	1A	FF	EE	00	00	08
0x0140	24	09	00	0E	03	90	06	4A	FD	4A	00	A1	0B	DF	00	46	F7	7C	00	44	A6	6A	FF	C9	94	0C	FF	89	FF	FD	08	24

while the latter is used in ARM processors of mobile devices like smartphones. Thus, it is more likely to encounter "Big Endian" encoding in A-GPS files.

Since our study is predominantly educational, for the sake of clarity we consider a Qualcomm-provided service for a "gpsOne" chipset with the first version of the binary file, containing an almanac and ephemerides only for GPS (file "xtra.bin" at URL "xtra1"). An example of the binary file header is presented in table 2.

4 GENERAL CONSIDERATIONS ON THE CRYPTOGRAPHY ATTACKS EXPLOITING DATA REDUNDANCY

It is established that the binary A-GPS files have proprietary formats but are not truly encrypted in the technical sense. However, the lack of information on the data structure of the files can be treated the same way as the encryption. This kind of encryption originates from the well-known paradigm: "security through obscurity." The paradigm itself is widely recognized as unreliable, and its implementations are considered bad practice. However, this circumstance makes the recovery of the data structure to be easier than in the case of proper classical encryption.

The data structure of the binary file is completely defined by the numeric values that fill it. By establishing matches between numeric values and their reference counterparts, one can determine the underlying data structure. Since the sought-for numeric values vary with time, we implement quasi-differential cryptanalysis to reveal change patterns within the data on different timescales. Contrary to the true differential cryptanalysis, this one is based on a partial quasi-known-plaintext attack instead of a chosen plaintext attack. The quasiknown-plaintext attack assumes that attacker still lacks the original plaintext but has some approximation at least with numerical values of the same magnitude and sign.

These approaches to cryptanalysis demand large amounts of cyphertexts with at least partially known differences of the respective plaintexts. The properties of A-GPS service fulfill the requirements since the underlying data on orbit elements changes in about half an hour. Therefore, one can assemble the required volume of cyphertexts with respective timestamps within a week or even a lesser timeframe.

5 ANALYSIS OF A-GPS BINARY FILES

The GNSS-positioning technology by design relies on timing, so the primary parameter is the timestamp of data origin. This timestamp is expressed in terms of GPS-week and GPS-day numbers (e.g. [6]) as well as seconds elapsed from some reference instance. Usually the GNSS-positioning

Table 3. File header (address 0×0000, big Endian).

Offset	Type	Content	Value	Comment
0 × 00	U1	Length of H1 block	0x01	–
0 × 01	U1	H1: Length of H2 block	0x1B	version 1
			0x34	versions 2, 3
0 × 02	…	Unmatched	…	…
0 × 0B	U4	H2: Length of file	0x00	0x000096DE – v.1
0 × 0C		0x00	0x00007E6C – v.2	
0 × 0D			0x00006347 – v.3	
0 × 0E				
0 × 0F	U2	H2: GPS Week since 1st epoch		
0 × 10		without rollover		
0 × 11	U4	H2: Actual seconds of GPS Week	Value changes	
0 × 12		for provided data, [ms]		every 30 minutes
0 × 13			and some seconds	
0 × 14				
0 × 15	U2	H2: GPS Week since 1st epoch		
0 × 16		without rollover		
0 × 17	U4	H2: Reference seconds of GPS Week		Value changes
0 × 18		for provided data, [ms]		Every hour
0 × 19				and some seconds
0 × 1A				
0 × 1B	…	Unmatched	…	…
0 × 1D	U1	Length of H3 block	0x01	
0 × 1E	U1	H3: Length of H4 block	0x00	
0 × 1F	U1	H4: Length of H5 block	0x25	
0 × 45	U1	H5: Length of H6 block	0x03	
0 × 46	U2	H5: GPS almanac length	0x03C2 32x31 = 962	
0 × 47				
0 × 48	U1	GPS PRN slot length	0x1F	Begin of almanac

operates on the timescale of milliseconds, so it is reasonable to also expect a data field holding the number of milliseconds.

At the initial stage we obtain seven triplets of binary files. The binary files of the first triplet are to be downloaded at a temporal distance of about 45 minutes, and the files of every next triplet are to be downloaded about 24–25 hours after last file of the previous triplet.

Next, we perform byte-to-byte comparison of the downloaded files via one of the hexadecimal viewers. The results reveal that WORD-variable occupying offsets 0×0F and 0×10 undergo changes between triplets, but not within a triplet (with rare exceptions). Moreover, these changes fit in a pattern of consequential daily increments. Assuming "Big Endian" format and referencing e.g. [6], one can deduce that the offsets 0×0F and 0×10 essentially hold the FULL GPS Week since 1st epoch. The same goes to the offsets 0×15 and 0×16.

Analyzing the DWORD gap (at offsets 0×11–0×14) between GPS Week timestamps one can see, that the value changes within each triplet. The computed difference between DWORD-values in consecutive files of a triplet corresponds to about 30 minutes expressed in milliseconds. The difference between said values in consecutive triplets confirms the guess by approximately corresponding to the number of milliseconds within a day. Thus, the DWORD-values of seven triplets form the sequence of milliseconds. Extrapolating this sequence backwards to zero, we deduce that the DWORD-value holds actual milliseconds of the GPS Week for the orbits data in the respective file. The partially deciphered header for the A-GPS binary file is presented in table 3. Since the distinctive blocks sometimes exhibit shifts in their offsets, we revised the layout and decoded bytes that contained relative offsets for the structures that followed.

To proceed with more refined analysis, one should use a larger volume of data, such as a month's worth of files with a temporary spacing of 30 minutes. Such level of detail is useful for decoding current and predicted ephemerides.

At the second stage of analysis we consider blocks of data that are almost constant for all files, downloaded within a day or a week. Our aim is to locate the offset and length of a byte sequence corresponding to the GPS almanac. Relying on the common sense, one can assume, that the almanac is represented as a set of uniform-sized blocks arranged in ascending order of the respective satellite numbers (GPS PRN designators, PRN$\in$ [1;32]). Moreover, each block should contain nine parameters describing orbital elements (see e.g. [7]) as the set: reference time, eccentricity, orbital inclination, rate of right ascension, a semimajor axis, argument of perigee, mean anomaly, clock correction and rate of clock correction. As one can see, orbital elements can be divided into two categories by a property of having a constant sign across all satellites. Thus, reference time, eccentricity, orbital inclination, rate of right ascension and semi-major axis constitute one category, while the remaining parameters form another category with varying sign across all satellites.

Using these categories, we can match orbital elements and data fields within the blocks via signtosign comparison. However, direct matching by value comparison is problematic due to issues of floating-point precision and the format of the element record. Consulting [8] on the topic of "SEM Almanac Description" we estimated the most likely byte lengths for the respective orbital elements.

After performing sign matching, we deduced that the first block of the satellite almanac starts at global offset 0×0049. This block corresponds to GPS PRN designator equal to one. The second block corresponds to PRN=2 and so forth. Thus, the size of the sought-for GPS PRN SLOT is 31 bytes. The revealed layout of data fields is provided in Table 4. For the purposes of sign matching we used fields of the longitude of ascending node $L\Omega$ and the argument of perigee ω. The first byte at each of these two respective fields has distinctive values of either 0×00 or 0×FF representing the sign of the 32-bit integer.

While analyzing the second version of A-GPS binary files we discovered that the header was extended and the offsets of the GPS PRN slots were described as an expression (ADDRESS 0×0073+0×001E*(PRN-1)). In contrast to the first version, the second one contains a GPS week number with respect to rollover procedure (i.e. modulo 1024).

6 THE OUTLINE OF THE GPS EPHEMERIS DATA LAYOUT

The bite-wise comparison of A-GPS binary files revealed repeating patterns of the length greater than the length of GPS almanac entries. We believe that the patterns correspond to ephemeris data, expressed in increments of respective Keplerian orbital elements and amplitudes of the harmonic correction terms(see e.g. [9]).

The respective data layout is provided in Table 5.

In turn, each data block has the following layout (see Table 6).

7 RESULTS AND DISCUSSION

The algorithm for reading the GPS almanac from an A-GPS binary file as well as its program implementation are straightforward once all offsets and record formats for orbital elements are established. The reference values for all orbital elements and values extracted from the corresponding binary file are compared in Table 7.

Considering Table 7, one can see that information on the reference GPS almanac helps quite a lot in revealing underlying parameters represented by the record fields of the binary A-GPS file. However, this problem is of moderate complexity since the orbit elements are updated at a slow pace in comparison to the ephemeris updates of A-GPS files.

Table 4. The block structure for GPS almanac (address 0×0049+0×001E*(PRN-1), big Endian).

Offset	Type	Content	Value	Comment
0×00	U1	PRN	0x01–0x20	
0×01	U1	Unmatched	0x00	Health ???
0×02	U2	e– Eccentricity	$\tilde{e}$	$e=$
0×03				$\tilde{e}\cdot$ 477E-7
0×04	U1	Unmatched		
0×05	I2	i – Orbital inclination,	$\tilde{i}$	$i=$
0×06		(deg)		180$\cdot$ (03 + $\tilde{i}\cdot$ 191$\cdot$E-6)
0×07	I2	$d\Omega/dt$ – Rate of right ascension W	$\tilde{\Omega}$	$d\Omega/dt=$
0×08		(deg/s)		180 $\cdot\tilde{\Omega}\cdot$ 364E-12
0×09	U4	A – Semi-major axis (km),	$\tilde{A}$	$A=$
0×0A				$(\tilde{A}\cdot 488\text{E-04})^2$
0×0B				
0×0C				
0×0D	I4	$L\tilde{\Omega}$ – Longitude of ascending node	$L\tilde{\Omega}$	$L\Omega=$
0×0E		on 00h.00min.00sec. base date,		180 $\cdot L\tilde{\Omega}\cdot$ 119E-7
0×0F		(deg)		
0×10				
0×11	I4	ω – Argument of perigee,	$\tilde{\omega}$	$\omega=$
0×12		(deg)		180 $\cdot\tilde{\omega}\cdot$ 119E-7
0×13				
0×14				
0×15	I4	m – Mean anomaly (deg),	$\tilde{m}$	$m=$
0×16				180 $\cdot\tilde{m}\cdot$ 119E-7
0×17				
0×18				
0×19	I2	$af0$ – Clock correction (sec),	$a\tilde{f}0$	$af0=$
0×1A				$a\tilde{f}0\cdot$954E-7
0×1B	I2	$af1$ – Rate of clock correction $af0$,	$a\tilde{f}1$	$af1=$
0×1C		(sec/sec)		$a\tilde{f}1\cdot$ 364E-12
1×1D	U2	Reference time		Full GPS week 1-st epoch
0×1E		without rollover		for 2 days ahead

Table 5. The layout for GPS ephemeris (address 0×0C54+0×049C*(PRN-1), big Endian).

Offset	Type	Content	Value	Comment
0×00	U1	Header	0x07	
0×01	U2	PRN Data	0x04	
0×02	block length		0x99	
0×03	U1	PRN	0x01–0x20	Start of the block
0×04+		Data block		
0×54∗k			$k \in [0;13]$	
...		Data block		
0×049B		Data block		

Decoding the ephemeris, especially the predicted ephemeris, is a more complex challenge. The prediction of orbit depends significantly on the employed numerical method and precision of initial data. Moreover, the long-term (7–24 days) predictions of the ephemeris always suffer from precision degradation and can become useless

Table 6. The data block structure for GPS phemeris (address 0×0C54+0×049C*(PRN-1)+0x04+0x54*k, big Endian).

Offset	Type	Content	Value	Comment
0×00	U4	Time, [ms]		0xFF – empty record
0×01				
0×02				
0×03				
0×04	U1	Unmatched		
…	…	…	…	…
0×2A	U4	Time, [ms]		0xFF – empty record
0×2B				
0×2C				
0×2D				
0×3E	U1	Unmatched		
…	…	…	…	…
0×53	U1	Unmatched		

Table 7. GPSalmanaccomparison(R-reference,GPSsecs=503808;A-assistedGPS-service,GPSsecs=309600) GPSWeek = 1060

SrcPRN *eidΩ/dtALΩωmaf* 0 *af* 1
R010.0092056.06475-4.45289E-726559.54867-18.9188943.53152130.34261-2.36511E-4-1.09139E-11
A010.0092156.06761-4.45536E-726528.96088-18.8856743.45509130.11378-2.36592E-4-1.09200E-11
R020.0195954.86312-4.53801E-726559.07559-23.10092-96.69964147.16695-3.70026E-4-7.27596E-12
A020.0196054.86431-4.54054E-726528.48835-23.06036-96.52987146.90858-3.70152E-4-7.28000E-12
R030.0026155.25725-4.61005E-726559.5134440.7251044.7677463.15913-5.62668E-5-7.27596E-12
A030.0026155.25900-4.61261E-726528.9256940.6536044.6891563.04825-5.62860E-5-7.28000E-12
R04N/AN/AN/AN/AN/AN/AN/AN/AN/A
R050.0058154.50743-4.69518E-726558.6226439.0231246.10250-65.59913-4.76837E-60.00000E0
A050.0058154.50817-4.69778E-726528.0359238.9546046.02155-65.48396-4.77000E-60.00000E0
R060.0017056.04209-4.43324E-726560.35393-19.39207-68.88250150.88559-1.59264E-4-1.09139E-11
A060.0017056.04492-4.43570E-726529.76521-19.35802-68.76156150.62068-1.59318E-4-1.09200E-11
R070.0132054.67875-4.55766E-726560.22810161.42729-138.6930290.09334-1.74522E-4-7.27596E-12
A070.0132154.67969-4.56019E-726529.63953161.14388-138.4495289.93517-1.74582E-4-7.28000E-12
R080.0048855.57277-4.53147E-726559.13598-80.14097-12.88825-92.00938-1.71661E-50.00000E0
A080.0048855.57495-4.53398E-726528.54867-80.00027-12.86562-91.84785-1.71720E-50.00000E0
R090.0017154.55378-4.51837E-726558.4414699.8369396.49580-75.38670-1.12534E-4-1.09139E-11
A090.0017154.55455-4.52088E-726527.8549599.6616596.32638-75.25435-1.12572E-4-1.09200E-11
R100.0052055.25622-4.60350E-726559.6543640.54601-156.0411815.35239-1.79291E-4-1.45519E-11
A100.0052055.25796-4.60606E-726529.0664540.47482-155.7672215.32543-1.79352E-4-1.45600E-11
R110.0162552.22812-4.78030E-726560.57537-45.51383113.3695381.62406-4.13895E-41.09139E-11
A110.0162652.22565-4.78296E-726529.98640-45.43392113.1704981.48076-4.14036E-41.09200E-11

Nevertheless, we plan to reveal the record format for the GPS ephemeris in A-GPS files of the gpsOne web service as well as to reveal complementary information on the ionosphere. Moreover, we are considering recovering almanac and ephemeris formats for GNSS other than GPS that are contained in the second version of the A-GPS file. The scope of our future work would also be concerned with file formats of other A-GPS services mentioned in this paper. Once all data formats are revealed, it is possible and plausible to create a program utility for data conversion between the formats. Among practical benefits we expect the increase of availability, versatility, and robustness of the A-GPS services, providing shorter satellite acquisition times.

8 CONCLUSION

In the presented study we considered the proprietary layout of a binary AGPS file for Qualcomm's gpsOne web service. Employing differential cryptanalysis in the form of quasi-known-plaintext attacks, we deduced the structures of the header and GPS almanac entries for each operational satellite. The comparison of the deciphered almanac with reference data showed good agreement on the values of the respective orbital elements. We also outlined the future study of more sophisticated GPS ephemeris entries along with ionosphere data in the A-GPS files of said binary format.

REFERENCES

[1] Qualcomm Press Release, Qualcomm Introduces gpsOneXTRA Assistance toExpand Capabilities of Standalone GPS, Feb 12, 2007 San Diego https://www.qualcomm.com/news/releases/2007/02/12/qualcomm-introduces-gpsonextra-assistance-expand-capabilities-standalone

[2] Nightwatch Cybersecurity, Advisory: Insecure Transmission of Qualcomm AssistedGPS Data [CVE-2016-5341] https://wwws.nightwatchcybersecurity.com/2016/12/05/cve-2016-5341/

[3] Nightwatch Cybersecurity, Advisory: Crashing Android devices with large AssistedGPS Data Files [CVE-2016-5348] https://wwws.nightwatchcybersecurity.com/2016/10/04/ advisory-cve-2016-5348-2/

[4] Zhang, W. (2018). New GNSS Navigation Messages for Inherent Fast TTFF and High Sensitivity-Underlying Theory Study and System Analysis. PhD Thesis on geomatics engineering, Calgary, Alberta https://prism.ucalgary.ca/bitstream/handle/1880/106628/ucalgary 2018 zhang wentao.pdf?sequence=1&isAllowed=y

[5] Qualcomm Developer Network forums (2019) Software. Qualcomm LTE for IoT.SDKgpsOneXTRA not working https://developer.qualcomm.com/comment/17055#comment-17055

[6] GPS Date Calendar http://navigationservices.agi.com/GNSSWeb/

[7] Information and Analysis Center for Positioning, Navigation and Timing, Korolyov,Russia https://www.glonass-iac.ru/en/GPS/ephemeris.php

[8] Official U.S. government information about the Global Positioning System (GPS) and related topics, Interface Control Document ICD-GPS-240 https://www.gps.gov/technical/icwg/ICD-GPS-870A.pdf

[9] Global Positioning Systems Directorate, System Engineering & Integration, Interface Specification IS-GPS-200, Navstar GPS Space Segment/Navigation User Interfaces https://www.gps.gov/technical/icwg/IS-GPS-200G.pdf

Proceedings of the 1st International Congress on Engineering Technologies – Kiwan & Banat (Eds)
 ISBN 978-0-367-77630-5

Traffic analysis of a software defined network

Uğur Özbek
Sakarya University, Sakarya, Turkey

Derya Yiltas-Kaplan
Istanbul University-Cerrahpasa, Istanbul, Turkey

Ahmet Zengin
Sakarya University, Sakarya, Turkey

Sema Ölmez
Hartlap Middle School, Onikisubat, Kahramanmaras, Turkey

ABSTRACT: A Software Defined Network is different from traditional networks because of its flexible and developable structure. This structure gives the researchers the opportunity to perform computational tests in their environments with realistic results. Software Defined Networks provide an extremely flexible environment for researchers. Network topology measurement, network traffic analysis, and network performance computations are performed on virtual networks created with the Mininet emulator. In this study, the Bellman-Ford routing algorithm was adapted into the routing module of the Floodlight controller. Then TCP traffic on a medium-sized network topology was generated and the traffic performance results were obtained in between the nodes even in different subnets. The implementation results were compared with the default routing algorithm of Floodlight, namely Dijkstra's algorithm.

Keywords: Benchmarking; Floodlight; Iperf; OpenFlow; Software Defined Network

1 INTRODUCTION

Traditional computer networks become complicated structures with their difficult management procedures and trials for scaling into today's requirements. Software Defined Network (SDN) aims to eliminate the weakness of the traditional network with its flexibility, central management, and open to research structure. With these properties, SDN is used in large data centers and large networks [1].

SDN architecture is mostly dependent on the efficiency of the control plane at which the controller is located. Having a controller dealing with network traffic successfully is important for making the control plane efficient. The controllers are located in the central part in an SDN environment. The main role of the typical controller is to manage all flows in the network through Southbound API to all switches as well as to maintain applications via the Northbound API. The routing operations in SDNs are performed by the controllers.

Floodlight controller [2] is one of the most popular controllers because it is Java-based, open-source, and supports both real and virtual OpenFlow compatible switches. Floodlight is a modular structure, so development and expansion operations can be made easily. Floodlight consists of four main bases: services, internal and external modules, and network applications. Status and events are exported via the services interface. The controller can operate without external modules,

DOI 10.1201/9781003178255-7

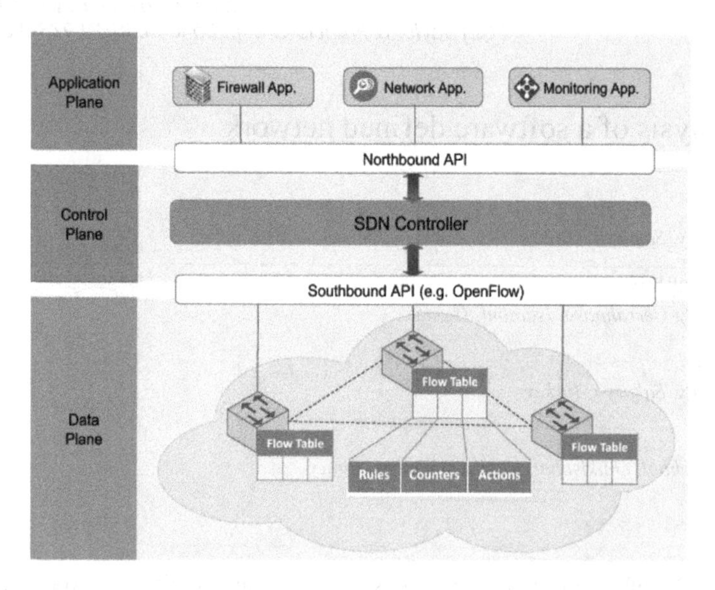

Figure 1. Basic SDN architecture [6].

but not without internal modules. Network applications are benefited to use Floodlight supported features through the REST API. Dijkstra's algorithm (DA) is an algorithm that allows us to find the shortest distance between any two nodes whose edges are values relative to a certain metric value on a weighted graph. DA is used as the default routing algorithm in the Floodlight controller. The Bellman-Ford algorithm (BFA) is the shortest single-source algorithm, so when you have a negative edge weight it can detect negative loops in a chart. The main difference between the two is that although DA only calculates positive weights, BFA can calculate both positive and negative weights. In this study, the network topology is constructed with Mininet [3] and then some comparison of the controller test results on the performance of DA and BFA are presented with four different scenarios on the same topology. An open-source traffic generator, Iperf [4] tool, is used during this test. DA's effects on the controller throughput is observed on large networks in which the main parameter is the distance between all the nodes. Two of the performed scenarios signified that BFA represents the controller more effectively. The rest of this study is planned as follows.

Some general information about SDN, literature studies on the routing algorithms and path computation in SDN are discussed in Section II. The implementation environment and test results are pointed out in Section III. Lastly, a conclusion and future works are given in Section IV. Section V contains an acknowledgement.

2 BACKGROUND

2.1 *Software defined network*

SDN is a flexible, dynamic, manageable, cost-effective, and adaptable architecture. Today, the increasing number of networked devices and the high demand for bandwidth make the management of the network more and more difficult. This architecture separates network control and routing functions so that network control can be directly programmable. Network applications and services are isolated from the underlying infrastructure [5].

Figure 1 shows a basic SDN architecture.

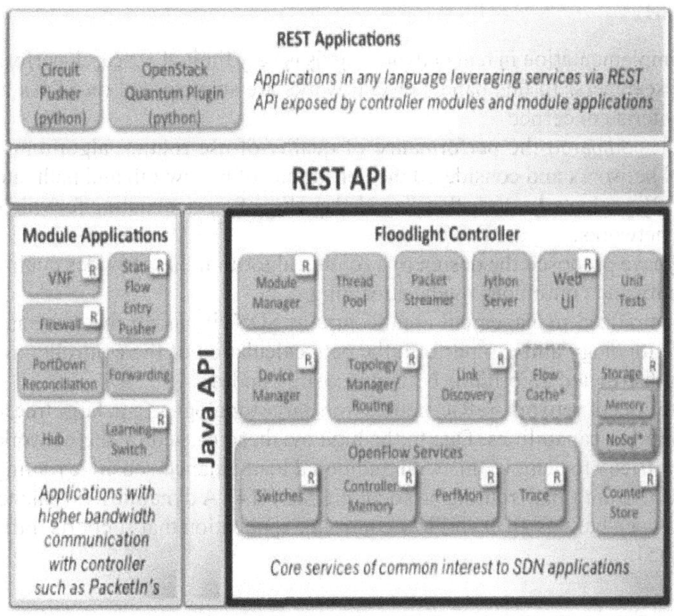

Figure 2. Floodlight controller structure [2].

The idea of programmable networks has gained significant momentum with the emergence of SDN approach. SDN promises to simplify network management significantly and allow innovation with network programmability [7].

In SDN architecture, the controller is a separate entity that can run complex algorithms to calculate the shortest path. As a result, the controller has a central view of the network and can dynamically redirect traffic demanded by an application or based on congestion, disconnection, or user policies.

The controller is like a GPS that tells you to follow the most effective route and dynamically change the route in case of an accident, road construction, or road closure.

2.2 *Floodlight controller*

Floodlight Controller, one of the best-known SDN controllers, supports OpenFlow-compatible virtual and physical switches [8].

The modular structure of the Floodlight Controller includes the device management module, topology module, learning switch, load balancer, Web Graphical User Interface (Web GUI) [9], and a counter module for statistics (see Figure 2) [8].

In Figure 2, the Floodlight Provider module, called the core module, processes I/O from network devices and converts incoming OpenFlow messages into Floodlight events. Topology Management is responsible for calculating the shortest path with DA. Link Discovery Provides link states using LLDP packets. The Forwarding Module provides end-to-end routing flow. The Device Manager maintains information about the nodes and storage resources on the network. Static Flow Entry is an application that loads a specific flow entry that contains match and transaction columns for a specific key that is enabled by default. Through the REST API, we can add, remove, and query flow entries [10].

2.3 *Related works*

Though SDN's implementation in real network stands is very limited, research performed for SDNs attracts many researchers. In this paper, several works are summarized on SDNs by considering the scope and criteria concerned.

Tomovic et al. compared the performance of quality-of-use routing algorithms applicable to large-scale SDN networks and considered the constraints of bandwidth and path latency [11].

Henni et al. [12] conducted a study that provided quality of service routing for video conferencing over OpenFlow networks.

Jin et al. [13] have proposed the design of a routing algorithm that provides quality of service to ensure high quality and continuous media streaming.

Due to the centralized management architecture in SDN, it has all the information about the existing network topology and can perform the path calculation of the controller as a simple task instead of using distributed algorithms between existing nodes.

It is possible to collect network traffic metrics from large topologies (e.g. large tree networks) with OpenFlow monitoring capabilities. Due to the high overhead caused by the network over density, it is sometimes difficult to achieve this with traditional load-balance-sensitive routing mechanisms such as cost-effective multipath routing protocol (ECMP) [14]. A dynamic load balancer with multipath support has been built as an OpenFlow controller application that supports traffic adaptations in the network [15].

2.4 *Methodology*

The routing process in the Floodlight Controller is performed by calculating the shortest path (minimum cost) with the hop number metric using DA. DA is considered to be the most effective way to calculate the shortest route in IP networks. However, routing is likely to become inefficient in large-scale networks where traffic is heavy [16]. This problem is the main motivation of the research done in this paper. In addition to DA, a state of the art algorithm called BFA is implemented into the Floodlight Routing Module. Briefly, two different routing technologies are obtained for comparison in particular for large-scale networks. One of them is DA, the default routing algorithm of the Floodlight controller, and the other is BFA, which is implemented to the Floodlight controller. After implementation, comparison tests are performed in consideration of performance characteristics such as throughput to depict the efficiency of the approaches.

Despite link states of DA being broadcast to all nodes in the network, BFA just deals with neighbors. Since BFA visits each node and calculates all relevant distances, it is possible to apply this algorithm to inputs with negative weights. On the other hand, DA uses a greedy approach, passing only once from each node. Therefore it cannot provide a solution for negative weights. Due to the possibility of existing negative cycles, BFA can detect the presence of a negative loop if it detects any changes in the stored values.

3 CONTROLLER PERFORMANCE EVALUATION

3.1 *Test configuration*

The test environment configuration is shown in Table 1. Network topology consists of 5 Gateway switches and 25 switches. 20 nodes are connected to per switch, so a total of 500 nodes are created. All of these virtual tools communicate via OpenFlow 1.3 protocol. Each switch IP assignment which is connected to the corresponding Gateway Switch has 5 separate subnets. In the test environment, Floodlight and Mininet were run on the same virtual machine.

As can be seen in Figure 3, traffic is generated between nodes in different subnets over the network topology created. It has been observed that routing occurs between the sample nodes.

Table 1. Experiment specifications.

Floodlight + Mininet Virtual Machine	
Ubuntu	14.04.5
Floodlight	V1.2
Mininet	V 2.3.0.d6
CPU	6 vCPU Intel Xeon E5530 2.40 GHz
RAM	16 GB RAM

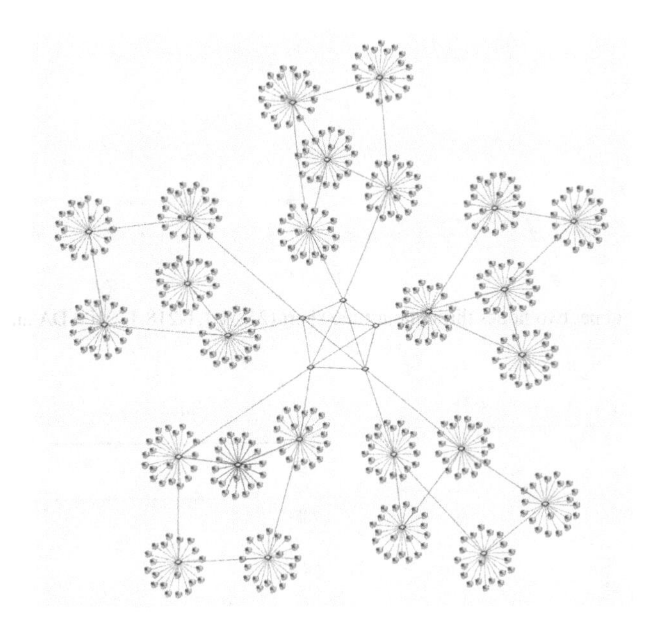

Figure 3. Network topology.

Table 2. Traffic generation scenario.

Scenario	Server Node IP	Client Node IP	Traffic Time (sec)	Sampling Rate (sec)
Scenario 1	218.1.1.1	218.1.2.100	25	1
Scenario 2	218.1.1.1	218.1.3.100	25	1
Scenario 3	218.1.1.1	218.1.4.100	25	1
Scenario 4	218.1.1.1	218.1.5.100	25	1

The scenarios applied are shown in Table 2. TCP traffic is generated by the Iperf tool between nodes located on different subnets. The traffic measurement time is 25 seconds and the sampling range is 1 second.

3.2 *Results*

As shown in Figure 4, the two routing algorithms have similar performances. More stability has been observed in DA after 15 seconds, so it has a better convergence.

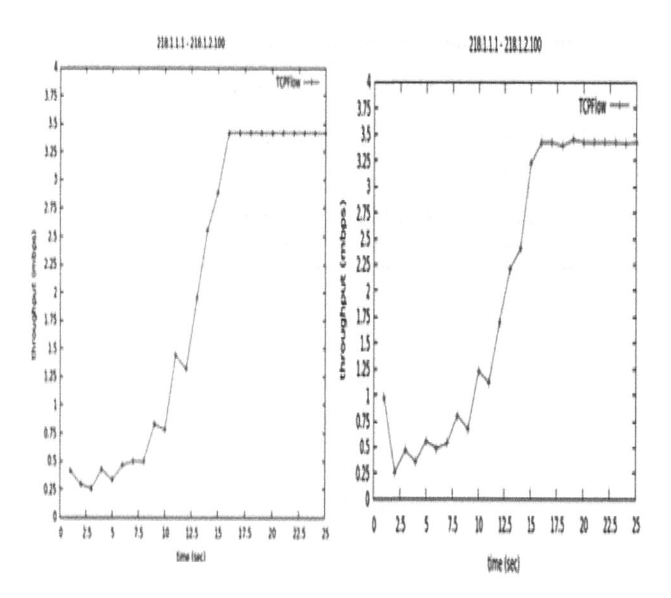

Figure 4. Different subnet two nodes throughput/time chart (218.1.1.1-218.1.2.100 DA and BFA).

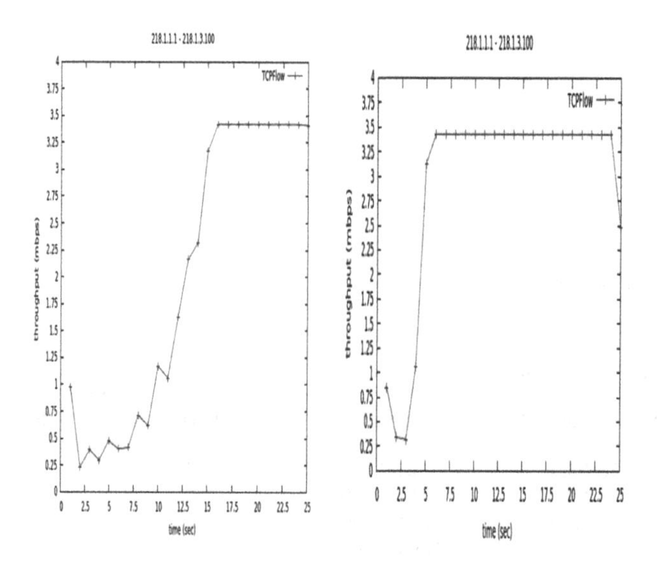

Figure 5. Different subnet two nodes throughput/time chart (218.1.1.1-218.1.3.100 DA and BFA).

As shown in Figure 5, BFA started to increase at 5 seconds and achieved a stable performance at 7 seconds, but DA achieved a stable performance at the end of 15 seconds. Therefore, BFA has a faster convergence.

As shown in Figure 6, the convergence times of the routing algorithms are very close to each other.

In Figure 7, BFA converged much earlier than DA, although the cost of routing was higher than other scenarios in the different subnet, and it took 5 seconds for the throughput value to increase and stabilize.

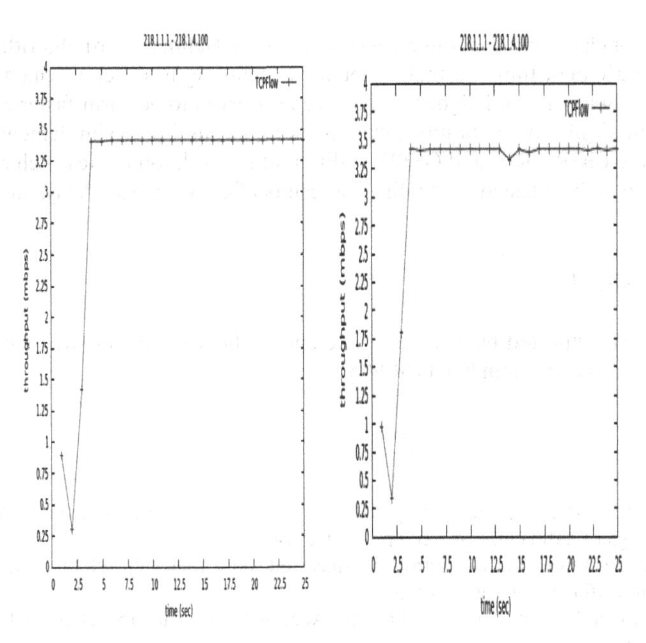

Figure 6. Different subnet two nodes throughput/time chart (218.1.1.1-218.1.4.100 DA and BFA).

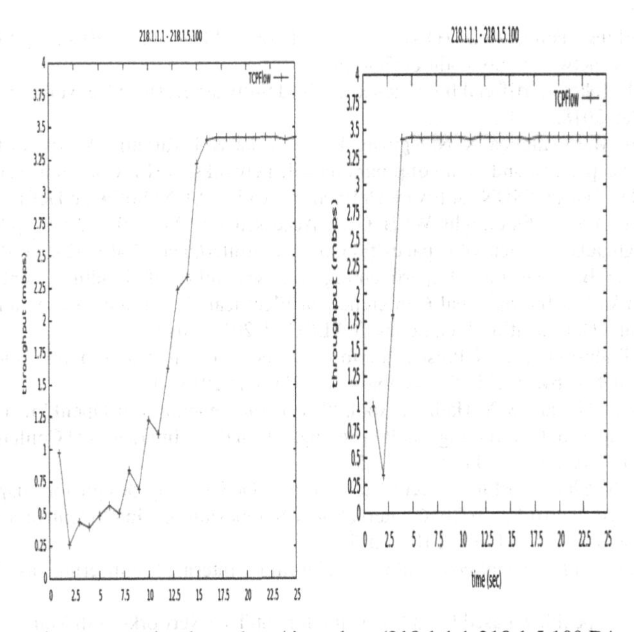

Figure 7. Different subnet two nodes throughput/time chart (218.1.1.1-218.1.5.100 DA and BFA).

4 CONCLUSION

In this study, we have implemented the comparative throughput tests between the Bellman-Ford and the Dijkstra's routing algorithms on the designed network topology. Due to the different path costs of these algorithms, different throughput results were obtained. As seen in Scenarios 2 and 4, the Bellman-Ford algorithm adapted to the situation more quickly and achieved a stable speed in less time. We conclude that the convergence time of the Bellman-Ford algorithm is faster than that of the Dijkstra's algorithm and that is obtained according to different characteristics of the network topology. Bellman-Ford algorithm has been observed to perform faster convergence and routing in medium-sized and link failure networks. In future studies, the implementation of routing protocols including EIGRP, RIP, and OSPF on different network topologies such as Tree, Fat Tree and Extended Stars will be tested. Then, the test results of convergence, delay, and efficiency will be compared.

ACKNOWLEDGMENT

This work has been supported by The Scientific and Technological Research Council of Turkey (TUBITAK) with the project number 116E905.

REFERENCES

[1] P. Göransson & C. Black, (Foreword by M. Davy), Software Defined Networks: A Comprehensive Approach, Morgan Kaufmann, Elsevier, pp. 1–20, 2014.
[2] Floodlight is an Open SDN Controller. Accessed on: Mar. 3, 2020. [Online]. Available: http://www.projectfloodlight.org/floodlight
[3] An Instant Virtual Network on your Laptop. Accessed on: Feb. 15, 2020. [Online]. Available: http://mininet.org/
[4] The ultimate speed test tool for TCP, UDP and SCTP. Accessed on: Feb. 25, 2020. [Online]. Available: https://iperf.fr
[5] Open Networking Foundation (ONF) Accessed on: Mar. 05, 2020. [Online]. Available: https://www.opennetworking.org/sdn-definition/
[6] M. Latah and L. Toker, "Artificial Intelligence Enabled Software Defined Networking: A Comprehensive Overview," Mar. 2018.
[7] B. A. A. Nunes, M. Mendonca, X. N. Nguyen, K. Obraczka, & T. Turletti, "A survey of software-defined networking: Past, present, and future of programmable networks," IEEE Commun. Surv. Tutorials, 2014.
[8] K. Gray & T. D. Nadeau, "SDN: Software Defined Networks," SDN: Software Defined Networks. 2017.
[9] R. Izard & H. Akcay, Floodlight WEB GUI. Accessed on: Mar. 05, 2020. [Online]. Available: https://floodlight.atlassian.net/wiki/spaces/floodlightcontroller/pages/40403023/Web+GUI
[10] S. Asadollahi & B. Goswami, "Experimenting with scalability of floodlight controller in software defined networks," in International Conference on Electrical, Electronics, Communication Computer Technologies and Optimization Techniques, ICEECCOT 2017, 2018.
[11] S. Tomovic, I. Radusinovic, & N. Prasad, "Performance comparison of QoS routing algorithms applicable to large-scale SDN networks," in Proceedings - EUROCON 2015, 2015.
[12] D. E. Henni, A. Ghomari, & Y. Hadjadj-Aoul, "Videoconferencing over OpenFlow networks: An optimization framework for QoS routing," in Proceedings - 15th IEEE International Conference on Computer and Information Technology, CIT 2015, 2015.
[13] X. Jin, H. Ju, S. Cho, B. Mun, C. Kim, & S. Han, "QoS routing design for adaptive streaming in Software Defined Network," in 2016 International Symposium on Intelligent Signal Processing and Communication Systems, ISPACS 2016, 2017.
[14] C. Hopps. Analysis of an equal-cost multi-path algorithm. Internet Engineering Task Force, RFC 2992, 2000.
[15] Y. Li & D. Pan, "OpenFlow based Load Balancing for Fat-Tree Networks with Multipath Support," Proc. 12th IEEE Int. Conf. Commun., 2013.
[16] Y. Sakumoto, H. Ohsaki, & M. Imase "Performance of Thorup's Shortest Path Algorithm for Large-Scale Network Simulation" IEICE Transactions on Communications, 2012 Volume E95.B Issue 5 pp. 1592–1601.

Proceedings of the 1st International Congress on Engineering Technologies – Kiwan & Banat (Eds)

Outage analysis of cooperative relay non orthogonal multiple access systems over Rician fading channels

Sari Khatalin & Hadeel Miqdadi
Department of Electrical Engineering, Jordan University of Science and Technology, Irbid, Jordan

ABSTRACT: Non-orthogonal multiple access (NOMA) is a promising technology for the fifth generation (5G) wireless networks. Integrating NOMA with cooperative relaying transmission technique assures remarkable improvement to the performance of the 5G networks in terms of outage probability, achievable rate and other important network performance metrics. In this paper, the outage probability of cooperative relaying system (CRS) integrated NOMA is investigated over Rician fading channels. Analytical expressions for the outage probability of the system mentioned above are obtained. Furthermore, corresponding expressions for Rayleigh fading channels are also presented in this paper as a special case of Rician fading channels. Some examples are presented here for illustration purposes. The analytical results obtained in this paper are all validated via simulation.

Keywords: Cooperative Relaying, Non Orthogonal Multiple Access, Successive Interference Cancellation, Outage probability Rician fading.

1 INTRODUCTION

The tremendous growth in the demands of wireless internet communications has led to the development in the multiplexing techniques [1]. Wireless communications systems allocate resources to different users and applications based on orthogonal multiple access (OMA) such as time division multiple access (TDMA), frequency division multiple access (FDMA) and code division multiple access (CDMA) [2]. The demand for higher spectral efficiency, massive connectivity and low latency is rapidly increasing in order to support the new applications in the fifth generation (5G) networks such as internet of things (IoT). OMA is not expected be able to support such demands [3]. Non orthogonal multiple access (NOMA) has shown to be the new multi user multiplexing technology that could support all of these requirements [4]. Unlike OMA techniques, NOMA is able to support data transmission for multiple users to transmit their data simultaneously by using the same resources of frequency, time and code, but the power levels for the signals are different from each other [1]. At the NOMA receiver, the strong symbol is decoded first, then successive interference cancellation (SIC) is utilized to decode the reminder of the symbols [5].

Integrating NOMA with cooperative relay transmission provides yet further improvement to the reliability and the coverage of the 5G networks. This scheme was proposed by [6] where the users with better channel conditions can be employed as relays to enhance the performance of the users with bad channel conditions.

A full-duplex cooperative NOMA system with dual users is proposed in [7] where a dedicated full-duplex relay assists the transmission of information to the weak channel condition user under the assumption of imperfect self-interference cancellation. The best relay selection has also been investigated in [8] for cooperative relay system (CRS) based NOMA (CRS-NOMA) to derive the average rate under independent Rayleigh fading channels. In [9] the authors considered relay selection strategies for cooperative NOMA where one base station communicates with two mobile users by the assistance of multiple relays to propose two stage relay selection with decode-and-forward (DF)

DOI 10.1201/9781003178255-8

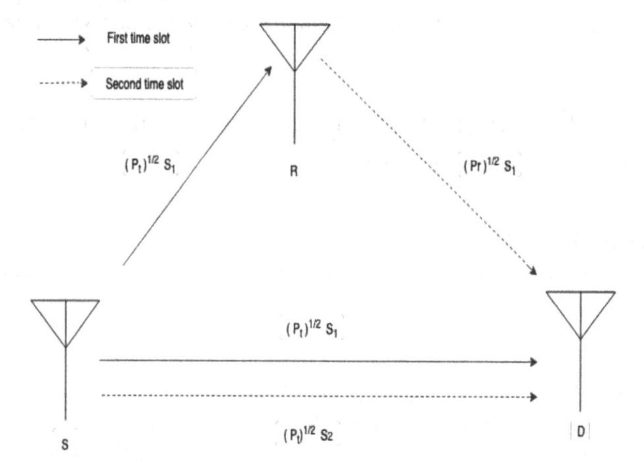

Figure 1. Proposed CRS-NOMA.

and amplify-and-forward (AF) relaying protocols. A downlink NOMA based cooperative relay system with a single relay over Nakagami-*m* fading channels considering both DF and AF protocols was proposed in [10]. A relay assisted CRS- NOMA scheme was proposed in [11] to calculate the outage performance, and the achievable rate over Rayleigh fading channels, where the destination can receive transmission in two time slots. Furthermore, the authors in [12] proposed a cooperative relay based NOMA scheme with the use of complex power allocation over Rician fading channels. In [13], the authors studied the achievable rate of the proposed system presented in [11] over Rician fading channels.

In this paper, the outage probability of the relay assisted CRS-NOMA system, which was proposed in [11] and [13], is studied over Rician fading channels. In this system, each node transmits symbols with its maximum power. Therefore, the complex power allocation utilized by the conventional CRS-NOMA system is avoided.

The rest of the work is organized as follows: description for the proposed CRS_NOMA system is presented in Section II. Section III presents the analytical analysis for the outage probability. Some numerical examples and the conclusion are presented in Section IV and V, respectively.

2 SYSTEM DESCRIPTION

Figure 1 illustrates a simple cooperative relay system, which consists of a source node, S, that communicates with the destination node, D, with the assistance of a fixed half duplex DF relay node, R. The cooperative Relay fully decodes, re-encodes and retransmits the message to the destination D [11],[13]. In addition, the source can also directly communicate with the destination. All links between the system nodes (i.e., S-R, S-D, and R-D links) are assumed to be available. The channel coefficients of S- R, S- D, and R- D links are denoted by h_{SR}, h_{RD} and h_{SD}, respectively, and are considered to be independent Rician Random variables, with average powers Ω_{SR}, Ω_{RD} and Ω_{SD}, respectively. Also, we assume that $\Omega_{\mathrm{SR}} > Omega_{\mathrm{SD}}$, because the path loss of the S-D link is normally smaller than that of the S-R link [11]. In this scheme, the destination will be able to receive two different signals in two time slots. In the first time slot, the source S transmits the symbol with power to both the relay and the destination, and therefore the received symbol at R in the first time slot can be presented as [13]

$$X_{SR} = \sqrt{P_t} h_{SR} S_1 + n_{SR}. \tag{1}$$

And the received symbol at D in the first time slot can be presented as

$$X_{SD} = \sqrt{P_t} h_{SD} S_1 + n_{SD}. \quad (2)$$

Where s_i denotes the the $i_t h$ data symbols with normalized power $E[|S_i|^2] = 1, P_t$ denotes the total transmit power, n_{SR} and n_{SD} denote the additive white Gaussian noise (AWGN) with zero mean and variance σ^2 . The received signal to noise ratio (SNR) for s_1 at the relay can be presented as:

$$\gamma_{SR}, s_1 = \frac{|h_{SR}|^2 p_t}{\sigma^2}. \quad (3)$$

And the SNR for s_1 at the destination can be presented as:

$$\gamma_{SD}, s_1 = \frac{|h_{SD}|^2 p_t}{\sigma^2}. \quad (4)$$

In the second time slot, R sends the decoded symbol s_1 to D and S transmits another symbol s_2 to D. Then, the received signal at the destination is given by

$$r_{RD} = h_{RD} \sqrt{\mathrm{P_R}} S_1 + h_{SD} \sqrt{\mathrm{P_t}} S_2 + n_{RD}, \quad (5)$$

where n_{RD} denotes the AWGN with zero mean and variance σ^2. It should be noted that, $E\{|h_{SD}|^2\} < E\{|h_{RD}|^2\}$, which means that the fading gain of the R-D link is greater than that for the S-D link due to the differences in the distances of the SD and R-D links, and this characteristic makes it easier to implement the CRS-NOMA principle at the second time slot. It is considered that the powers at the source and the relay are equal (i.e., $P_t = P_R = P$) Applying SIC, the destination initially decodes $s_1$while treating S_2 as a noise, then this signal is removed to decode s_2 [5]. Then, the received SNR for s_1 at the destination can be written as:

$$\gamma_{RD}, s_1 = \frac{\frac{|h_{RD}|^2 P}{\sigma^2}}{\frac{|h_{SD}|^2 P}{\sigma^2} + 1}. \quad (6)$$

And the received SNR for s_2 at the destination can be written as:

$$\gamma_{SD}, s_2 = \frac{|h_{SD}|^2 P}{\sigma^2}. \quad (7)$$

3 OUTAGE PROBABILITY ANALYSIS

Expressions for the outage probability of the proposed CRS-NOMA system will be derived in this section over Rician fading channels The outage probability is defined as the probability that the instantaneous SNR γ falls below a pre-defined threshold γ_{th} The probability density function (PDF) γ for Rician fading is presented in [14–15] as

$$p_r(\gamma) = \frac{(k+1)}{\Omega} \exp\left(-k - \frac{\gamma(k+1)}{\Omega}\right) I_o\left(2\sqrt{\frac{\gamma k (k+1)}{\Omega}}\right), \quad \gamma \geq 0. \quad (8)$$

And the cumulative distribution function (CDF) of γ is presented in [14–15] as

$$\mathrm{F}(\gamma) = 1 - \mathrm{Q}_1\left(\sqrt{2\mathrm{k}}, \sqrt{\frac{2(\mathrm{k}+1)\gamma}{\Omega}}\right), \quad (9)$$

where Ω denotes the average SNR equals E[$|\gamma|^2$], k is the Rician factor, $I_o(x)$ is the modified Bessel function with first kind and zero order defined [16, Eq. (8.445)] and Q_1(a,b) is the first order Marcum Q function defined as [17]

$$Q_m(a,b)=\sum_{l=0}^{\infty}\sum_{k=0}^{m+l-1}\frac{a^{2l}b^{2k}}{l!k!2^{l+k}}e^{-\frac{a^2+b^2}{2}}. \tag{10}$$

The outage probability is derived for both s_1 and s_2. We first obtain outage probability for s_1, P_{o,S_1}. s_1 goes into outage if it failed to arrive at the destination, and this occurs when either the direct or the indirect link is in outage. So we define { $\gamma_{SD}, s_1 < \gamma_{th}$} as the event $s_1$cannot be detected, and so the outage for the direct link, $F_{SD,s1}(\gamma_{th})$ is defined by the CDF of the SNR, γ_{SD}, s_1 as

$$\begin{aligned}F_{SD,s1}(\gamma_{th})&=p_r(\gamma_{SD},s_1<\gamma_{th})\\&=1-Q_1\left(\sqrt{2k_{SD}},\sqrt{\frac{2(k_{SD}+1)\gamma_{th}}{\rho\Omega_{SD}}}\right),\end{aligned} \tag{11}$$

where $\rho=\frac{P}{\sigma^2}$ denotes the transmit SNR.

As for the indirect link, the outage probability for the s_1, $F_{D,s1}(\gamma_{th})$ can be derived by following the same process presented in [18, Eq. (2)] First of all, we obtain the equivalent γ for the indirect link presented in Eq. (6) as $\gamma_D, s_1=\frac{\gamma_{RD},s_1}{\gamma_{SD},s_1+1}$ Now, let x $=\rho|h_{RD}|^2$, y $=\rho|h_{SD}|^2$and z $=\gamma_D, s_1$ [11]. The outage probability is obtained as [18, Eq. (2)]

$$F_{D,S_1}(z)=\int_0^{\infty}F_x(z(y+1))f_Y(y)\,dy. \tag{12}$$

Defining x = z (y+1), and inserting Eq. (8) and Eq. (9) in Eq. (12) yields

$$\begin{aligned}F_{D,S_1}(z)=&\int_0^{\infty}\left(1-Q_1\left(\sqrt{2k_{RD}},\sqrt{\frac{2(k_{RD}+1)Z(y+1)}{\rho\Omega_{RD}}}\right)\right.\\&\times\frac{(k_{SD}+1)}{\rho\Omega_{SD}}\exp\left(-k_{SD}-\frac{y(k_{SD}+1)}{\rho\Omega_{SD}}\right)I_o\left(2\sqrt{\frac{yk_{SD}(k_{SD}+1)}{\rho\Omega_{SD}}}\right)dy.\end{aligned} \tag{13}$$

Eq. (13) can be presented as

$$\begin{aligned}F_{D,S_1}(z)=&\int_0^{\infty}\frac{(k_{SD}+1)}{\rho\Omega_{SD}}\exp\left(-k_{SD}-\frac{(k_{SD}+1)y}{\rho\Omega_{SD}}\right)I_o\left(2\sqrt{\frac{k_{SD}(k_{SD}+1)y}{\rho\Omega_{SD}}}\right)dy\\&-\int_0^{\infty}Q_1\left(\sqrt{2k_{RD}},\sqrt{\frac{2(k_{RD}+1)z(y+1)}{\rho\Omega_{RD}}}\right)\frac{(k_{SD}+1)}{\rho\Omega_{SD}}\exp\left(-k_{SD}-\frac{(k_{SD}+1)y}{\rho\Omega_{SD}}\right)\\&\times I_o\left(2\sqrt{\frac{k_{SD}(k_{SD}+1)y}{\rho\Omega_{SD}}}\right)dy.\end{aligned} \tag{14}$$

Evaluating the first integral in Eq. (14), which is equal to one, replacing $Q_1(a,b)$ in Eq. (14) with it's infinite series representation as shown in Eq. (10), evaluating the second in Eq. (14) by utilizing [19, Eq. (3.15.2.5)] and using the power series in [16, Eq. (1.111)] yields

$$F_{D,s1}(z) = 1 - \frac{(k_{SD}+1)}{\rho\Omega_{SD}} \exp\left(-k_{SD} - k_{RD} - \frac{z\,(k_{RD}+1)}{\rho\Omega_{RD}}\right) \sum_{l=0}^{\infty}\sum_{r=0}^{l} \frac{(k_{RD})^{l}\left(\frac{z(k_{RD}+1)}{\rho\Omega_{RD}}\right)^{r}}{l!k!}$$
$$\times \sum_{i=0}^{r} \binom{r}{i} \frac{\Gamma(i+1)}{\left(\frac{(k_{RD}+1)z}{\rho\Omega_{RD}} + \frac{(k_{SD}+1)}{\rho\Omega_{SD}}\right)^{i+1}} {}_1F_1\left(i+1; 1; \frac{\frac{k_{SD}(k_{SD}+1)}{\rho\Omega_{SD}}}{\left(\frac{(k_{RD}+1)z}{\rho\Omega_{RD}} + \frac{(k_{SD}+1)}{\rho\Omega_{SD}}\right)}\right), \tag{15}$$

where ${}_1F_1$(a; b; x) is the Confluent hypergeometric function defined by the power series in [14 , (3.37)].

The end-to-end (e2e) SNR for s_1 during the indirect link is $\gamma_{e2e}, s_1 = \min\{\gamma_{SR,s1}, \gamma_{D,s1}\}$, and the e2e outage probability for the symbol s_1 can be written as [20]

$$F_{e2e,s_1}(\gamma_{e2e}, s_1) = F_{D,s1}(\gamma_{th}) + F_{SR,s1}(\gamma_{th}) - F_{D,s1}(\gamma_{th})\, F_{SR,s1}(\gamma_{th}), \tag{16}$$

where

$$F_{SR,s1}(\gamma_{th}) = 1 - Q_1\left(\sqrt{2k_{SR}}, \sqrt{\frac{2\,(k_{SR}+1)\,\gamma_{th}}{\rho\Omega_{SR}}}\right). \tag{17}$$

The outage probability for the s_1 is expressed as:

$$P_o, S_1 = F_{SD,s1}(\gamma_{th}) \times F_{e2e,s1}(\gamma_{th}). \tag{18}$$

The symbol s_2 goes into outage if s_2 cannot be decoded properly during the transmission through the S-D link. Furthermore, if the indirect link goes in outage (i.e., based on the concept of SIC, D detects the symbol $s_1$first, and then removes it from the received signal and after that detects the symbol s_2 Therefore, if the destination failed to detect $s_1$correctly, it will not be able to detect s_2). The outage probability for s_2 can be expressed as [21]

$$\begin{aligned} P_o, S_2 &= p_r(\gamma < \gamma_{th}) \\ &= 1 - p_r(\gamma_{D,s1} \geq \gamma_{th}, \gamma_{D,s2} \geq \gamma_{th}) \\ &= 1 - p_r(\gamma_{D,s1} \geq \gamma_{th}) p_r(\gamma_{D,s2} \geq \gamma_{th}) \\ &= 1 - (1 - p_r(\gamma_{D,s1} < \gamma_{th}))(1 - p_r(\gamma_{D,s2} < \gamma_{th})) \\ &= 1 - (1 - F_{D,s1}(\gamma_{\text{th}}))\,(1 - F_{sD,s2}(\gamma_{\text{th}})), \end{aligned} \tag{19}$$

where $P_r(\gamma_{SD,S2} < \gamma_{th})$ can be calculated from Eq.(9) as

$$\begin{aligned} F_{\text{SD,S2}}(\gamma_{\text{th}}) &= \text{P}_\text{r}(\gamma_{\text{SD,S2}} < \gamma_{\text{th}}) \\ &= 1 - \text{Q}_1\left(\sqrt{2\text{k}_\text{SD}}, \sqrt{\frac{2\,(\text{k}_\text{SD}+1)\,\gamma_{\text{th}}}{\rho\Omega_{\text{SD}}}}\right) \end{aligned} \tag{20}$$

Inserting Eq. (15) and Eq. (20) into Eq. (19) yields the expression of outage probability for s_2 Setting $k = 0$ (k_{SD}, k_{SR}, k_{RD}) in Eq. (18) and Eq. (19), yields the expressions of outage probability for s_1 and $s_2$over Rayleigh fading, respectively, which are in agreement with the results in [11]

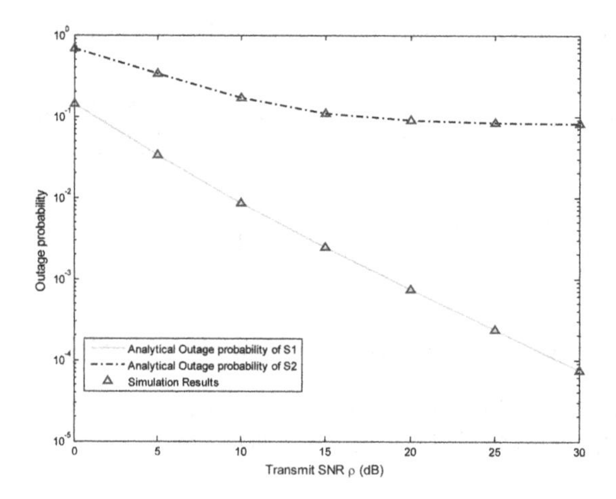

Figure 2. Outage probability vs transmit SNR, ρ over Rician fading channels for γ_{th}= 0 dB, k_{SD}= 0.5, k_{SR} = 0.55, k_{RD}= 0.7, Ω_{SD} =1, Ω_{SR} = 10, Ω_{RD} =10.

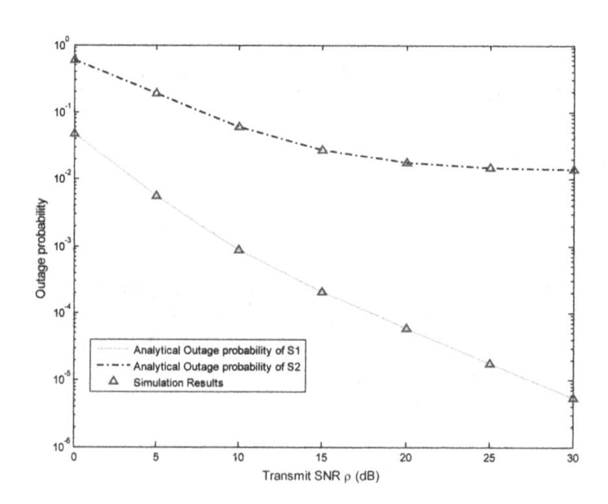

Figure 3. Outage probability vs transmit SNR, ρ over Rician fading channels for γ_{th}= 0 dB, k_{SD}= 2, k_{SR} = 2, k_{RD}= 5, Ω_{SD} =1, Ω_{SR} = 10, Ω_{RD} =10.

4 NUMERICAL RESULTS

We present in this section some examples to illustrate the main results obtained here. All results are verified via simulations. Figures show that the simulation results are matched very well with the analytical results.

Figure 2 plots the outage probability versus ρ for k_{SD}= 0.5, k_{SR} = 0.55, k_{RD}= 0.7, Ω_{SD} =1, Ω_{SR} = 10, Ω_{RD} =10, and the threshold SNR γ_{th}= 0 dB. It can be seen from the figure that as ρ increases P_o, s_1, and P_o, s_2 decrease. It can also be seen the decrease in P_o, s_1, is more obvious than in the case of P_o, s_2

In Figure 3, the values of k are selected as k_{SD}= 2, k_{SR} = 2, k_{RD}= 5 and the values for Ω are set as , Ω_{SD} =1, Ω_{SR} = 10, Ω_{RD} =10, and γ_{th}= 0 dB. Comparing Figure 3 with Figure 2, it is obvious that increasing k will decrease the outage probability meaning that the outage performance is improved with increasing k

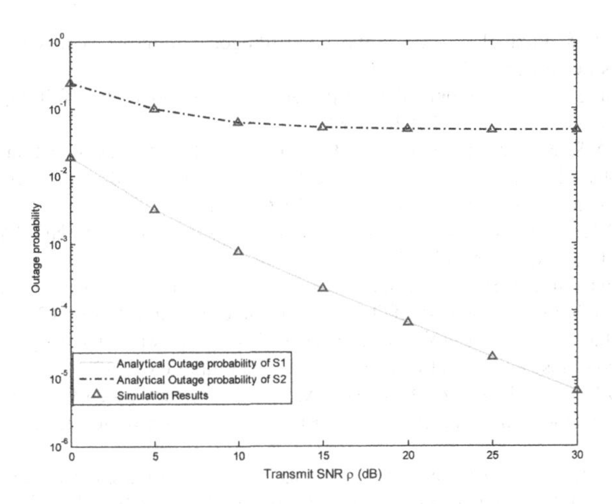

Figure 4. Outage probability vs transmit SNR, ρ over Rician fading channels for γ_{th}= 0 dB, k_{SD} =0.5, k_{SR} = 0.55, k_{RD} =0.7, Ω_{SD} =3, Ω_{SR} = 12, Ω_{RD} =10.

Figure 4 plots the outage probability against ρ for the same values of k as in Figure 2 and Ω_{SD} = 3, Ω_{SR} = 12, Ω_{RD} =15. Comparing Figure 4 with Figure 2, It can be noted that as the value of Ω increases, both s_1 and s_2 achieve better outage probability, and so the outage performance of the system is improved We can notice from all figures that both P_o, s_1, and P_o, s_2 improve as the values of ρ or $k$or Ω increase. However, the improvement is more noticeable in the case of P_o, s_1 than that of P_o, s_2. Furthermore, the improvement for P_o, s_2 declines when ρ gets to 15 dB and beyond, which means increasing ρ beyond 15 dB will not improve P_o, s_2. However, improving the conditions of the channels and/or average power may help in the improvement for P_o, s_2

5 CONCLUSIONS

The outage probability of CRS-NOMA system was investigated in this paper over Rician fading channels. Closed form expressions for P_o, s_1, and P_o, s_2 were derived. It was noticed that increasing the values of ρ or $k$or Ω will improve P_o, s_1. Furthermore, P_o, s_2 will improve as the values of k or Ω increase, but the improvement for P_o, s_1 is more pronounced than P_o, s_2. Moreover, increasing ρ beyond 15 dB will not help improving P_o, s_2 much. This means that improving P_o, s_2 will depend on the links average powers and channels conditions

REFERENCES

[1] Do, D.-T.; Le, C.-B. Application of NOMA in Wireless System with Wireless Power Transfer Scheme: Outage and Ergodic Capacity Performance Analysis. Sensors 2018, 18, 3501.

[2] Z. Han, T. Himsoon, W. P. Siriwongpairat, & K. J. R. Liu, "Resource allocation for multiuser cooperative OFDM networks: Who helps whom and how to cooperate," IEEE Trans. Veh. Technol., vol. 58, no. 5, pp. 2376–2391, Jun. 2009.

[3] K. M. Rabie B. Adebisi E. H. Yousif H. Gacanin A. & M. Tonello "A comparison between orthogonal and non-orthogonal multiple access in cooperative relaying power line communication systems" IEEE Access vol. 5 pp. 10118–10129 2017.

[4] Y. Saito, Y. Kishiyama, A. Benjebbour, T. Nakamura, A. Li, & K. Higuchi, "Non-orthogonal multiple access (NOMA) for cellular future radio access," in Proc. IEEE Vehicular Technology Conference, Dresden, Germany, Jun. 2013.

[5] S. M. R. Islam, N. Avazov, O. A. Dobre, & K. S. Kwak, "Power-domain non-orthogonal multiple access (NOMA) in 5G systems: Potentials and challenges," arXiv preprint arXiv:1608.05783, 2016.

[6] Ding Z, Peng M, & Poor HV. Cooperative non-orthogonal multiple access in 5G systems. IEEE Communications Letters. 2015;19(8):1462–5.
[7] Zhong C & Zhang Z. Non-orthogonal multiple access with cooperative full-duplex relaying. IEEE Communications Letters. 2016;20(12):2478–81.
[8] Kim J-B, Song MS, & Lee I-H, editors. Achievable rate of best relay selection for non-orthogonal multiple access-based cooperative relaying systems. 2016 international conference on information and communication technology convergence (ICTC); 2016: IEEE.
[9] Yang Z, Ding Z, Wu Y, & Fan P. Novel relay selection strategies for cooperative NOMA. IEEE Transactions on Vehicular Technology. 2017;66(11):10114–23.
[10] Wan D, Wen M, Ji F, Liu Y, & Huang Y. Cooperative NOMA systems with partial channel state information over Nakagami-m fading channels. IEEE Transactions on Communications. 2017;66(3):947–58.
[11] Zhang J, Dai L, Jiao R, Li X, & Liu Y. Performance analysis of relay assisted cooperative non-orthogonal multiple access systems. submitted to IEEE Wireless Commun Lett. 2017.
[12] Jiao R, Dai L, Zhang J, MacKenzie R, & Hao M. On the performance of NOMA-based cooperative relaying systems over Rician fading channels. IEEE Transactions on Vehicular Technology. 2017;66(12): 11409–13.
[13] P. K. Jha & D. S. Kumar, "Achievable rate analysis of relay assisted cooperative NOMA over Rician fading channels," in *Proc. Int. Conf. Recent Adv. Inf. Technol. (RAIT)*, Dhanbad, India, Mar. 2018, pp. 1–5.
[14] M. K. Simon & M.-S. Alouini, Digital Communications over Fading Channels: A Unified Approach to Performance Analysis. New York: John Wiley & Son, Inc, 2005.
[15] Wannaree Wongtrairat & Pornchai Supnithi, "New Simple Form for PDF and MGF of Rician Fading Distribution," in 2011 Symp. Intelligent Sig. Processing And Comm. Sys.(ISPACS), pp. 1–4, Dec. 7–9, 2011.
[16] I. S. Gradshteyn & I. M. Ryzhik, Table of Integrals, Series, and Products, 6th ed. San Diego, CA: Academic, 2000.
[17] Sofotasios PC, Tsiftsis TA, Brychkov YA, Freear S, Valkama M, & Karagiannidis GK. Analytic expressions and bounds for special functions and applications in communication theory.
[18] J.Zhang,L.Dai,Y.Zhang, & Z.Wang,"Unified performance analysis of mixed radio frequency/free-space optical dual-hop transmission systems," IEEE/OSA J. Lightw.
[19] P. Prudnikov, Y. A. Brychkov, & O. I. Marichev, Integrals and Series. New York: Gordon and Breach, 1992, vol. 4.
[20] K. P. Peppas, "Dual-hop relaying communications with cochannel inter- ference over $\eta - \mu$ fading channels," IEEE Trans. Veh. Technol., vol. 62, no. 8, pp. 4110–4116, Oct. 2013.
[21] Z. Ding, Z. Yang, P. Fan, & H. V. Poor, "On the performance of non-orthogonal multiple access in 5G systems with randomly deployed users," IEEE Signal Process. Lett., vol. 21, no. 12, pp. 1501–1505, Dec. 2014.

Proceedings of the 1st International Congress on Engineering Technologies – Kiwan & Banat (Eds)

A miniaturized ultra-wideband Wilkinson power divider using non-uniform coplanar waveguide

Heba H. Jaradat, & Nihad I. Dib
Electrical Engineering Department, Jordan University of Science and Technology, Irbid, Jordan

Khair A. Al Shamaileh
Electrical and Computer Engineering Department, Purdue University Northwest, Hammond, IN, USA

ABSTRACT: In this paper, the design of an ultra-wideband Wilkinson power divider (WPD) using non-uniform coplanar waveguide (CPW) is introduced. The main idea is to replace the conventional CPW by a non-uniform one that achieves the ultra-wideband operation. To generate the non-uniform design, the even and odd mode circuits are analyzed, and the Fourier series expansion is embedded in an optimization process. Then, three resistors are incorporated to improve the output port's isolation over the entire band. A Full-wave simulation is performed using HFSS. The new divider has a compact size, enhanced bandwidth (0.2 GHz-14 GHz), and isolation better than –10 dB.

Keywords: Coplanar waveguide (CPW), Non-uniform coplanar waveguide, Ultra-wideband (UWB), Wilkinson power divider (WPD).

1 INTRODUCTION

Power dividers and combiners are essential parts in wireless communications systems which are used to implement different devices such as antenna array, array networks, power amplifiers, etc. [1]. Wilkinson power divider (WPD) is one of the most well-known power dividers. It is characterized by ports matching, isolation between output ports and the narrow bandwidth (BW). In the literature, different techniques were proposed in order to enhance the bandwidth of the conventional power divider and improve the isolation over a wide range of frequencies such as tapered transmission lines [2], trapezoidal and triangular-shaped resonators [3] and band-pass transformers [4]. However, enhancing the performance costs means increasing the utilization area which is considered a major drawback of the mentioned techniques. Un-equal split WPDs were discussed in [5, 6] based on microstrip technology. In [7], an ultra-wideband WPD was incorporated to design energy harvesting circuits.

In this paper, the conventional design approach of coplanar waveguide (CPW) is modified to achieve a new power divider with an extended bandwidth and enhanced isolation. The design approach depends mainly on replacing the conventional coplanar waveguide with non-uniform one and incorporating three isolation resistors based on an optimization process. The rest of the paper is organized as follows: in section II, the design procedure is highlighted, and in section III an ultra-wideband WPD is designed, simulated and analyzed.

2 DESIGN PROCEDURE

In this section, the design of an ultra-wideband (UWB) CPW-based WPD is presented. In this approach, the design is proposed to work in a frequency range from 1 GHz to 13 GHz. Figure 1 shows the layout of the proposed divider and its equivalent even/odd mode circuits. To generate a

DOI 10.1201/9781003178255-9

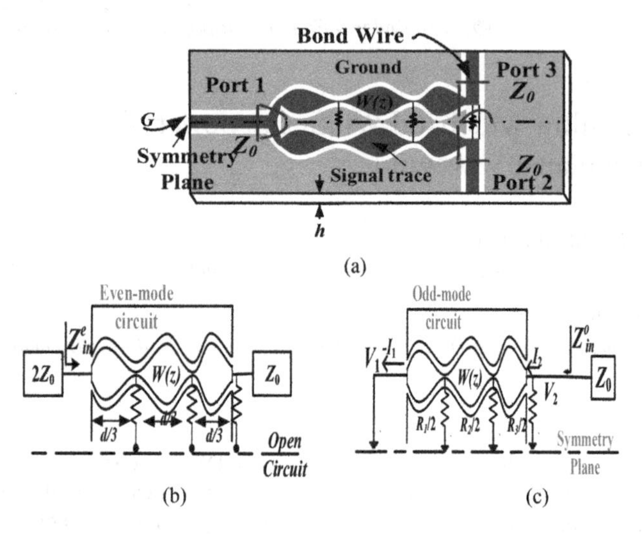

Figure 1. (a) The configuration of the ultra-wideband CPW-based WPD, (b) the equivalent even-mode circuit and (c) the equivalent odd-mode circuit.

non-uniform CPW operating in the UWB range, the non-uniform CPW transformer is subdivided into K short segments.

Then, the center conductor is computed based on the Fourier series expansion which is expressed as follows [8]:

$$\ln\left(W(z)/W_{ref}\right) = \sum_{n=0}^{N} a_n \cos\left(\frac{2\pi nz}{d}\right) + \sum_{n=1}^{N} b_n \sin\left(\frac{2\pi nz}{d}\right) \tag{1}$$

where W_{ref} is the reference line width, N is the number of series coefficients (N=5 is used here) and d is the length of the non-uniform CPW which is selected arbitrarily. In this context, d is chosen to be 20 mm. The even mode analysis is used to determine the Fourier coefficients related to the UWB-design by carrying out an optimization process. The adopted Error function in the optimization process is intended to minimize the reflection coefficients of the input port at the target frequencies as follows [8]:

$$\text{Error} = \sqrt{\sum_{j=1}^{M} \left|\Gamma_{in}^{e}(f_j)\right|^2} \tag{2}$$

where f_j are the operating frequencies, M is their number and $\Gamma_{in}^{e}(f_j)$is the even mode's reflection coefficient at the input port which is written as follows [8]:

$$\Gamma_{in}^{e}(f_j) = \frac{Z_{in}^{e}(f_j) - 2Z_o}{Z_{in}^{e}(f_j) + 2Z_o} \tag{3}$$

where Z_o=50 Ω and

$$Z_{in}^{e} = \frac{A(f_j) + B(f_j)}{C(f_j)Z_o + D(f_j)} \tag{4}$$

The total $ABCD$ matrix of the non-uniform CPW is given by [8]:

$$\begin{bmatrix} A & B \\ C & D \end{bmatrix} = \prod_{i=1}^{K} \begin{bmatrix} A_i & B_i \\ C_i & D_i \end{bmatrix} \tag{5}$$

Table 1. The UWB non-uniform cpw fourier coefficients.

a_0	a_1	a_2	a_3	a_4	a_5
0.1853	−0.0215	−0.0153	−0.0709	−0.0706	−0.0069
–	b_1	b_2	b_3	b_4	b_5
–	−0.2881	−0.1525	−0.1233	0.1088	−0.1970

K is set to be 50.The *ABCD* parameters of the ith section are expressed as [8]:

$$\begin{bmatrix} A_i & B_i \\ C_i & D_i \end{bmatrix} = \begin{bmatrix} \cos(\Delta\theta) & jZ((i-0.5)\Delta z)\sin(\Delta\theta) \\ j\frac{\sin(\Delta\theta)}{Z((i-0.5)\Delta z)} & \cos(\Delta\theta) \end{bmatrix} \tag{6}$$

where $\Delta\theta$ is the electrical length of each segment and $Z((i-0.5)\Delta z)$ is the characteristic impedance at the center of each segment.

On the other hand, to determine the values of the optimum resistors, another optimization process is performed based on the odd mode circuit. Figure 1 (c) shows the odd-mode circuit which consists of non-uniform CPW divided into three equal sections separated by three resistors of $R_1/2$, $R_2/2$ and $R_3/2$. The whole ABCD matrix of the odd-mode circuit can be calculated as follows [8]:

$$\begin{aligned} ABCD_{total} &= ABCD_{section\,1} \times ABCD_{R_1/2} \\ &\times ABCD_{section\,2} \times ABCD_{R_2/2} \times ABCD_{section\,3} \times ABCD_{R_3/2} \end{aligned} \tag{7}$$

The relation between V_1, I_1, V_2, and I_2 can be written as follows [8]:

$$\begin{bmatrix} V_1 \\ -I_1 \end{bmatrix} = \begin{bmatrix} A & B \\ C & D \end{bmatrix}_{total} \begin{bmatrix} V_2 \\ -I_2 \end{bmatrix} \tag{8}$$

Accordingly, the input impedance and the reflection coefficient of the odd mode circuits are calculated as:

$$Z_{in}^{o} = \left.\frac{V_2}{I_2}\right|_{V_1=0} = \frac{B}{A} \tag{9}$$

$$\Gamma_{in}^{o}(f_j) = \frac{Z_{in}^{o}(f_j) - Z_o}{Z_{in}^{o}(f_j) + Z_o} \tag{10}$$

Consequently, to determine the value of the optimum resistors that achieve the optimum output ports matching, the adopted Error function in the optimization process is intended to minimize the reflection coefficients at the output ports simultaneously at the target frequencies as follows:

$$\text{Error} = \sqrt{\sum_{j=1}^{M} \left|\Gamma_{in}^{o}(f_j)\right|^2} \tag{11}$$

In this regard, the operating frequencies are from 1 GHz to 13 GHz with a step size of 1 GHz. This equation can be solved using an optimization code where the values of the three resistors are the optimized variables.

3 DESIGN EXAMPLE AND RESULTS

Based on this procedure and fixing the ground-to-ground spacing ($W+2G=3.596$ mm), an UWB WPD is designed considering an FR-4 substrate with 4.4 relative permittivity and 1.5 mm thickness. The obtained Fourier coefficients are listed in Table 1. The length of the proposed non-uniform

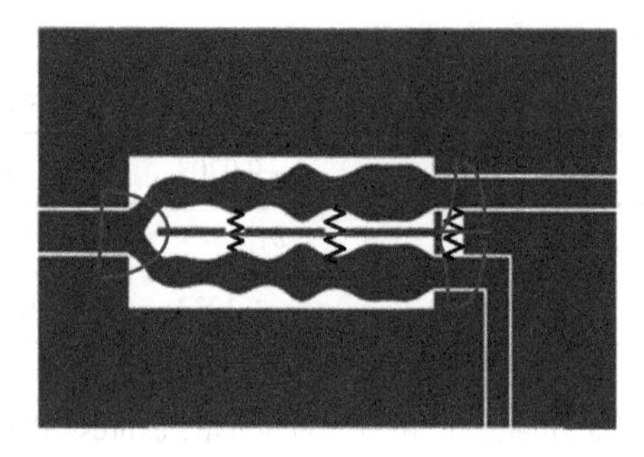

Figure 2. The layout of the proposed UWB WPD.

Table 2. The physical dimensions of the designed WPD. (All dimensions are in mm).

Feed Lines		Non-uniform CPW			
Trace width	**Gap width**	W_{min}	W_{max}	d	**Gap width**
2	0.216	0.15	2.81	20	0.579

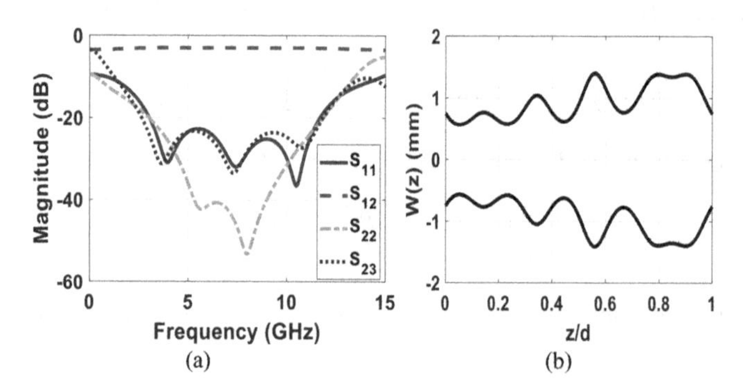

Figure 3. (a) The analytical scattering parameters, (b) the profile width of the UWB non-uniform CPW transformer.

CPW section is set to be 20 mm. In addition, the obtained values of the optimum resistors are R_1 = 357.566 Ω, R_2 = 283.6 Ω and R_3 = 120 Ω. Figure 2 shows the layout of the UWB power divider. It is paramount to point out that CPW technology requires adding bond wires at the discontinuities and junctions to avoid the occurance of the slot-line mode as shown in Figure 2. Table 2 lists the physical dimensions of the proposed non-uniform CPW. The analytical results and the profile width of the proposed divider are shown in Figure 3. Then, the ultra-wideband performance is demonstrated by performing the full-wave simulation. Figure 4 depicts the simulation results of the proposed divider. The input/output ports matching and the output ports isolation are better than −10 dB within the range of 0.2–14 GHz. The transmission parameters are within −4 ±1 dB in the frequency range of 1–10 GHz.

4 CONCLUSION

In this work, an ultra-wideband WPD based on non-uniform CPW was designed and simulated. The wide-band operation was achieved by replacing the uniform CPW with a non-uniform one whereas the isolation was enhanced through adding three resistors separated by equal distance along the

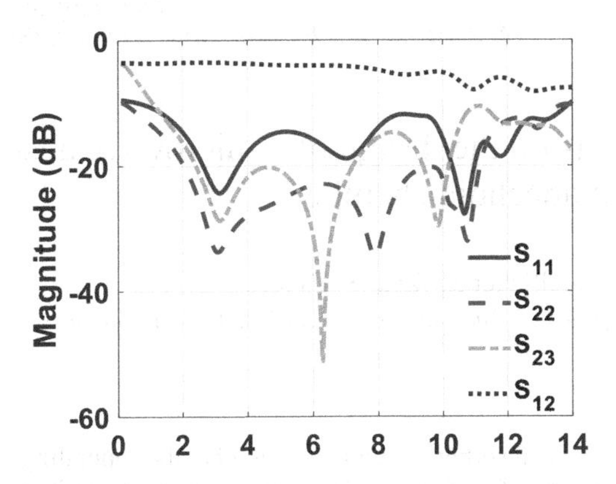

Figure 4. The simulated (HFSS) results of the proposed ultra-wideband divider.

transmission line. To obtain the parameters of the non-uniform CPW, an optimization process was carried out based on simplified mathematical formulation for the odd and even mode circuits. A wide range of operation extending from 1 GHz to 10 GHz was achieved with a small utilization area.

REFERENCES

[1] D. Pozar, Microwave engineering, 3rd ed., Wiley, 2005.

[2] H. Habibi & H. Miar Naimi, "Taper transmission line UWB Wilkinson power divider analysis and design," *International Journal of Electronics*, vol. 106, no. 9, pp. 1332–1343, Sep. 2019.

[3] S.A. Zonouri & M. Hayati, "A compact ultra-wideband Wilkinson power divider based on trapezoidal and triangular-shaped resonators with harmonics suppression," *Microelectronics Journal*, vol. 89, pp. 23–29, July 2019.

[4] X. Wang, Z.Ma, T. Xie, M. Ohira, CP. Chen, & G. Lu, "Synthesis theory of ultra-wideband bandpass transformer and its Wilkinson power divider application with perfect in-band reflection/isolation," *IEEE Transactions on Microwave Theory and Techniques*, vol. 67, no. 8, pp. 3377–3390, July 2019.

[5] V. Ravi, D. Sharma, K. Purnima, P.V Naidu, A.R Raja, A. Kumar, A.P. Singh, & P. Raveendra. "Design of compact wideband unequal wilkinson power divider with improved isolation," *International Journal of Engineering & Technology*, vol. 7, no. 3, pp.1063–1066, 2018.

[6] S. Saleh, W. Ismail, I.S. Zainal Abidin, M.H. Jamaluddin, S.A. Al-Gailani, A.S. Alzoubi, & M.H. Bataineh, "Nonuniform compact Ultra-Wide Band Wilkinson power divider with different unequal split ratios," *Journal of Electromagnetic Waves and Applications*, vol. 34, no. 2, pp.154–167.

[7] O. Kasar, M. Kahriman, & M.A. Gozel, "Application of ultra wideband RF energy harvesting by using multisection Wilkinson power combiner," *International Journal of RF and Microwave Computer Aided Engineering*, vol. 29, no. 1, p.e21600, 2019.

[8] H. Jaradat, N. Dib, & K. Al Shamaileh, "Miniaturized multi'frequency Wilkinson power dividers based on nonuniform coplanar waveguide," *International Journal of RF and Microwave ComputerAided Engineering*, vol. 29, no. 6, May 2019.

Proceedings of the 1st International Congress on Engineering Technologies – Kiwan & Banat (Eds)
 ISBN 978-0-367-77630-5

Multi variant effects on the design of a microwave absorber and performance of an anechoic chamber

Raneem AlShair, Waleed AlSaket & Yanal S. Faouri
Electrical Engineering Department, The University of Jordan, Amman, Jordan

ABSTRACT: A microwave absorber for an anechoic chamber operating at RF ranges has been designed to study the effects of each variable on the reflection coefficient $\boldsymbol{S}_{11}$. The observed variables are the absorber shape, the effect of full and partial coating, pyramidal height and base size, number of pyramids, excitation source location, the thickness of the underneath perfect electric conductor (PEC) layer and the electrical properties of the absorber coating. The proposed electromagnetic absorbers are investigated using the high-frequency structure simulator (HFSS). The simulated reflection coefficient $\boldsymbol{S}_{11}$ result for each effect is plotted and its average over the entire band is calculated. The RF range from 1–6.5 GHz is considered to achieve better return loss (RL = $|\boldsymbol{S}_{11}|$) $\geq$ 10 dB. The achieved average $\boldsymbol{S}_{11}$ is -26.86 dB. These absorbers are a necessity in conducting experiments related to antennas to measure their characteristics, especially the radiation pattern and gain.

Keywords: Microwave absorber, Reflection Coefficient S_{11}, Pyramidal absorber, EM absorber coating, polyethylene, Anechoic chamber, Reverberation room, Radiation pattern, Antenna measurement setup.

1 INTRODUCTION

An anechoic chamber has been a major issue throughout history. Many researchers have been studying and trying to design a chamber in the radio frequency range for radiation pattern measurements. Some of them focused on the simulation using different techniques, other researchers have been conducting research about the absorption materials and the conductive layer on the interior walls of the anechoic chamber. Antenna radiation pattern measurements either the co-polarized or cross-polarized patterns with a smaller error is required for all antenna applications such as ultra-wideband with [1] or without [2–3] rejection band, MIMO antennas [4], frequency agility, polarization diversity [5–6] and pattern reconfigurable antenna [7] etc.

A study on the microwave absorber design that shows that there are many parameters that can affect the performance of the absorber has been reported in [8]. The most affected condition of the absorber performance is the shape of the absorber walls. There are many models for absorber design such as cubic, truncated pyramid, wedge, oblique wedge and pyramidal. Pyramidal and wedge showed the best results. For pyramidal absorbers, there are eight polygonal based pyramidal shapes, i.e. triangular, square or tetragonal, pentagonal, hexagonal, heptagonal, octagonal, nonagonal and decagonal. A Pyramidal shape with a square base has shown the best result for the reflectivity parameter in the reported simulations and analysis [8–9].

Most anechoic chambers are made from ferrite coating [10] since the ferrite can absorb electromagnetic waves with high absorption for the range 1 to 10 GHz if it is composed with other materials and be efficient for lower frequencies from 30 MHz to 1 GHz. An alternative material for microwave absorbers from agricultural waste called biomaterial or bio-based material that can be used to fabricate the absorbers has been reported in [11]. Biomaterial has been verified as a good candidate for microwave absorbers to replace the conventional absorber material in terms of

 DOI 10.1201/9781003178255-10

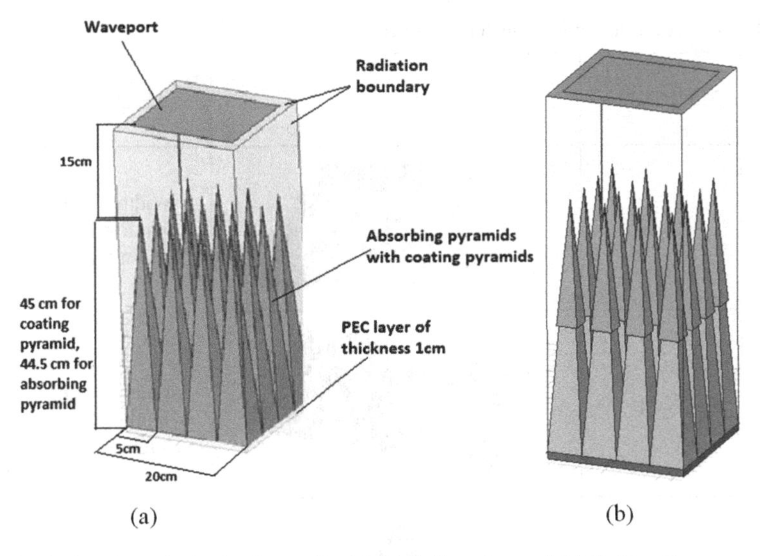

Figure 1. The designed anechoic chamber wall with (a) Full coating and (b) Half coating.

good reflectivity, low cost, minimal industrial waste and environmental problems. Other studies on biomaterials were on rice husk [12]. Then, the addition of other organic and agricultural materials to improve the results further such as tire rubber dust and sugarcane baggage is reported in [13].

A new electromagnetic absorbing material developed from carbon fiber loaded epoxy foam for anechoic chamber application has been presented in [14]. A description of the anechoic walls and radiating elements as equivalent models is reported in [15] to reduce the degrees of freedom of the numerical problem and the required computational resources required by the modeling of anechoic chambers.

This paper is a guide to understanding how an anechoic chamber will be affected under the influence of each parameter change. This will inform the selection of proper, cost-effective coatings that will obtain good experimental results once manufactured. The effect of each parameter, either related to the absorber properties, such as permittivity, conductivity and the percentage of the coating, or related to the chamber shape and dimensions, such as base size, height, the thickness of the PEC layer and the number of the absorbers are investigated, and their effect on S11 is plotted and compared. A parametric study for each parameter is conducted to reach the optimum design. The anechoic chamber structure with the pyramidal absorbers' detailed dimensions is outlined in Section II. In Section III, the simulated reflection coefficient results and their explanations are viewed. In Section IV, the conclusion is given.

2 MICROWAVE CHAMBER STRUCTURE

The designed chamber with the assigned parameters is shown in Figure 1(a). An example of a coating layer is shown in Figure 1(b), where the four sides of the surrounding box and the area around the excitation waveguide ports are assigned as a radiation boundary. A comprehensive parametric analysis was conducted in all dimensions to obtain the optimum dimensions. The shape of the absorbers was tested first, and these shapes are oblique, wedge and pyramidal as summarized in Table 1. The chamber has a base made of a PEC layer to block all the electromagnetic radiation from the surroundings. The designed 4 × 4 chamber wall overall size is 20 × 20 × 45cm^3. The excitation source is placed above the pyramids and the radiating wave is directed towards the pyramidal absorbers. The size of the port is calculated using Eq. (1) to ensure only the 1 GHz frequency is transmitted, this required the port to be of a rectangular shape with the length of the minimum edge is 15 cm.

Table 1. The shapes of different investigated absorbers.

Model Name	Model Shape	Description
Oblique		Width = 5 cm Depth = 10 cm Height = 35 cm
Wedge		Width = 5 cm Depth = 10 cm Height = 35 cm
Pyramidal		Width = 5 cm Depth = 5 cm Height = 35 cm

$$f_c = \frac{1}{2a} \cdot \frac{c}{\sqrt{\epsilon_{reff}}} \tag{1}$$

where 'c' is the speed of light, 'a' the narrower edge, fc is the cutoff frequency and ϵ_{reff} is the effective dielectric constant.

3 SIMULATION RESULTS AND PARAMETRIC STUDY

The designed anechoic chamber shown before was simulated using HFSS to test the performance of the microwave absorbers. The absorber shape made with polyethylene foam with characteristics $\epsilon_r = 2.25, \mu_r = 1 and \sigma = 0 S/m$ was tested first. The considered shapes are wedge, oblique and pyramidal shapes. The average S_{11} calculated over the range 1–6.5 GHz for these shapes and without utilizing any type of coatings are −19 dB, -16.25 dB and -17.3 dB respectively as shown in Figure 2. The pyramidal shape gives the best average S_{11} after inserting the coating, so it is selected for further studies. The selected pyramid is coated with absorber material of characteristics as $\epsilon_r = 2.5, \mu_r = 50 and \sigma = 0.04 S/m$. The average S_{11} curves with different coating levels are shown in Figure 3. The achieved results for a coating percentage of 25%, 50%, 75% and full coating are -16.76 dB, -21.94 dB, -20.63 dB and -24.11 dB respectively. The complete foam coating will give the best absorption for the monitored range.

In Figure 4, the pyramid height with a full coating is investigated with a fixed base size of $5 \times 5cm^2$. The average S_{11} enhances with increased height as it changed from -19.56 dB, -21.87 dB, -23.93 dB to -26.01 dB with heights of 25 cm, 35 cm, 40 cm and 45 cm respectively. Then for a fixed height of 45 cm and different squared base sizes with an edge side of 4 cm, 5 cm and 6 cm produced an average S_{11} of -24.89 dB, -26.01 dB and -23.35 dB respectively. So, a pyramid of base size $5 \times 5cm^2$ and height of 45 cm will be a good agreement together. Another combination

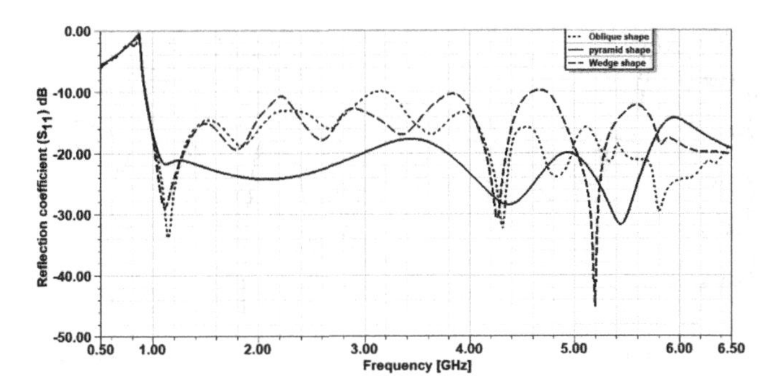

Figure 2. The impact of different absorber shapes on S_{11}.

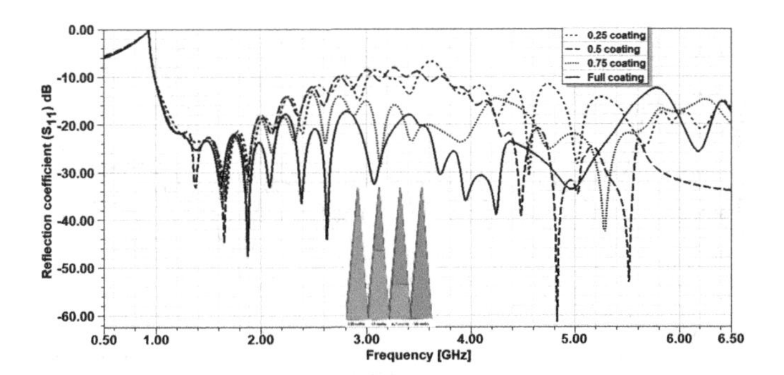

Figure 3. Variation of S_{11} vs frequency due to different coating percentage.

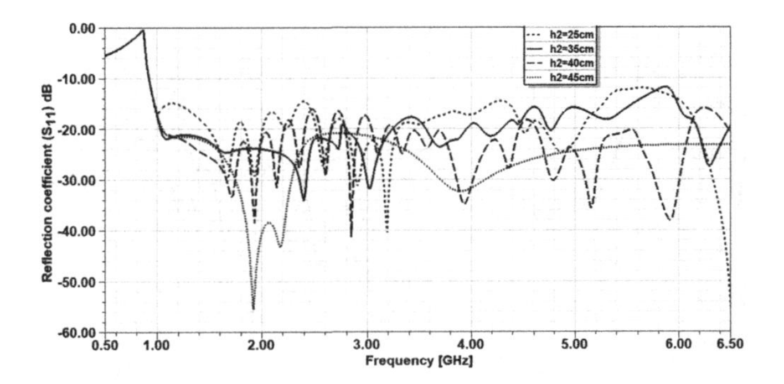

Figure 4. The impact of pyramid height on S_{11}.

with a good agreement is a base size of $4 \times 4cm^2$ with a height of 35 cm that gives an average S_{11} of -26.86 dB as shown in Figure 5.

To extend the designed chamber, the number of pyramidal absorbers has increased to different combinations as shown in Figure 6. Results show that squared chamber walls will give better results since for $4 \times 4cm^2$, $5 \times 5cm^2$ and $6 \times 6cm^2$ the average S_{11} is -26.86 dB, -24.3 dB and -2.04 dB respectively.

The location of the waveguide port is then investigated to determine the best location by considering various height values of 7.5 cm (0.25λ), 15 cm (0.5λ) where the far-field region starts, 30 cm (λ) and 90 cm (3λ) and the simulated average S_{11} are -27.37 dB, -26.86 dB, -23.03 dB and -25.16 dB respectively as shown in Figure 7. So, being closer to the excitation source to the chamber wall within a distance 0.25–0.5 λ will give better average S_{11}. The layer beneath the pyramidal

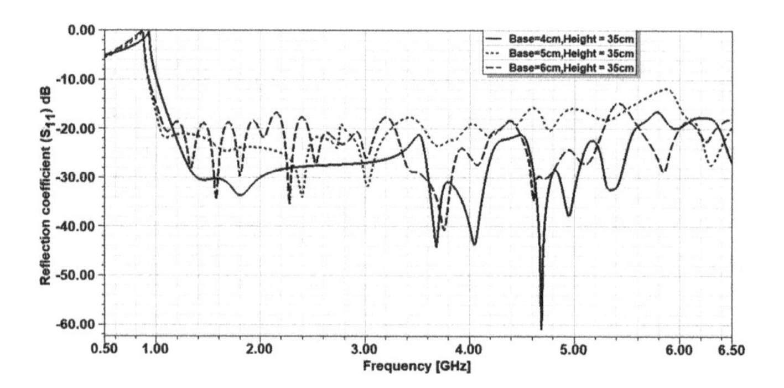

Figure 5. Variation of S_{11} vs frequency due to the variation of the pyramid base dimensions.

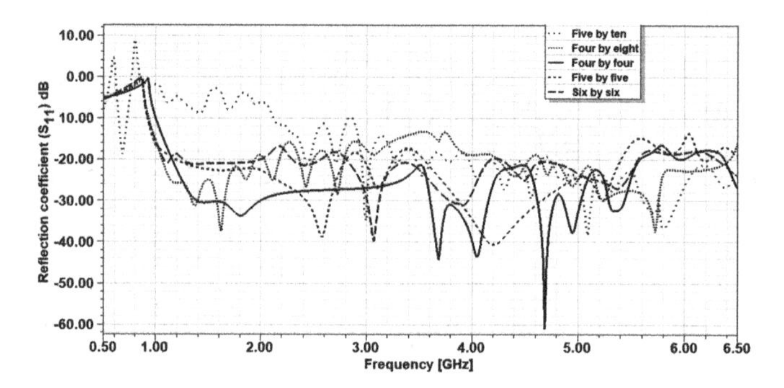

Figure 6. The effect of the number of pyramids on S_{11} results.

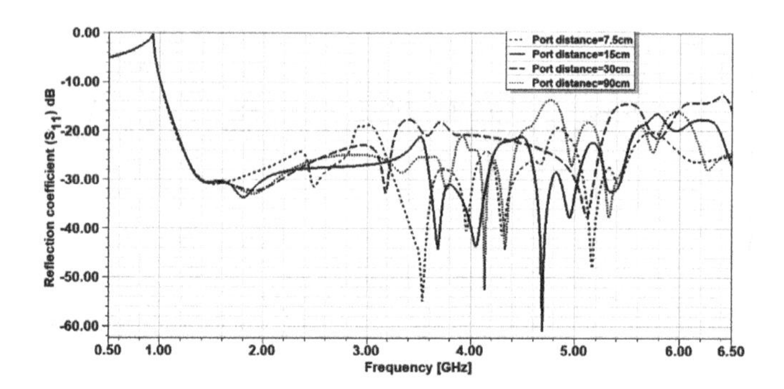

Figure 7. The effect of the port location on S_{11}.

absorbers must be made of PEC to prevent any unwanted signal from the outer environment and this layer thickness is studied for a small thickness of 0.5 mm, 1 mm, 1.5 mm and 2 mm as shown in Figure 8 and the achieved average S_{11} are -23.86 dB, -22.77 dB, -24.15 dB, and -25.114 dB for the considered thicknesses respectively.

It is better for the pyramids to have a sharper vertex since when the top vertex is removed the average S_{11} for the considered range is dropped to -13.15 dB as shown in Figure 9.

The electrical properties of the microwave absorber coating, such as permittivity and conductivity, are also examined, and their results are plotted in Figure 10 for the effect of the dielectric constant and tabulated in Table 2 for the effect of the conductivity. A material with varying dielectric constant values of 2.5, 2.9 and 3.3 achieves an average S_{11} of -26.86 dB, -24.4 dB and -22.89 dB

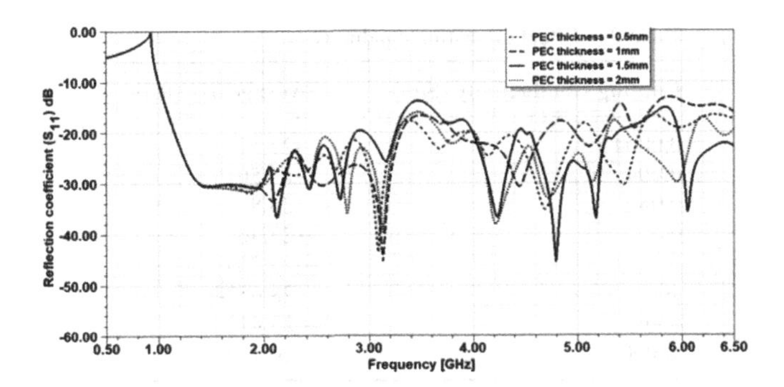

Figure 8. The variation of S_{11} due to different PEC layer thickness.

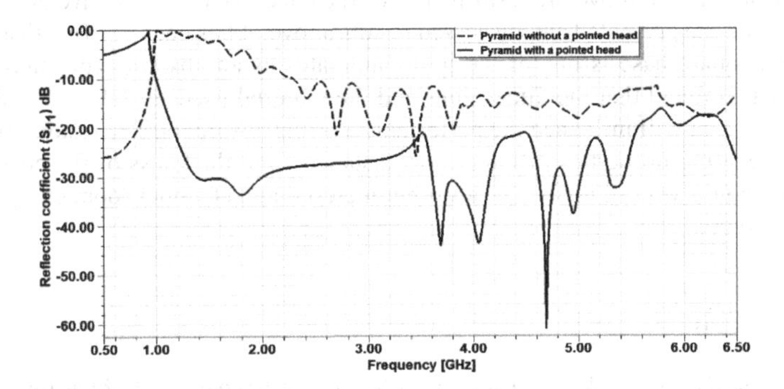

Figure 9. The effect of the pyramid vertex on S_{11}.

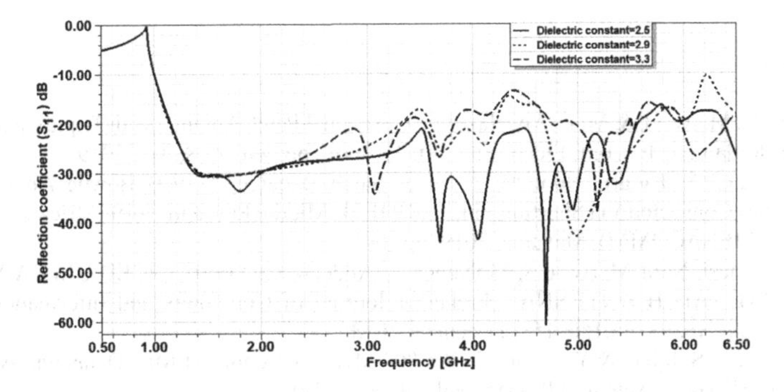

Figure 10. The effect of the dielectric constant of the pyramidal absorber on S_{11}.

in turn while changing conductivity will also affect the absorber characteristics in a non-clear way. Very low conductivity of 0.04 S/m is used through this conducted study.

It is worth mentioning that this chamber is not strictly used for testing antennas; it is also necessary for microwave circuits such as amplifiers, mixers, and tunable filters [16] for measuring their results in a low noise environment.

4 CONCLUSION

A guide to designing an anechoic chamber for microwave frequencies is presented. After conducting the intensive simulation, the effect of several parameters and design factors that influenced the

Table 2. Effect of coating conductivity.

Coating Conductivity (S/m)	Average S_{11} (dB)
0.0004	−27.5754
0.004	−24.5614
0.04	−26.8615
0.4	−27.469
4	−23.7291
40	−28.6574
400	−23.4179
40000	−20.1375

reflectivity performance and overall performance of the anechoic chamber have been studied and analyzed. Based on the presented and discussed results above, it has been verified that a pyramidal absorber with a square base is one of the best candidates as an absorber for microwave signal absorbers. It is also found that the dimensions and the material used for fabricated absorbers are the critical factors that will influence the performance of the absorbers. The results so far indicate that for certain coating properties different port distances, the thickness of the coating material and PEC layer, and the overall volume of absorber material and chamber room can give distinctly performing absorbers.

ACKNOWLEDGEMENT

The authors would like to thank the Deanship of Scientific Research (DSR) at the University of Jordan for providing facilities in conducting this research for the year 2019–2020.

REFERENCES

[1] Y. S. Faouri, N. M. Awad, & M. K. Abdelazeez, "Hexagonal Patch Antenna with Triple Band Rejections," 2019 IEEE Jordan Int. Jt. Conf. Electr. Eng. Inf. Technol., pp. 446–448, Apr. 2019.

[2] Ali Salim, Yanal S. Faouri, Noor M. Awad & Saleh Baqaleb, "UWB Bowtie Antenna with Two Ground Planes Connected Via Shorting Pin," 2019 IEEE Middle East and North Africa Communications Conference (MENACOMM), Manama, 2019, pp. 1–4.

[3] Yanal S. Al-Faouri, Noor M. Awad & Mohamed K. Abdelazeez, "ENHANCED ULTRA-WIDE BAND HEXAGONAL PATCH ANTENNA," Jordanian Journal of Computers and Information Technology (JJCIT), Vol. 04, No. 03, pp.150–158, December 2018

[4] M. S. Sharawi, S. S. Iqbal, & Y. S. Faouri, "An 800 MHz 2×1 compact MIMO antenna system for LTE handsets," IEEE Trans. Antennas Propag., vol. 59, no. 8, 2011.

[5] Malek Y. Al-mallah, Zaid I. Saleh & Yanal S. Faouri, "Two Feed Reconfigurable Microstrip Patch Antenna for Polarization Diversity," 2019 IEEE Jordan International Joint Conference on Electrical Engineering and Information Technology (JEEIT), Amman, 2019.

[6] Zaid I. Saleh, Malek Y. Al-mallah & Yanal S. Faouri, "Polarization Reconfigurable Microstrip Slotted Antenna with Two Opposite Feeds," Global Power, Energy and Communication Conference- IEEE GPECOM 2019, 12–15 June 2019.

[7] S. Hamdan, A. Nofal, R. A. Nawaiseh & Y. Faouri, "Microstrip antenna for radiation pattern reconfigurability," 2017 IEEE Jordan Conference on Applied Electrical Engineering and Computing Technologies, AEECT 2017 (2018), pp.1–5, January 2018, 11–13 October 2017.

[8] H. Nornikman, P.J Soh, & A.A.H Azremi, "Modelling simulation stage of pyramidal and wedge microwave absorber design," 4th International Conference on Electromagnetic Near Field Characterization and Imaging. June 2009.

[9] H. Nornikman, P.J Soh, & A.A.H Azremi, "Performance of Different Polygonal Microwave Absorber Designs Using Novel Material," The 2009 International Symposium on Antennas and Propagation (ISAP 2009) October 20–23, 2009, Bangkok, THAILAND.

[10] K. Shimada, T. Hayashi, & M. Tokuda, "Fully compact anechoic chamber using the pyramidal ferrite absorber for immunity test," IEEE Int. Symp. Electromagn. Compat., vol. 1, pp. 225–230, 2000.
[11] H. Nornikman, P. J. Soh, A. A. H. Azremi, F. H. Wee, & M. F. Malek, "Investigation of an Agricultural Waste as an Alternative Material for Microwave Absorbers," Progress In Electromagnetics Research Symposium Proceedings, Moscow, Russia, August 18–21, 2009. Pp. 1287–1291.
[12] H. Nornikman, P. J. Soh, F. Malek, A. A. H. Azremi, F. H. Wee, & R. B. Ahmad, "Microwave wedge absorber design using rice husk - An evaluation on placement variation," 2010 Asia-Pacific Symp. Electromagn. Compat. APEMC 2010, no. May, pp. 916–919, 2010.
[13] F. Malek et al., "Rubber tire dust-rice husk pyramidal microwave absorber," Prog. Electromagn. Res., vol. 117, no. April, pp. 449–477, 2011.
[14] Chloé Méjean, Laura Pometcu, Ratiba Benzerga, Ala Sharaiha, Claire Le Paven-Thivet, Mathieu Badard, & Philippe Pouliguen, "Electromagnetic absorber composite made of carbon fibers loaded epoxy foam for anechoic chamber application," Materials Science and Engineering: B, Volume 220, June 2017, Pages 59–65.
[15] Silvano Chialina, Matteo Cicuttin, Lorenzo Codecasa, Giovanni Solari, Ruben Specogna, & Francesco Trevisan "Modeling of Anechoic Chambers with Equivalent Materials and Equivalent Sources," IEEE Transactions on Electromagnetic Compatibility. (2016). Volume: 58, Issue: 4, Aug. 2016.
[16] Yanal S. Faouri, Hanin Sharif, Leena Smadi & Hani O. Jamleh, "LOW PASS AND QUAD BAND PASS TUNABLE FILTER BASED ON STUB RESONATORS TECHNIQUE," Jordanian Journal of Computers and Information Technology (JJCIT), Vol. 05, No. 02, pp.124–136, August 20.

Proceedings of the 1st International Congress on Engineering Technologies – Kiwan & Banat (Eds)
 ISBN 978-0-367-77630-5

Automatic detection of acute lymphoblastic leukemia using machine learning

Lamis R. Bany Issa & Areen K. Al-Bashir
Department of Biomedical Engineering, Jordan University of Science and Technology, Irbid, Jordan

ABSTRACT: The identification of acute leukemia blast cells in colored microscopic images is a challenging task. Usually, visual assessment for microscopic images of blood samples is performed. However, considering the quick advances in utilizing different image processing techniques,rapid and more accurate assessment can be achieved. Therefore, this paper proposes an enhanced automatic method to detect Acute Lymphoblastic Leukemia (ALL) utilizing microscopic blood sample images.

Our proposed methodology includes Color-based segmentation using the K-means clustering technique with morphological feature extraction algorithms and cell classification. The proposed method was tested on blood microscopic images from the ALL-IDB1 database, University of Milano, Italy. ALL_IDB1 is comprised of 108 blood cell images (healthy and leukemia) in which the lymphocytes are labeled by expert oncologists. The accuracy achieved was 97.2%. This relatively high accuracy suggests that the colour-based method for segmentation was more efficient and acceptable compared to the traditional thresholding methods and the methods that do not consider the overlapped cells.

Keywords: Acute Leukaemia, White blood cells (WBC), K-means, Neural Network (NN).

1 INTRODUCTION

Leukemia is defined as a cancer of the blood-forming tissue. It is usually detected from the white blood cells (WBCs); however, it can begin in other blood cell types as well. There are various types of leukemia, which are split based on whether the leukemia is rapid-growing (acute) or slower-growing (chronic), and whether it begins in myeloid cells (myeloid leukemia) or lymphoid cells (lymphoblastic leukemia) [1]. These types can also be classified into different subcategories. Specialists generally split leukemia into these four major groups [2] as shown in Table 1.

The visual analysis of blood samples is a significant step in the detection of leukemia, which is accomplished by analyzing WBCs. Therefore, accurate classification of blood cells is an important factor. Automated systems can accelerate this progress and increase the uniformity and accuracy of the detection. Accordingly, there is a regular need for a powerful, simple and stable method to analyze, count, and classify blood cells.

Reviewing the literature, WBCs segmentation using microscopic images was the very first step in most proposed methods. However, there are different segmentation methods utilized. A study by

Table 1. Major groups of leukemia.

Cell Type	Acute	Chronic
Lymphoid cell	Acute Lymphoblastic Leukemia (ALL)	Chronic Lymphoblastic Leukemia (CLL)
Myeloid cell	Acute Myeloid Leukemia (AML)	Chronic Myeloid Leukemia (CML)

 DOI 10.1201/9781003178255-11

Joshi et al used Otsu's automatic thresholding method for segregating the blood cell together with arithmetic algorithms for WBC segmentation. The K-nearest neighbor (KNN) classifier was used to classify cells into normal cells and blast cells giving overall accuracy of 93%. [3].

Other methods used for WBC segmentation based on the support vector machine (SVM) classifier with k-means clustering methods depending on some cells' features like geometry and texture to detect leukemia cells. The algorithm accomplished an acceptable performance for diagnosing Acute Myeloid Leukemia (AML) only, and its accuracy reaches 96% [4]. For the multi-SVM classifier, an accuracy of 87% has been achieved [4]. Other studies worked on full blast cells, which consist of cytoplasm and the nucleus region and was segmented by utilizing a combination of bright stretching and partial contrast techniques. They showed a better result because techniques used for image enhancement to extract the nucleus region in the WBC image sample and applying the same threshold value on both ALL and AML images. [5].

Nevertheless, the problem of overlapping cells in blood microscopic images is also one of the most limiting factors of accuracy. Hence, several algorithms are presented to solve it. Most methods are based on splitting and merging cells at the edges, as proposed in [6][7]. Other methods proposed a suggestion based on morphological operations, thresholding, and watershed mechanisms [8][9]. The drawback of these methods is the incomplete localization of the cells of interest, including the nucleus and cytoplasm as a whole [8][9].

A study by Huey Nee Lim et. al. [10] used a morphological operation together with a Color-based segmentation algorithm to segment blood cells from microscopic blood sample images. Their work resulted in reduced over-segmentation and accuracy reached 100% in some images. Only acute myeloid leukemia (AML) is considered in their study.

The authors in [11] segmented abnormal white blood cells in blood cell images using k-means, fuzzy c-means (FCM) and moving k-means (MKM) clustering. They made a comparison between three clustering techniques. The study showed that moving k-means clustering achieved the best performance for WBC segmentation. Based on the segmentation performance of 100 acute leukemia images, the MKM clustering algorithm has proven to be the best in producing a fully segmented blast area with a sensitivity value of 86.18% as compared to segmentation results provided by KM and FCM algorithms with sensitivity values of 84.04% and 85.78%, respectively.

Some studies used the changes in the features of the cells that related to geometry and statistical parameters such as mean and standard deviation to separate white blood cells from other blood components [12]. Chaitali Raje et. Al. [13] also utilized geometrical features such as area and perimeter of WBC for the prediction of Leukaemia. Both studies used the processing tool LabVIEW for feature extraction of WBCs, however, they did not solve the problem of overlapped cells [12] [13].

As discussed above, there are many proposed methods to segment leukemia cells but some methods have inaccurate results and the others have a hidden process with a long processing time. As an improvement on the previous work in [10], which gave good accuracy and solved the problem of overlapped cells but only detected the AML type of leukemia, this paper will propose an enhanced method to detect ALL utilizing microscopic blood sample images. It uses Color-based segmentation using the K-means clustering algorithm together with the changes of the WBC features such as geometrical, textural, and statistical features to detect lymphoblastic cells. This method is expected to be more efficient and acceptable in contrast with the thresholding as in [8].

2 PROPOSED METHODOLOGY

In this paper, the Color-based segmentation method is applied for WBC segmentation from other blood components in microscopic images. Different features such as texture features and geometric features are extracted from the WBCs. Finally, a neural network (NN) is applied for the WBCs classification based on the extracted features into healthy and leukemic cells. The overall working principle is depicted in Figure 1.

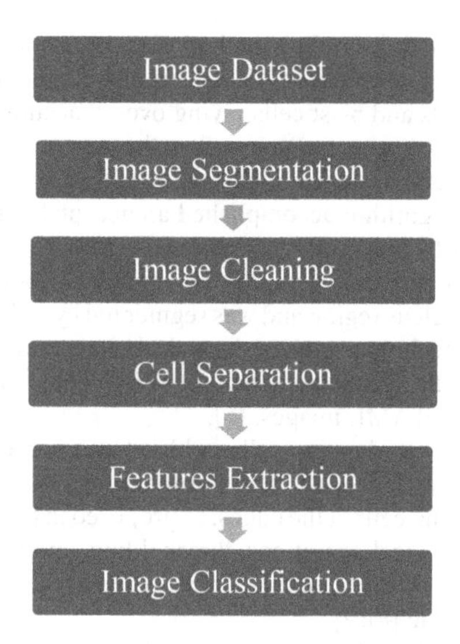

Figure 1. The overall proposed method of Lymphoblastic cells segmentation and classification.

2.1 *Image dataset*

Microscopic blood cell images from the ALL-IDB1 database were used to test our proposed algorithm. Dr. Fabio Scotti, University of Milano, Italy, collected the dataset. The database is comprised of blood sample images of healthy and leukemia cases. The database ALL IDB contained two datasets ALL_IDB1 and ALL_IDB2. An optical laboratory microscope attached to a Canon PowerShot G5 camera took the JPG blood microscopic images. All images have a resolution of 2592 × 1944 and 24-bit color depth.

ALL_IDB1 were only used in this method, it is comprised of 108 blood cells images (59 healthy and 49 leukemia). The lymphoblasts were labeled by expert oncologists resulting in 510 elements [14].

2.2 *Image segmentation*

Colour based segmentation using the K-means clustering algorithm was used to provide a preliminary segmentation of the image. Kmeans clustering succeeded as a preliminary segmentation stage to split the image pixels depending on the similarity of color components and spatial coordinates into K clusters

The L*a*b* color space is derived from the CIE XYZ tristimulus values. It is comprised of a luminosity layer 'L*' a chromaticity-layer 'a*' showing the position of the color on the red-green axis, and a chromaticity-layer 'b*' showing where the color falls on the direction of the blue-yellow axis. The color information exists in the 'a*' and 'b*' layers. Each pixel in the Lab color space was categorized into any of the four clusters. The clusters depend on computing the Euclidean distance between the pixel and each color index using K-means clustering. In this way each pixel of the whole image will be labeled to a specific color based on the minimum distance from each index that supples a color discrimination feature depending on vision [15]

Hence, Color-based clustering segmentation is accomplished for extracting the nuclei of WBCs. The 'CIE L_A_B_color' structure is used as in [12]. Each pixel of an object has two values (a* and b*). According to these two values, each pixel is categorized into four clusters, to segment the whole image into four parts, i.e., nucleus, cytoplasm, other cells, and background.

2.3 *Image cleaning*

The small particles in the image were eliminated using the area open function to eliminate objects with an area less than 400. Then, eclectic median filtering was applied to remove the noise accumulative in the images to eliminate the stain artifacts. The eclectic median filter with a size of 12*12 and open area function were applied to maintain the details of the cells.

2.4 *Cell separation*

The overlapped cells may give misclassification results because it considered as one cell with abnormal features. To separate the overlapping cells in blood cells images, the image gradient was produced from the resultant image of the image cleaning stage, and then the watershed transform is performed on the gradient image.

As known WBCs consists of five different types and ALL type of leukemia can only be initiated in the lymphocyte cells. Normal lymphocytes have a condensed and spherical nucleus, with a scanty bluish cytoplasm[16]. Features like area and circularity (roundness) were used to extract only the lymphocyte from the other type of WBCs. From the literature, the roundness threshold of the lymphocytes is almost 0.8 [17]. The image only consists of lymphocyte and lymphoblast cells. The other cells were rejected.

2.5 *Feature extraction*

The nuclear structure of cancer cells is subjected to changes that result in a large, dark and irregularly shaped nucleus. These morphological features have been considered the gold standard for diagnosing cancer.

Generally, normal cells have a regular and oval shape with a smooth appearance that is spherical, whereas cancer cells are mostly irregular and significantly vary in size. Decreased adhesion in cancer cells can lead to deranged cell propagation and smaller cell-to-cell contacts. On the contrary, normal cells will develop as a uniform layer of cells with many strong connections between neighboring cells [18].

In our proposed method, a set of geometrical, textural, and statistical features of lymphocytes were extracted. The extracted features from the segmented cells are used to classify lymphocytes into healthy and leukemic cells:

- The Geometric Features are comprised of geometrical parameters such as area, perimeter, equivalent diameter, circularity, solidity, major & minor axis length and eccentricity.
- Texture Features are comprised of correlation, homogeneity, entropy, and energy.
- Statistical Features are comprised of the mean, standard deviation, variance, and skewness.

2.6 *Classification*

In this step, a shallow neural network [19] was used to classify the lymphocytes into healthy or leukemic cells The cells are classified based on their morphological features. After descriptors extraction the class assign to the lymphocytes was used as input for the neural classifier

The segmented lymphocytes were given as an input for the NN with two layers to classify these lymphocytes according to the labeled lymphocytes from the database. The test was applied and the system was validated on the ALL-IDB1 database.

The algorithm was run using MATLAB version 2017a on a conventional PC with the processer core i5 and 4 Giga RAM. The processing time of all (108) images was 25 minutes including reading, segmentation, feature extraction, and classification (the neural network takes just a few seconds)

Figure 2 shows the results of each stage of our proposed method. Figure 2(A) shows the input original image, which was then segmented as seen in figure 2(B) The result of image cleaning after the segmentation stage is shown in figure 2(C) The cells are still overlapped which may allow misclassification. Enhanced watershed transform was applied to separate the cells as shown in figure 2(D). Finally the individual separated lymphocytes are shown in figure 2(E).

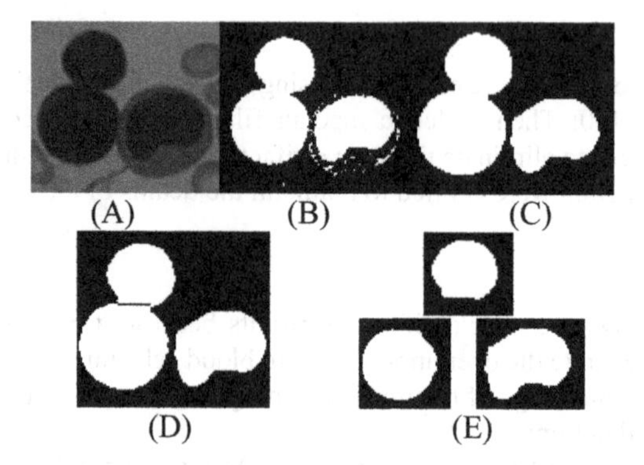

Figure 2. The results of cells segmentation using the proposed method. (A): The original image, (B) the primary segmented WBCs using Colour Based K-means Clustering, (C): The cleaned image. (D): watershed transform result. (E): the segmented lymphocytic cells

Table 2. Results of the classifier.

Data Set	Accuracy (%)	Precision (%)	Sensitivity (%)	Specificity (%)
Training	97.4	97.7	97.7	96.9
Testing	100	100	100	100
Overall system	97.2	98.3	96.6	98

3 RESULTS

The performance of our proposed algorithm has been evaluated by calculating its accuracy, sensitivity and specificity using eq's (1), (2) and (3) respectively. Our proposed algorithm achieved relatively high accuracy with values of 974%, 100% and 972% for training, test and the overall algorithm respectively. Sensitivity reached 977%, 100%, 96.6%, and the specificity reached 96.9%, 100%, and 98% for training, test and the overall algorithm respectively, see Table 2.

$$Accuracy = \frac{TP + TN}{TP + TN + FP + FN} \tag{1}$$

$$Sensitivity = \frac{TP}{TP + FN} \tag{2}$$

$$Specificity = \frac{TN}{TN + FP} \tag{3}$$

Where TP is true positives, TN is true negatives, FP is false positives, and FN is false negatives. The algorithm confusion matrixes are shown in figure 3 below.

4 DISCUSSION

Our algorithm provided an automated learning technique to classify the WBCs as healthy or unhealthy cells. Colorbased segmentation using the K-means clustering algorithm was used together with WBCs features detection. Our algorithm discusses the ALL detection that is not mentioned in [10] The proposed method for segmentation is more efficient and acceptable in contrast with the thresholding method used in [8] and in contrast with the methods that did not consider the

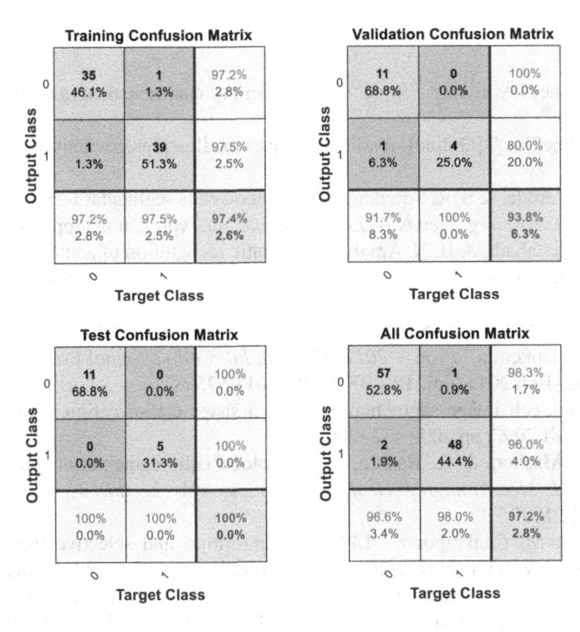

Figure 3. Confusion matrix.

overlapped cells [8][9] Nevertheless, our proposed algorithm achieved higher accuracy comparing to other algorithms presented in [20][21] and [22] In these algorithms authors used different pre-processing techniques followed by K-mean clustering and SVM for classification.

The algorithm successfully recognized the lymphoblastic cells from all of the cells that were introduced as the input. The classes feed-forward the neural classifier and it achieves an exact accuracy with less complication compared to the classifiers that depend on nearest neighbors [1]

The system achieved a relatively high accuracy that is shown with the detailed information in the confusion matrix [Figure 3] that resulted from the classification stage using NN.

5 CONCLUSION AND FUTURE WORKS

This research successfully detects the blood cancer cells using microscopic image samples. The lymphocyte cells are segmented based on the Colorbased segmentation method, then Kmeans clustering together with enhanced watershed transform is applied to separate the overlapped cells, after that a classification depending on the cells morphological features using a neural network was performed. The system is designed to segment the lymphoblast cells. The algorithm successfully recognized the lymphoblast cells from all of the cells that were introduced as the input for the classifier The test images were applied, and the system was validated on the database. The system has achieved the proposed purpose with relatively high accuracy, acceptable processing time on a personal computer with acceptable specifications. In the future additional research will be developed to the extent of the proposed method to be applied in the detection of other types of leukemia and to classify its subtypes.

ACKNOWLEDGMENT

We are so grateful to Dr. Fabio Scotti, associate professor at the University of Milan, Crema, Italy, for providing us with the database utilized to inspect our algorithm.

REFERENCES

[1] "Leukemia." [Online]. Available: https://www.cancer.org/cancer/leukemia.html. [Accessed: 21-Feb-2020].

[2] "Leukemia | MedlinePlus." [Online]. Available: https://medlineplus.gov/leukemia.html. [Accessed: 21-Feb-2020].

[3] M. D. Joshi, A. H. Karode, & S. R. Suralkar, "White blood cells segmentation and classification to detect acute leukemia," *Int. J. Emerg. Trends Technol. Comput. Sci.*, vol. 2, no. 3, pp. 147–151, 2013.

[4] F. Kazemi, T. A. Najafabadi, & B. N. Araabi, "Automatic recognition of acute myelogenous leukemia in blood microscopic images using k-means clustering and support vector machine," *J. Med. Signals Sens.*, vol. 6, no. 3, p. 183, 2016.

[5] N. H. A. Halim, M. Y. Mashor, A. S. Abdul Nasir, N. R. Mokhtar, & H. Rosline, "Nucleus segmentation technique for acute leukemia," *Proc. – 2011 IEEE 7th Int. Colloq. Signal Process. Its Appl. CSPA 2011*, no. March, pp. 192–197, 2011, doi: 10.1109/CSPA.2011.5759871.

[6] S. Shah, "Automatic cell image segmentation using a shape-classification model," *Proc. IAPR Conf. Mach. Vis. Appl. MVA 2007*, pp. 428–432, 2007.

[7] L. H. Nee, M. Y. Mashor, & R. Hassan, "White blood cell segmentation for acute leukemia bone marrow images," *J. Med. Imaging Heal. Informatics*, vol. 2, no. 3, pp. 278–284, 2012, doi: 10.1166/jmihi.2012.1099.

[8] R. Bhattacharjee & M. Chakraborty, "LPG-PCA algorithm and selective thresholding based automated method: ALL & AML blast cells detection and counting," in *2012 International Conference on Communications, Devices and Intelligent Systems (CODIS)*, 2012, pp. 109–112.

[9] H. T. Madhloom, S. A. Kareem, H. Ariffin, A. A. Zaidan, H. O. Alanazi, & B. B. Zaidan, "An automated white blood cell nucleus localization and segmentation using image arithmetic and automatic threshold," *J. Appl. Sci.*, vol. 10, no. 11, pp. 959–966, 2010.

[10] H. N. Lim, M. Y. Mashor, N. Z. Supardi, & R. Hassan, "Color and morphological based techniques on white blood cells segmentation," in *2015 2nd International Conference on Biomedical Engineering (ICoBE)*, 2015, pp. 1–5.

[11] N. H. Harun, A. S. A. Nasir, M. Y. Mashor, & R. Hassan, "Unsupervised segmentation technique for acute leukemia cells using clustering algorithms," *Int. J. Comput. Electr. Autom. Control Inf. Eng.*, vol. 9, no. 1, pp. 253–259, 2015.

[12] S. Khobragade, D. D. Mor, & C. Y. Patil, "Detection of Leukemia in Microscopic White Blood Cell Images," *Proc. – 2015 Int. Conf. Inf. Process.*, pp. 435–440, 2015.

[13] C. Raje & J. Rangole, "Detection of Leukemia in Microscopic Images Using Image Processing," *2014 Int. Conf. Commun. Signal Process.*, pp. 255–259, 2014.

[14] R. D. Labati, V. Piuri, & F. Scotti, "All-IDB: The acute lymphoblastic leukemia image database for image processing," in *2011 18th IEEE International Conference on Image Processing*, 2011, pp. 2045–2048.

[15] A. Chitade, S. K.-I. J. of E. S. and, and undefined 2010, "Colour based image segmentation using k-means clustering."

[16] "The Histology Guide | Blood." [Online]. Available: https://www.histology.leeds.ac.uk/blood/blood_wbc.php. [Accessed: 21-Feb-2020].

[17] C. T. Basima & J. R. Panicker, "Enhanced Leucocyte Classification for Leukaemia Detection," pp. 65–71, 2016.

[18] "Cancer Cells vs Normal Cells | Technology Networks." [Online]. Available: https://www.technologynetworks.com/cancer-research/articles/cancer-cells-vs-normal-cells-307366. [Accessed: 21-Feb-2020].

[19] "Shallow Networks for Pattern Recognition, Clustering and Time Series - MATLAB & Simulink." [Online]. Available: https://www.mathworks.com/help/deeplearning/gs/shallow-networks-for-pattern-recognition-clustering-and-time-series.html. [Accessed: 21-Feb-2020].

[20] M. M. Amin, S. Kermani, A. Talebi, & M. G. Oghli, "Recognition of acute lymphoblastic leukemia cells in microscopic images using k-means clustering and support vector machine classifier," *J. Med. Signals Sens.*, vol. 5, no. 1, p. 49, 2015.

[21] L. Putzu, G. Caocci, & C. Di Ruberto, "Leucocyte classification for leukaemia detection using image processing techniques," *Artif. Intell. Med.*, vol. 62, no. 3, pp. 179–191, 2014.

[22] N. Chatap & S. Shibu, "Analysis of blood samples for counting leukemia cells using Support vector machine and nearest neighbour," *IOSR J. Comput. Eng.*, vol. 16, no. 5, pp. 79–87, 2014.

Proceedings of the 1st International Congress on Engineering Technologies – Kiwan & Banat (Eds)

Simulation and synthesis of homogeneous drug encapsulating nanoparticles

Ruba Khnouf, Ala'a Migdade & Esra' Shawwa
Department of Biomedical Engineering, Jordan University of Science and Technology, Irbid, Jordan

ABSTRACT: Polymer-based Nanocapsules have proven to be an attractive vehicle for targeted cancer therapies, however drug encapsulation using polymeric materials has been difficult in terms of maintaining a homogeneous capsule size distribution while maintaining a small size. We employed a microfluidic system to synthesize homogenous polymeric nanoparticles (NPs) with controlled characteristics. We also demonstrated the effect of polymer concentration, flow rate ratio (FRR), and the flow rate of the aqueous phase on the NPs size. We finally simulated similar conditions to better explain the observations The optimum concentration was found to be in the range of 6 – 9 mg'ml. It was also found that higher FRR resulted in larger NPs size because of reduced mixing, and that low FRR (0.05) with high flow rates (100 µl/ min) resulted in the smallest NP size

Keywords: microfluidic, nano medicine, hydrodynamic focusing, drug encapsulation

1 INTRODUCTION

Cancer is a group of diseases caused by an abnormal growth of tissue forming clusters of abnormal cells, whether benign or malignant. Slowing the abnormal growth or eradicating the solid tumors is accomplished using different approaches including surgery, radiotherapy, and chemotherapy. However, these methods do not precisely target abnormal cells, and could harm normal ones. One of the recent solutions for the aforementioned problem is the development of nanoparticles (NPs) for cancer therapy and diagnosis, opening opportunities to cope with the dilemma of targeting only the abnormal cells to enhance cancer treatment [1]. There are several bulk methods for NP synthesis involving nanoprecipitation, self-assembly methods, and the top-down method. However, these techniques are complex and produce NPs with a wide size distribution, making controlled, targeted drug delivery less efficient and more challenging. Microfluid-based techniques are expected to be a promising methodology for the synthesis of drug loaded NPs with precisely controlling the characteristics of NPs including their size, composition, surface modification, and mechanical properties [2].

Although it has been shown that regulated mixing through the application of microfluidic devices can reduce the size of nanocapsules and the variation in the size, the exact mechanism through which this is achieved is still unclear and requires optimization [3]. Liu et al. have found that the characteristics of the nanocapsules are dependent on the concentration of the polymer, its chemistry, molecular weight, and on the microfluidic channel design including channel length, geometry, flow rate of the channel and flow rate ratio (FRR) [4]. Karnik et al used a hydrodynamic focusing design to control and tune the mixing of the aqueous and polymeric phase in order to tune the size and polydispersity index of polylactic co-glycolic acid and polyethylene glycol co-polymer and concluded that efficient, rapid mixing would result in more homogenous, smaller nanocapsule [5]. The same group used the same mixing mode to synthesize different formulations of the same copolymer, producing 45 different nanocapsule types in a high throughput format [6].

DOI 10.1201/9781003178255-12

Hydrodynamic focusing has been used to achieve fast and efficient mixing on the micro and nano scale for decades and has been one of the simplest methods to achieve mixing in laminar flow dominant environments like the one described by Karnik et al. [7].

Xu et al. compared the a two phase hydrodynamic focusing device with a herringbone structure microfluidic device, the latter is usually applied in microfluidics to generate three dimensional twisting flows that enhance the efficiency of mixing, they concluded that because of the randomness in herringbone mixers, the properties of the NPs generated using this method are more similar to bulk mixing than when using regular hydrodynamic focusing. [8]

In this work, we focus on the synthesis of Methcarylic acid copolymers, also known as Eudragit® polymers nanoparticles using microfluidic devices and understanding the effect of polymer concentration and the hydrodynamic parameters on the structure of the nanoparticles.

2 EXPERIMENTATION

2.1 *Materials*

EUDRAGIT s100 polymer, Ethanol, Deionized water, and poly(dimethyl)siloxane(Sylgard 184, Dow Corning).

2.2 *Device design*

A device based on T-type configuration with three inlets and one outlet was used to carry out homogeneous microfluidic mixing. The overall dimensions of the device are 5mm(L) × 500μm (W) × 300μm (H) [3]. To introduce relatively high pressure drops and hence decrease the mixing time a throttle of 160μm diameter was introduced just after the T-Junction and the mixing angle which is defined as the angle between the inlet and the horizontal was set to 30° because it affects the mixing length. The micromold was created using CNC machining (figure 1) and the microstructures were created out of polydimethylsiloxane (PDMS). In brief, PDMS was mixed with its curing agent (ratio 10:1) and cast onto the fabricated mold, degassed using a vacuum chamber connected to a vacuum pump and placed in an isothermal oven for curing. The PDMS device is then peeled off the mold, and inlets and outlets are punched into it, subjected to plasma treatment using a high voltage generator to increase its surface tension and hydrophilicity, and finally the mixers are sealed using a standard glass microscope slide.

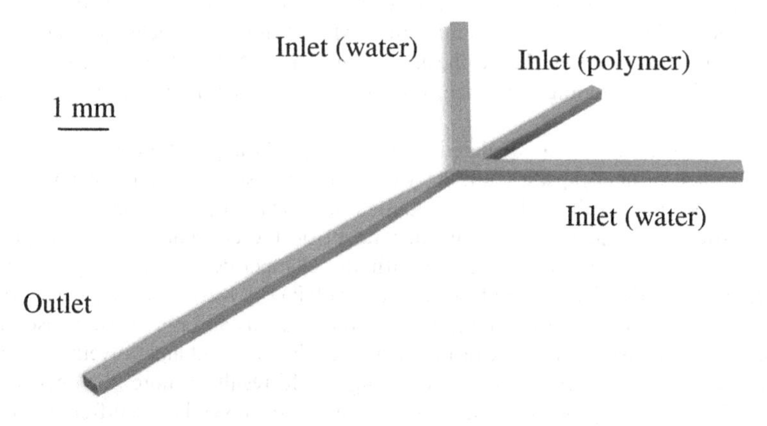

Figure 1. Photograph of the setup.

2.3 *Synthesis of nanoparticles in the microfluidic system*

Deionized water was injected to the side channels of the microfluidic device using a syringe pump (LEGATO SY RINGE PUMPS). Stock solution of different concentrations of eudragit s100 was dissolved in ethanol and pumped in the middle inlet using another syringe pump (NEXT ADVANCE syringe pump) with different flow rates and flow rate ratios, so that the NPs spontaneously formed upon mixing of the polymer/ethanol solution with water . The NP samples were collected from the outlet of the microfluidic device then the collected sample was left with the cap open overnight to evaporate the ethanol as shown in figure 2.

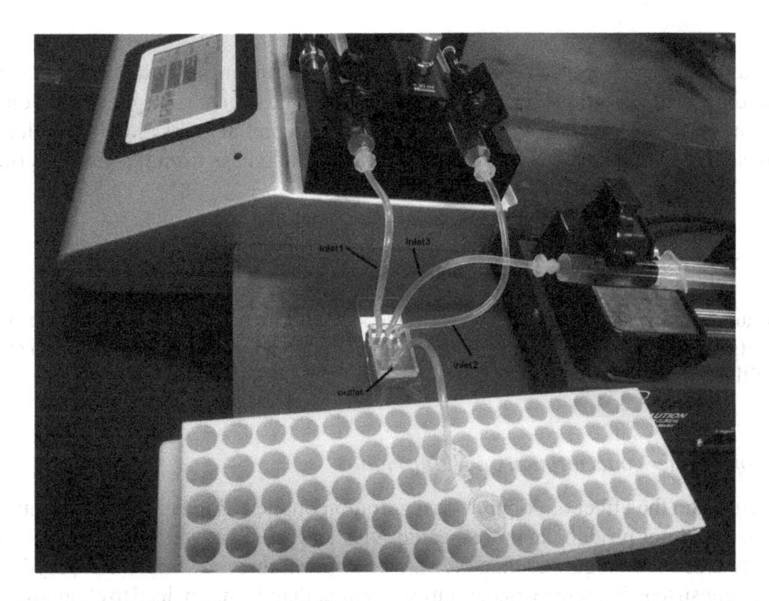

Figure 2. Photograph of the mold.

2.4 *Characterization of NPs*

Different parameters which were expected to affect the NPs' properties were evaluated by investigating the characteristics of the diluted NPs samples such as the size and size distribution using Malvern Zetasizer Nano series dynamic light scattering (DLS) system.

2.5 *Simulation setup and parameters*

The design was done in 2-D using AutoCAD software 2019, and for analysis it was imported to COMSOL Multiphysics software as shown in figure 1. 2-D microchannels were simulated numerically using COMSOL Multi-physics software (V 5.1). From the microfluidics module, a laminar, two phase flow physics interface was selected for fluid flow, while transport of diluted species physics interface was selected to evaluate the mixing in the microchannels. To solve the problem, a time dependent solution was chosen. The study time was set for 10 minutes. Water and ethanol were considered for analysis. The density for water and ethanol were 1000kg/m^3 and 789kg/m^3 respectively, whereas the viscosity was considered as 0.00089 Pa.s and 0.001095 Pa.s. The diffusion coefficient used was $1 \times 10 - 9m^2/s$. Then, the input flow rates for inlets at two flow rate ratios were specified for simulation, as shown in table 1. The channel walls were subjected to the no slip boundary condition. Furthermore, the concentration was specified at side inlets where water was pumped as 0mol/m^3 and 50mol/m^3 at the middle inlet where the polymer solution was pumped. The transport mechanism was determined to be through convection. First order elements

Table 1. Input flow rates used in simulation at 0 5 and 0 8 flow rates.

FRR	DI water [μL/min]	Polymeric solution [μL/min]
0.05	10	0.5
0.05	100	5
0.8	10	8
0.8	100	80

for velocity and pressure were chosen. Although extremely fine mesh slows down the simulation, it was created for better solutions because the accuracy of the results strongly depend on the mesh resolution. The convergence of the solution is sensitive to the initial guess. So to solve this problem, the equations were solved in sequence by updating the initial guess from the previous run.

3 RESULTS AND DISCUSSION

A microfluidic-based system was used to produce size and homogeneity controllable NPs, a few factors affecting the NPs size were evaluated including concentration, flow rate ratio (FRR), and flow rate of the polymer and water [2], [9], [10].

3.1 *Effect of polymer concentration on NP formation*

First, we focused on the effect of polymer concentration on the NP formation. Concentrations less than 4mg/ml could not produce monodispersed NPS and we observed several peaks and a high polydispersity index (PDI) (greater than 0.75), which meant that the sample was polydisperse and did not have a consistent diameter. Concentrations greater than 14mg/ml lead to clogging the mixing channel because of agglomeration, where the polymer clusters as it is hydrophobic and forms large clumps that block the channel. While the concentrations ranging from 5mg/ml to 8mg/ml produced a small NP size and almost the same PDI regardless of the flow rate and flow rate ratio as shown in table 2.

Table 2. Mean diameter (nm) and pdi of different polymer concentrations

Concentration(mg/ml)	Mean Diameter(nm)	PDI
2	132.3	0.375
3	1.04E4	0.76
4	93.58	0.557
5	85.3	0.296
6	58.82	0.146
7	60.17	0.194
8	96.45	0.129
9	119.8	0.187
10	141.2	0.292
12	151.5	0.316
14	176.3	0.222
16	221.4	0.325
18	323.5	0.715
20	349.5	0.683

3.2 *Effect of flow rate and Flow Rate Ratio (FRR)*

In order to establish an efficient hydrodynamic focusing and the control of the interfacial mixing between the polymeric solution and the deionized water, stable flow rate ratio must be optimized. So, the effect of flow rate and flow rate ratio (FRR) was investigated by changing the flow FRR of the polymeric solution to DI-water from 0.05 to 1 and the flow rate of DI-water was fixed at 10, 50, and 100 μl/min. We observed a dramatic size reduction from 183.8 nm to 62.44 nm at FRR of 0.05 and larger NPs were produced at FRR greater than 0.1.as well as, smaller NPs were formed by increasing the flow rate of DI-water for FRRs less than 0.1 as shown in fig 3.

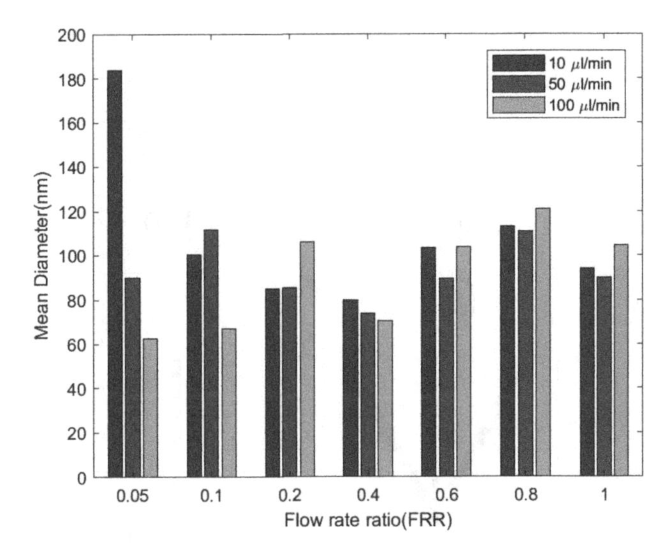

Figure 3. Effect of the flow rate ratio NP size obtained experimentally.

3.3 *Simulation analysis*

Figure 4 shows the results of keeping the FRR constant at 0.8 while the flow rate of DI- water was changed. Increasing the flow rate from 10 μl/min to 100 μl/min results in an increasing in the hydrodynamic focusing, shown in the length of the focused yellow line which is extended at the higher flow rate. This leads to efficient mixing between the polymeric solution and deionized water. Through fixing the flow rate of aqueous solution and varying the flow rate of the organic phase of different flow rate ratios (0.05 and 0.8), figure 5 illustrates an increase in the width of the focus point of the organic phase that occurred as a result of the higher FRR, shown by the formation of the green region, and that implies that higher flow rates produce larger NPs. Because of the increased width, the simulation time was longer at higher FRR, as shown in table 3.

It has been hypothesized that the homogeneity of NPs is dependent on the mixing efficiency while maintaining laminar flow, hydrodynamic focusing achieves that by having two interfacial surfaces between water and the polymer allowing laminar flow to mix the two solutions where the hydrophobic polymer would start to nucleate and form small spherical nanoparticles [5]. As far as the authors know, this is the first work to use Methcarylic acid copolymers to form nanocapsules using microfluidics and one of the first to explain the phenomenon with simulations.

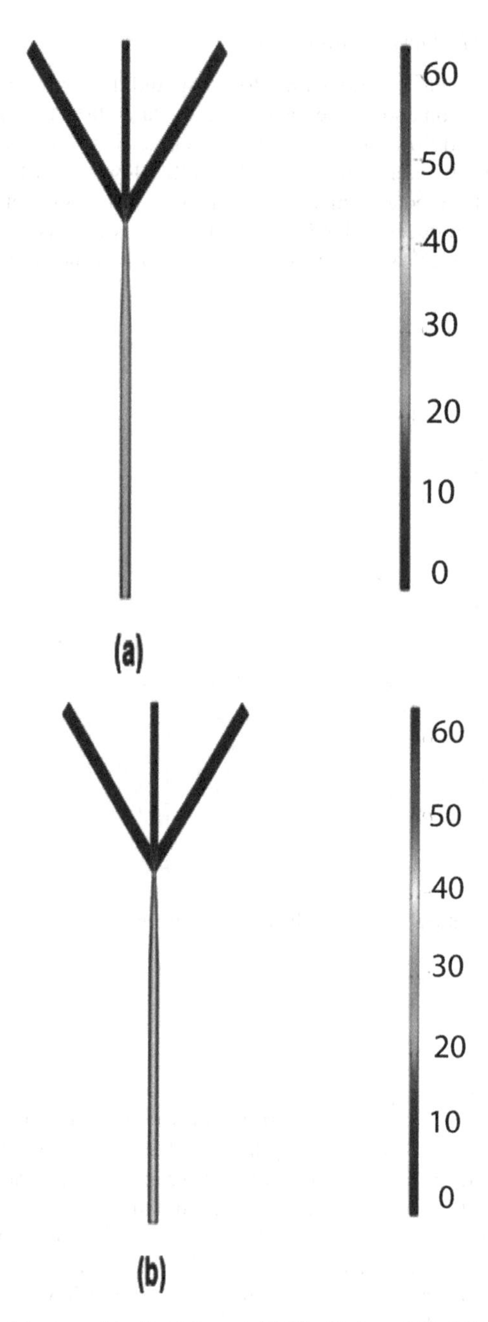

Figure 4. Hydrodynamic focusing of the fluid flow at (a) 10 μl/min and (b) 100 μl/min, respectively.

4 CONCLUSION

Nanoparticles have a big advantage in the field of cancer therapy because of their specific cancer cell targeting and reduction of toxicity to normal cells. Furthermore, characteristics of the NPs such as size play a crucial role in their pharmaceutical efficacy. This has led to the optimization of the Methcarylic acid copolymers concentration, where the concentration between 6 and 9 mg/ml

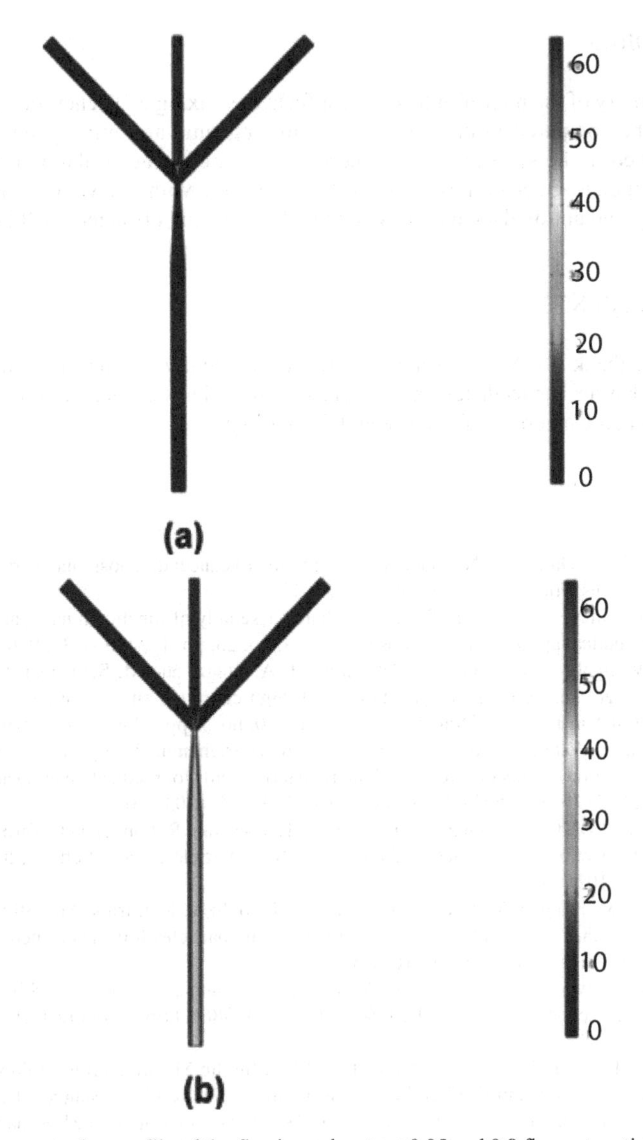

Figure 5. The concentration profile of the flowing solvents at 0.05 and 0.8 flow rate ratios.

Table 3. Input flow rates used in simulation at 0.5 and 0.8 flow rates.

FRR	Simulation time (miutes)
0.05	39
0.8	120

resulted in the smallest polydispersity index indicating a more homogenous NP size. FRR, and flow rate were also optimized both experimentally and through simulation, where it has been shown that a high FRR results in reduced mixing causing larger PDIs and NP sizes, while high flow rates (100 μl/ min) with small FRR (0.05) achieved the smallest NP diameter.

5 FUTURE WORKS

Because the geometry of the microfluidic device affects the mixing efficiency, other microchannel designs need to be compared in order to get less mixing time and mixing length. The device will then be applied in the synthesis of polymeric nanoparticles encapsulating the plant-derived polyphenolic compound Quercetin as a model drug molecule with known anti-cancer as well as anti-inflammatory and antioxidant activities and the drug loading efficiency will be evaluated.

ACKNOWLEDGEMENTS

We would like to thank Dr. Suhair Sunnuqrot from the Faculty of Pharmacy at Al Zaytoonah University of Jordan for her feedback and advice. This work has been supported by the Deanship of Research at Jordan University of Science and Technology.

REFERENCES

[1] A. S. Thakor and S. S. Gambhir, "Nanooncology: the future of cancer diagnosis and therapy," CA: a cancer journal for clinicians, vol. 63, no. 6, pp. 395–418, 2013.

[2] Q. Feng, J. Sun, and X. Jiang, "Microfluidics-mediated assembly of functional nanoparticles for cancer-related pharmaceutical applications," Nanoscale, vol. 8, no. 25, pp. 12, 430–443, 2016.

[3] N. Lababidi, V. Sigal, A. Koenneke, K. Schwarzkopf, A. Manz, and M. Schneider, "Microfluidics as tool to prepare size-tunable plga nanopar- ticles with high curcumin encapsulation for efficient mucus penetration," Beilstein Journal of Nanotechnology, vol. 10, no. 1, pp. 2280–2293, 2019.

[4] K. Liu, Z. Zhu, X. Wang, D. Gonçalves, B. Zhang, A. Hierlemann, P. Hunziker, Microfluidics-based single-step preparation of injection-ready polymeric nanosystems for medical imaging and drug delivery, Nanoscale. 7 (2015) 16983–16993. https://doi.org/10.1039/C5NR03543K.

[5] R. Karnik, F. Gu, P. Basto, C. Cannizzaro, L. Dean, W. Kyei-Manu, R. Langer, O.C. Farokhzad, Microfluidic Platform for Controlled Synthesis of Polymeric Nanoparticles, Nano Lett. 8 (2008) 2906–2912. https://doi.org/10.1021/nl801736q.

[6] P.M. Valencia, E.M. Pridgen, M. Rhee, R. Langer, O.C. Farokhzad, R. Karnik, Microfluidic Platform for Combinatorial Synthesis and Optimization of Targeted Nanoparticles forCancer Therapy, ACS Nano. 7 (2013) 10671–10680. https://doi.org/10.1021/nn403370e.

[7] J.B. Knight, A. Vishwanath, J.P. Brody, R.H. Austin, Hydrodynamic Focusing on a Silicon Chip: Mixing Nanoliters in Microseconds, Phys. Rev. Lett. 80 (1998) 3863–3866. https://doi.org/10.1103/PhysRevLett.80.3863.

[8] Z. Xu, C. Lu, J. Riordon, D. Sinton, M.G. Moffitt, Microfluidic Manufacturing of Polymeric Nanoparticles: Comparing Flow Control of Multiscale Structure in Single-Phase Staggered Herringbone and Two-Phase Reactors, Langmuir. 32 (2016) 12781–12789. https://doi.org/10.1021/acs.langmuir.6b03243.

[9] D. Gobby, P. Angeli, and A. Gavriilidis, "Mixing characteristics of t-type microfluidic mixers," Journal of Micromechanics and microengineering, vol. 11, no. 2, p. 126, 2001.

[10] M. Maeki, Y. Fujishima, Y. Sato, T. Yasui, N. Kaji, A. Ishida, H. Tani, Y. Baba, H. Harashima, and M. Tokeshi, "Understanding the formation mechanism of lipid nanoparticles in microfluidic devices with chaotic micromixers," PloS one, vol. 12, no. 11, 2017.

Proceedings of the 1st International Congress on Engineering Technologies – Kiwan & Banat (Eds)
 ISBN 978-0-367-77630-5

Image based microfluidic mixing evaluation technique

Ruba Khnouf, Areen Al Bashir, Ala'a Migdade, Esra'a Alshawa & Arwa Sheyab
Department of Biomedical Engineering, Jordan University of Science and Technology, Irbid, Jordan

ABSTRACT: Micromixers are considered an integral part in biomedical microfluidic applications because of important chemical and biological processes that require rapid and efficient mixing. In this work, flow regimes in throttle-T and square wave type micromixers are studied. The dependence of mixing efficiency on the geometry design is examined. Passive mixing methods with different flow rates at the inlets are studied experimentally using an imaging based technique. Moreover, a finite element model is used in order to study the mixing process in 500 μm and 400 μm – wide throttle and square wave microchannels, respectively. The results illustrated the effects of both flow rate and channel geometry on the mixing efficiency. Experimental studies at different flow rates are reported showing the same mixing tendency.

Keywords: Microfluidics, Micromixers, Microchannels, Creeping flow, Diffusive transport, Image analysis

1 INTRODUCTION

Microfluidics is the innovation that permits controlling and processing small quantities of fluid utilizing channels with length scales in the micrometer range [1]. In recent years, Lab-on-a-Chip (LoC) technologies have dramatically highlighted the development of microsystems, mostly for chemical, biological, and medical applications. They have many characteristics that save time and cost due to their miniaturized size, which makes them portable devices. They implement quick analysis using low amounts of samples and expend little reagents, lowering the waste generated. Ward and Hugh show a rapidly rising growth in the number of micromixer publications, highlighting the fact that the scientific community is expanding in this area of research [2]. However, one of the challenges in microfluidic devices for LoC or Micro Total Analysis System (μ TAS) applications is mixing [3]. In LoC devices, mixing is important for sample dilution, chemical and biological reactions. Micromixers are considered as an integral part in microfluidic devices. Due to their small sizes, the velocity of fluid flow is very small [4], viscous forces will significantly dominate the fluid properties rather than inertial forces [5], therefore the Reynolds number will be very small, indicating that the fluid flow is laminar. This nature of flow prevents achieving sufficient mixing in a microfluidic system in a short time. Furthermore, mixing in microfluidic devices is usually accomplished by using some limited volume, which significantly improves the diffusion and advection effects. This diffusion property can induce homogeneous mixing instead of heterogeneous mixing triggered by convective micromixers usually operating in larger microchannels under higher Reynolds number flows. Micromixers are generally classified as active and passive mixers [6]. Active micromixers require power from outside inputs. Passive micromixers are achieved by modifying fluid channels' structure or configuration [7], and are not managed externally by a source. Ansari and Kim report that restructuring the flow using special channel designs minimizes the diffusion path and maximizes the contact surface area to enhance molecular diffusion and advection for efficient mixing [8]. In addition, passive micromixers involve more efficient manufacturing than active micromixers, and can be easily integrated into more complex LoC devices. This paper, flow regimes in throttle-T and square wave type micromixers are studied

DOI 10.1201/9781003178255-13

in terms of efficiency at different flow rates using both finite element analysis and experiments based on imaging techniques.

2 EXPERIMENTATION

2.1 *Device design*

There are two different geometries for mixing efficiency evaluation, a throttle-T micromixer and a square wave micromixer. The devices have two inlets and one outlet. A throttle-T micromixer is based on T-type configuration, and the overall dimensions of the device are 5mm (L) × 500 μm (W) × 300 μm (H). In addition, to introduce relatively high-pressure drops and hence decreasing the mixing time a throttle of 160μm was introduced just after the T-Junction and the mixing angle, which is defined as the angle between the inlet and the horizontal, was set to 30° because it affects the mixing length. The square wave microchannel has a total length, width, and height of 11.5mm400 μm and 200 μ respectively. Micromolds were created using CNC machining and the microstructures were created out of polydimethylsiloxane (PDMS). In brief, PDMS was mixed with its curing agent (ratio 10:1) and cast onto the fabricated molds, degassed using a vacuum chamber connected to a vacuum pump and placed in an isothermal oven for curing. The PDMS devices are then peeled off the molds, then inlets and outlets are punched into it, subjected to plasma treatment using a high voltage generator to increase their surface tension and hydrophilicity, finally the mixers sealed using a standard glass microscope slide.

2.2 *Microfluidic device operation*

The dyes (yellow and blue food coloring) were driven through the inlets at different flow rates, figure 1. First, we mixed yellow and blue dyes to get green and then we injected the yellow, blue and the mixed for purposes of thresholding in image processing at an arbitrary flow rate (500 μl/min) using a syringe pump (LEGATO SYRINGE PUMPS). After that, the blue and yellow were injected through inlets at flow rates comprising small, medium and large ranges. The flow rates were 22nl/ min, 2μ*l*/ min, 3μ*l*/ min, 5 μ*l*/ min and 10μ*l*/ min.

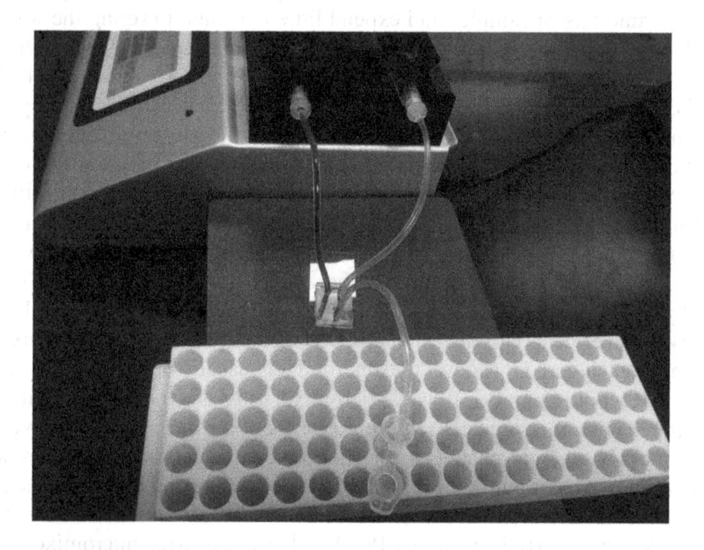

Figure 1. Microfluidic mixing device setup.

2.3 *Video recording*

The flow of dyes was recorded using the LaxcoLMS-S2 Series microscope (Carl Zeiss Microscopy GmbH). A digital camera with a color CMOS sensor was connected to the microscope and recorded videos at resolution $2560(H) \times 1920$ (V) pixels.

2.4 *Image processing*

In order to evaluate the mixing between the two colors, the following steps were followed, figure 2. In short, calculations of the device's volume and the time of flow rate, afterwards the videos were read and divided into multiple frames, and for each frame the images were cropped, sharpened, and the amount of green was quantified in 4 different locations using Matlab.

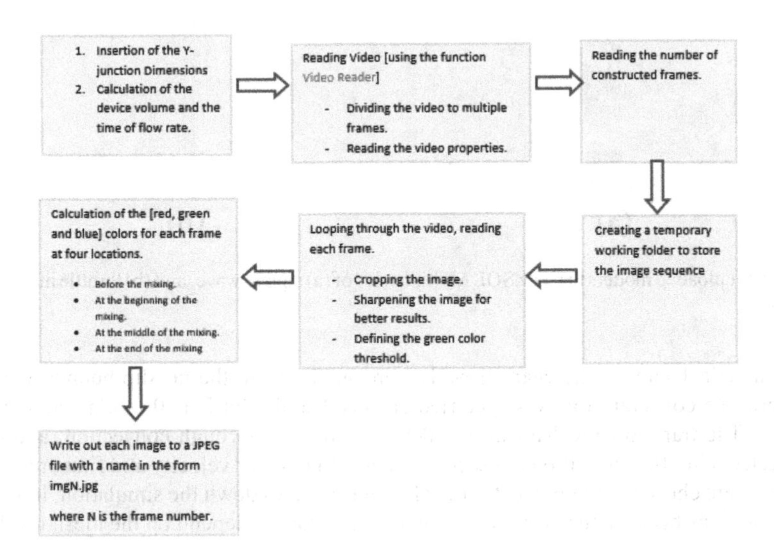

Figure 2. Block diagram for intensity measurement using image processing.

3 NUMERICAL STUDY

Mixing of fluids flowing through micro-channels was done using two different geometries: Square wave and throttle.

3.1 *Geometry modelling*

The designs were done in 2-D using AutoCAD software 2019, and for analysis, they were imported to COMSOL Multiphysics software, as shown in the figure 3.

3.2 *Simulation*

2-D microchannels were simulated numerically using COM-SOL Multi-physics software (V 5.1). From microfluidic module, creeping flow physics interface was selected for fluid flow, while transport of diluted species physics interface was selected to evaluate the mixing in the microchannels. To solve the problem, stationary solution was picked out. Water was considered for analysis. The density and viscosity for water were 1000 kg/m^3 and 0.001 Pa.s, respectively. The diffusion coefficient was taken as 1×10^{-9} m^2/s. Then, the input parameters for inlets were specified for simulation,

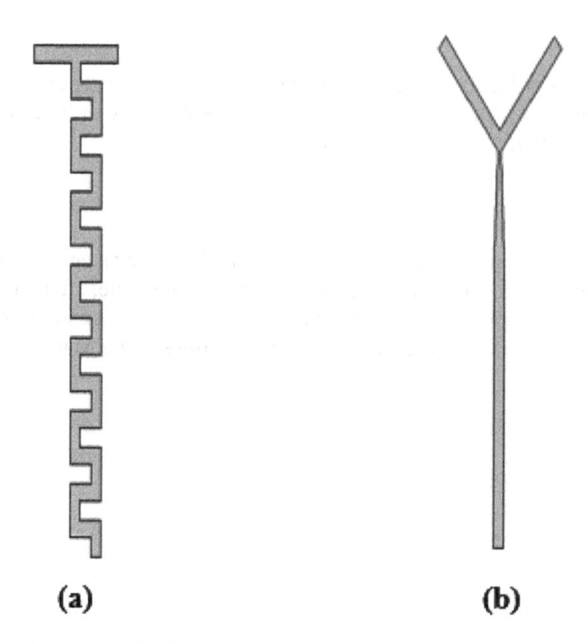

Figure 3. 2-D uploaded models to COMSOL Multiphysics of (a) square wave, and (b) throttle micro-channels, respectively.

as shown in table 1 and 2. The channel walls were subjected to the no slip boundary condition. Furthermore, the concentration was specified at inlet 1 and inlet 2 as 0 mol/m^3 and 50mol/m^3, respectively. The transport mechanism was determined to be through convection, and the walls were subjected to no flux conditions. A second order elements for velocity and first order elements for pressure were chosen. Although extremely fine mesh slows down the simulation, it was created for better solutions because the accuracy of the results strongly depend on the mesh resolution.

Table 1. Input parameters used for simulation of square wave micro-channel

Flow rate [μl/min]	Re	Entrance length[m]
0.022	$1.22*10^{-3}$	$1.63*10^{-8}$
2	0.11	$1.49*10^{-6}$
3	0.17	$2.23*10^{-6}$
5	0.28	$3.71*10^{-6}$
10	0.56	$7.43*10^{-6}$

Table 2. Input parameters used for simulation of throttle micro-channel.

Flow rate [μl/min]	Re	Entrance length[m]
0.022	$1.14*10^{-3}$	$2.14*10^{-8}$
24	0.10	$1.95*10^{-6}$
3	0.16	$2.93*10^{-6}$
5	0.26	$4.88*10^{-6}$
10	0.52	$9.76*10^{-6}$

4 RESULTS AND DISCUSSIONS

The plot in figure 4 finds the velocity of the streamline at different downstream positions for all input flow rates. It gives some information about the fluid flow along the microchannel. The flow is uncoupled, so its viscosity does not depend on the concentration. The colors in the figure indicate the respective flow values.

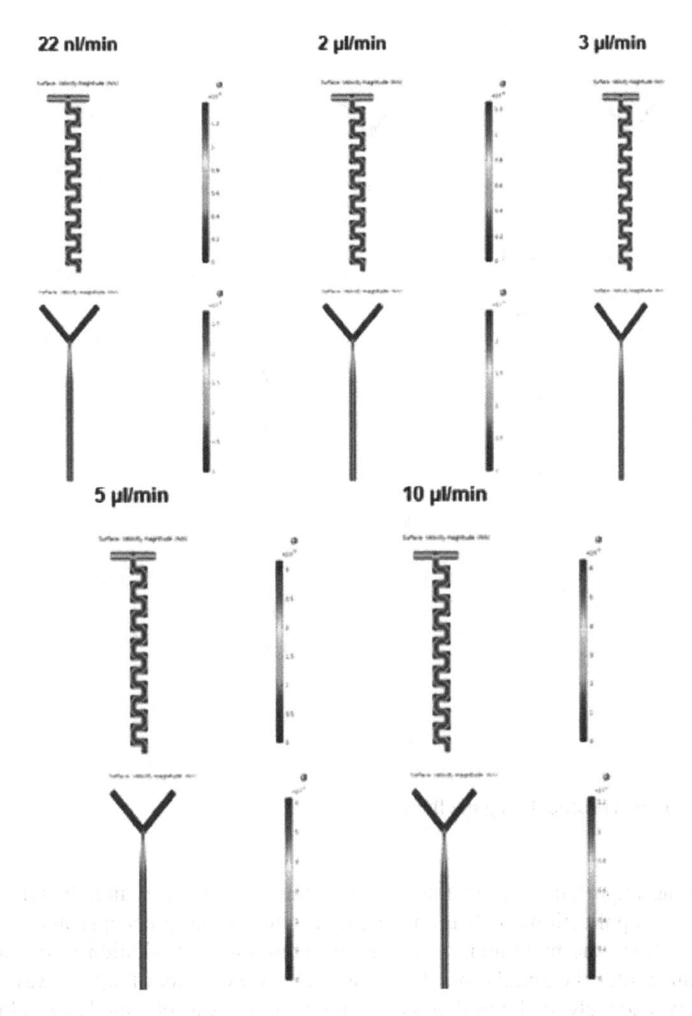

Figure 4. Streamline velocity during the fluid flow.

Figure 5 provides the simulation result for the concentration of the species along the microchannels at different flow rates. Square wave microchannel has better mixing than throttle microchannels at all flow rates. This means that when a fluid flows around a blind corner, the change in direction of the flow ultimately leads to a secondary area of flow perpendicular to the fluid flow in the channel and a good mixing will result over a wide range of Reynolds numbers.

Concentration profile at the exit of the channel is illustrated in figure 6. These graphs have been plotted using the data obtained from above simulation. From these figures, it is clear that among different flow rates, 22 nl/min has a complete mixed concentration profile at the outlet. Since the flow is laminar, as the volumetric flow rate decreases, there is more time for the solutions to diffuse into each other. This ascertains that the flow is completely laminar and flow is diffusion dependent

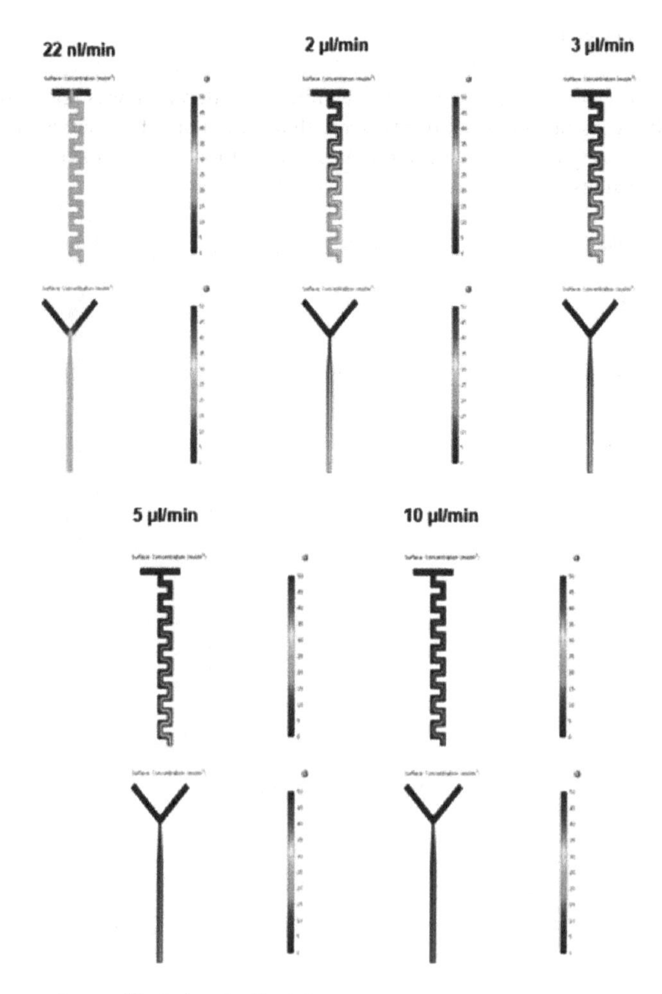

Figure 5. Concentration profile during the fluid flow.

it also stressed the importance of creating micro disturbances to guarantee better mixing when mixing is needed in applications such as enzymatic reactions or drug encapsulation.

Table 3 and 4 shows the maximum, minimum concentration, in addition to the concentration difference obtained after the simulation along the width at the outlet of square wave and throttle microchannels, respectively. It shows that the complete mixing was obtained at 22 nl/min for both designs, where the difference in the concentration found to be 0 mol/m^3. Other flow rates have concentration differences that are higher in throttle than square wave design, which indicates that the square wave design, have better mixing than the throttle design.

To estimate the validity of the numerical results, the mixing behavior was studied by intensity quantification at four different positions of the microchannel using confocal microscope. Figure 7 and 8 show the corresponding white light microscopic images at 22 nl/min and 10 µl/min for square wave and throttle microchannels, respectively. It is observed that almost complete mixing was done at 22 nl/min of both designs, which is similar to the concentration profile resulted from the simulation (see table V and VI) which indicates the mixing by the percentage of the intensity of the color green.

Table 5 and 6 represent the mixing percentage estimated from the intensity information of the micromixing images at different flow rates for square wave and throttle microchannels, respectively.

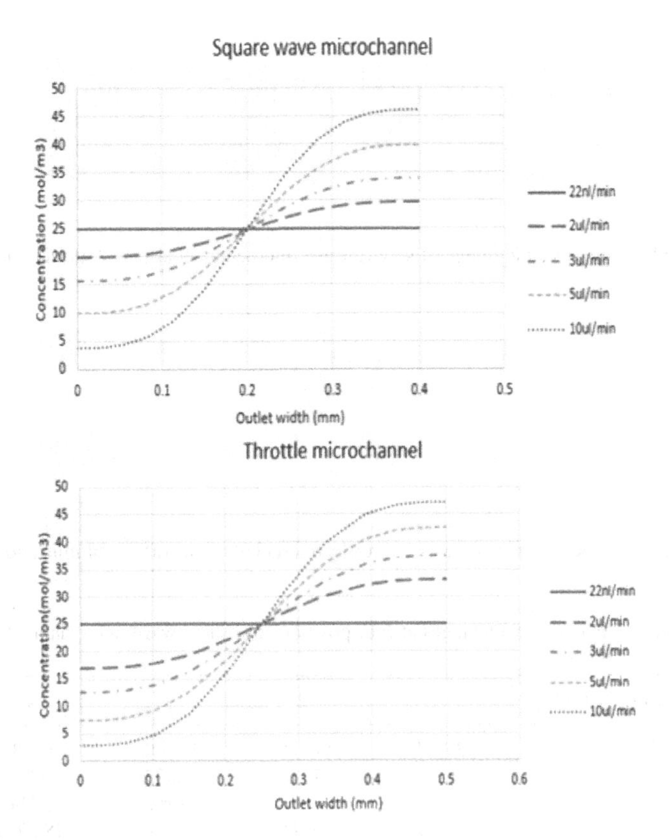

Figure 6. Concentration profile at the outlet.

Table 3. Maximum, minimum, and difference of square wave micro-channel concentration at the outlet.

Flow rate [μl/min]	Maximum	Minimum	Difference
0.022	24.87	24.87	0
2	29.66	20.05	9.61
3	33.93	15.78	18.15
5	39.86	9.88	29.98
10	46.09	3.71	42.38

Table 4. Maximum, minimum, and difference of throttle micro-channel concentration at the outlet.

Flow rate [μl/min]	Maximum	Minimum	Difference
0.022	25	25	0
2	33.03	16.99	16.04
3	37.42	12.61	24.81
5	42.52	7.52	35
10	47.10	2.96	44.14

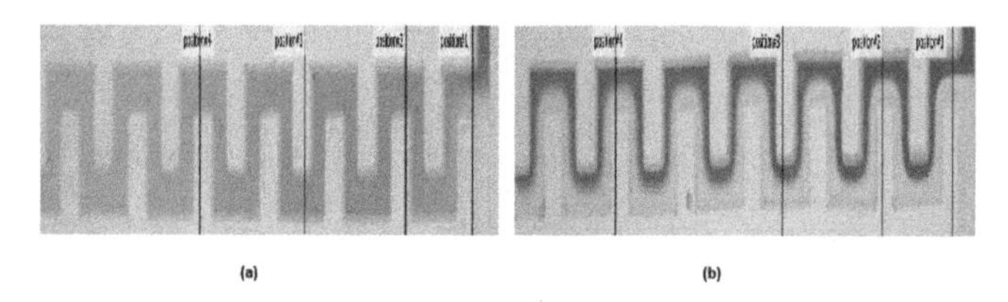

Figure 7. White light microscopic image of square wave microchannel at (a) 22 nl/min and (b) 10 μl/min, respectively.

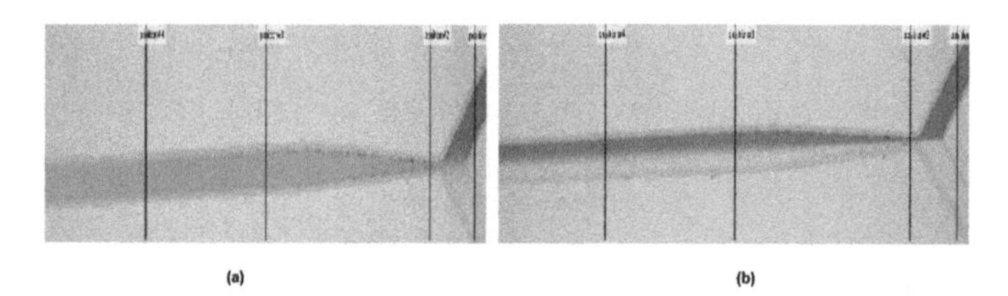

Figure 8. White light microscopic image of throttle microchannel at (a) 22 nl/min and (b) 10 μl/min, respectively.

Table 5. Intensity percentages at four positions of square wave microchannel at different flow rates.

Flow rate [μl/min]	1	2	3	4
0.022	37.5 %	90.6%	96.7%	96.7%
2	7%	23%	33.33%	60%
3	7%	13.7%	30%	34%
5	7%	21.4%	22.2%	28.57%
10	7.6%	13.8%	14.8%	20%

Table 6. Intensity percentages at four positions of throttle microchannel at different flow rates.

Flow rate [μl/ min]	1	2	3	4
0.022	0%	55%	73%	77%
2	0%	23%	27%	31%
3	0%	6.45%	17.3%	20%
5	0%	13.6%	14.2%	15%
10	0%	7.1%	10%	14.28%

From the results, it is shown that the highest percentage was at 22 nl/min for both designs. As a compar- ison, the square wave microchannel has higher percentage at different flow rates, which is compatible with the simulation result. This also provides an affordable method to evaluate microfluidic mixing where more expensive fluorescent approaches are traditionally used.

5 CONCLUSION

Two types of passive micromixers are presented here to demonstrate the differences in the mixing within the mixing chamber between them. These types can improve mixing at low Reynolds numbers. The mixing mechanism highly depends on the geometry of the microchannel. To predict the performance of the devices, 2D numerical models were developed. Simulation results showed that mixing in square wave design is better than that in throttle design, which was confirmed by experimentally measuring the color intensity in the channels. These results and calculations are critical for designing micromixers for various applications.

ACKNOWLEDGEMENTS

This work has been supported by the Deanship of Research at Jordan University of Science and Technology.

REFERENCES

[1] G. M. Whitesides, "The origins and the future of microfluidics," *Nature*, vol. 442, no. 7101, pp. 368–373, 2006.

[2] K. Ward and Z. H. Fan, "Mixing in microfluidic devices and enhancement methods," *Journal of Micromechanics and Microengineering*, vol. 25, no. 9, p. 094001, 2015.

[3] A. Manz, N. Graber, and H. a'. Widmer, "Miniaturized total chemical analysis systems: a novel concept for chemical sensing," *Sensors and actuators B: Chemical*, vol. 1, no. 1-6, pp. 244–248, 1990.

[4] N.-T. Nguyen and S. T. Wereley, "Fundamentals and applications of microfluidics artech house inc," *Boston, MA, USA*, pp. 354–369, 2002.

[5] J. P. Brody, P. Yager, R. E. Goldstein, and R. H. Austin, "Biotechnology at low reynolds numbers," *Biophysical journal*, vol. 71, no. 6, pp. 3430–3441, 1996.

[6] V. Hessel, H. Lö we, and F. Schönfeld, "Micromixers—a review on passive and active mixing principles," *Chemical Engineering Science*, vol. 60, no. 8–9, pp. 2479–2501, 2005.

[7] C. Saikat, M. Sharath, M. Srujana, K. Narayan, and P. P. Kumar, "Modelling and analysis of microfluidic micromixer for lab-on-a-chip (loc) application," in 2015 Annual IEEE India Conference (INDICON). IEEE, 2015, pp. 1–6.

[8] M. A. Ansari and K.-Y. Kim, "Mixing performance of unbalanced split and recombine micomixers with circular and rhombic sub-channels," Chemical Engineering Journal, vol. 162, no. 2, pp. 760–767, 2010.

Proceedings of the 1st International Congress on Engineering Technologies – Kiwan & Banat (Eds)
 ISBN 978-0-367-77630-5

A nano/micro-sphere excited by a subwavelength laser for wireless communications

Amer Abu Arisheh
Electrical Engineering Department, Jordan University of Science and Technology, Irbid, Jordan

Said Mikki
Department of ECECS, University of New Haven, West Haven, CT, USA

Nihad Dib
Electrical Engineering Department, Jordan University of Science and Technology, Irbid, Jordan

ABSTRACT: The problem of exciting a nanosphere by an external light has been extensively studied in the literature. In this paper, a far-field simulation and analysis is provided for the problem of a nano/micro-sphere excited by a subwavelength laser (800 nm wavelength) as a part of a wireless communication system. The simulation is done using a full-wave finite-element method based multiphysics solver. Nonparaxial Gaussian beam formulation, which is based on plane wave expansion, is utilized to simulate an accurate subwavelength laser. Evanescent fields are included in the expansion which help in simulating the tightly focused Gaussian beam. The subwavelength laser beam is accurate as it drops to exactly 1/e of its peak at the spot radius distance. The studied sphere is swept from 100 nm to 1000 nm while the subwavelength laser spot radius is swept from 600 nm to 800 nm. The relationship between ratio of scattered power to absorbed power and ratio of subwavelength laser spot radius to sphere radius is discussed in this work. Far-field antenna quantities are calculated and a nano/micro-scale spherical antenna is optimized for optimum sphere radius and subwavelength laser spot radius.

Keywords: Optical antennas, micro-sphere, subwavelength laser, nanotechnology, nano-communications.

1 INTRODUCTION

Nanoantennas are efficient devices that are extensively studied and applied in many areas including optical microscopy, photovoltaics, lasing, near-field nano-optics and quantum communications [1-5]. The physics of a nanoantenna is different from the classical RF antenna due to the appreciable skin depth in nanoantennas and due to the surface plasmons that appear when the nanoantenna is excited by an optical wave [6, 7]. Nanoantennas have been extensively studied when excited by a plane wave. On the other hand, some other works used a focused laser to excite a nanoantenna such as in [7-9]. However, to the best of our knowledge, a fully-fledged transmitting nanoantenna using a subwavelength laser was not simulated to be a part of a wireless communications system except for our recent work [10, 11] where a nano-dipole and a two nanosphere antenna excited by a fixed spot radius subwavelength laser is designed and optimized for wireless communications purposes. The related simulation model aspects and the postprocessing expressions were explained in detail. In this paper, an extension to our work is achieved using the latest version of COMSOL Multiphysics (version 5.5) [12] by studying the basic idea of exciting a single sphere by a subwavelength laser

 DOI 10.1201/9781003178255-14

with 800 nm wavelength for wireless communications purposes. The latest version of COMSOL helped in enhancing our work since evanescent fields are included in the expansion which help in simulating a tightly focused Gaussian beam, where the spot radius is smaller than the wavelength, propagating away from the focus. The laser beam is accurate such that it drops to exactly 1/e of its peak at the spot radius distance unlike our previous work. The spot radius of the subwavelength laser is swept from 600 nm to 800 nm. The sphere radius is swept from 100 nm to 1000 nm and is not confined to the nano-range as it also extends to a part of the micro-range. Far-field antenna quantities are calculated and a nano/micro-scale spherical antenna is optimized for optimum sphere radius and subwavelength laser spot radius. The paper is organized as follows: Section II discusses the specifications of the computational model including geometry, material and physics. In Section III, far-field expressions to be used in assessing the introduced nano/micro-spherical antenna are stated. Section IV provides and discusses the numerical results of the far-field quantities corresponding to the parametric sweep over sphere radius and subwavelength laser spot radius. Section IV concludes the work and provides some possible extensions.

2 SPECIFICATIONS OF THE COMPUTATIONAL MODEL

The main simulation stages of our computational model are geometry, material, physics, mesh and parametric sweep. These are followed by postprocessing expressions that will be evaluated in order to determine the optimum antenna. These were discussed extensively in our previous work [11]. However, here only the key points are highlighted and we focus on the changes we made in the computational model.

The simulated sphere material is taken from a well-known gold material for easier comparison with other models [13]. The sphere is excited by a background field using the nanparaxial Gaussian beam formulation that is based on plane wave expansion and is an exact solution of Helmholtz equation in contrast to the paraxial wave equation which is a solution for the paraxial approximation of the Helmholtz equation. The paraxial wave equation is not suitable for our study since the laser must have a spot radius below the wavelength in order to partially illuminate the nano/micro-sphere while the paraxial wave formula is not accurate when the spot radius is equal to or smaller than the wavelength. The simulated laser is an 800 nm wavelength subwavelength laser propagating in the +x direction and it is polarized in the z-direction with an electric field of 1 V/m in the point of focus. Evanescent fields are included in the expansion to help in simulating a tightly focused Gaussian beam, where the spot radius is smaller than the wavelength, propagating away from the focus [12]. The generated laser field is accurate and it drops to exactly 1/e of its peak at the spot radius distance. The sphere is located at the center of the laser beam and the symmetry of the sphere is utilized to reduce the computational burden. Hence, only a quarter of the sphere is considered and the total field is inferred from the proper symmetry relations.

The spot radius of the laser beam is swept from 600 nm to 800 nm with 50 nm steps. The sphere radius is swept from 100 nm radius to 1000 nm radius with 20 nm steps which covers most of the nano-range and a part of the micro-range. The ratio of spot radius to sphere radius generally varies for different combinations in the parametric sweep while some cases have the same ratio. Figure 1 shows a logarithmic plot of the ratio in order to give some sense of the cases that have an equal spot radius to sphere radius ratio.

Four cases of the parametric sweep are considered in Figure 2 where laser with spot radii 600 nm, 800 nm are shown illuminating 600 nm and 1000 nm radius spheres at x=0 (yz plane). Note that the electric field drops to exactly 1/e of its peak value at exactly the spot radius distance. Laser stages while propagating in the +x-direction are shown in Figure 3 for the same cases of Figure 2. These stages shows how the laser is alternating negatively and positively and focusing gradually in the center and then propagating away from the focus. It is shown that around three laser stages exist in the 600 nm radius sphere at the xz-plane (y=0) and around five laser stages exist in the larger sphere (1000 nm radius) at the same plane.

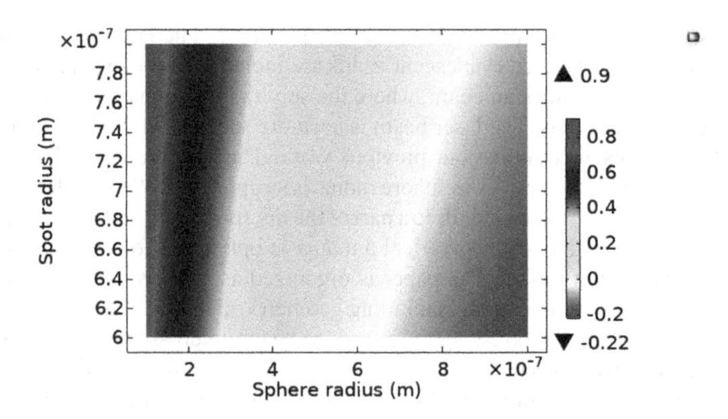

Figure 1. Ratio of subwavelength laser spot radius to sphere radius (logarithmic scale is used to accentuate the results). Values below zero indicate cases where laser is focused to the sphere.

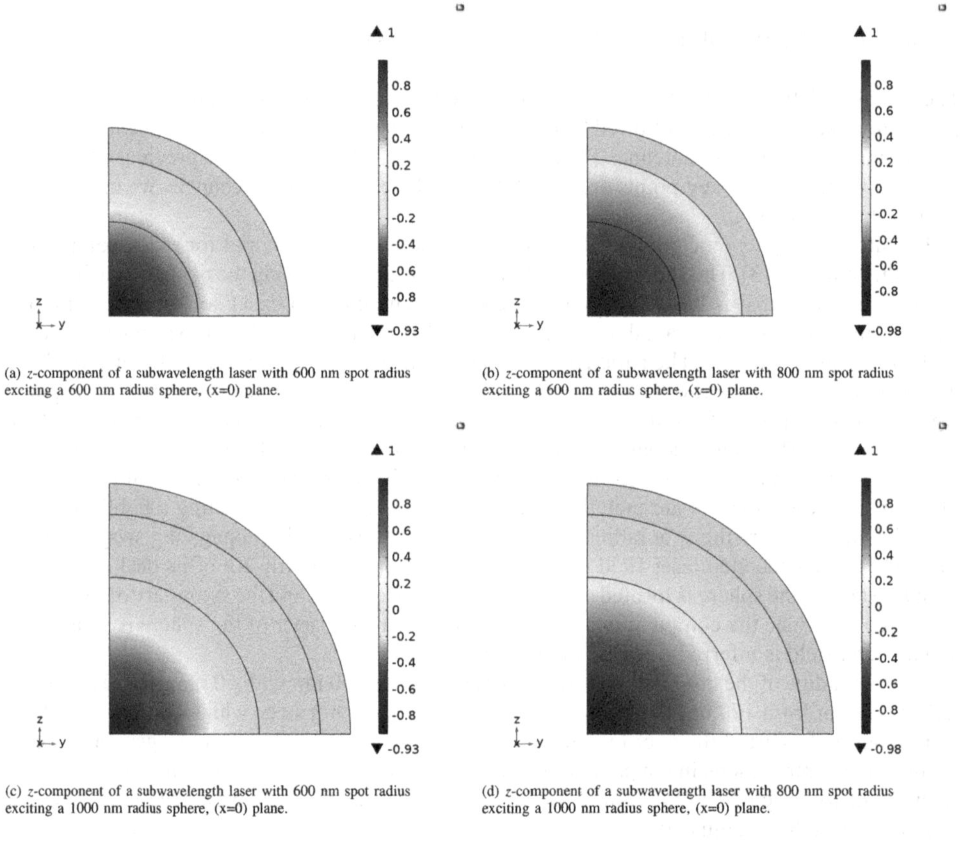

(a) z-component of a subwavelength laser with 600 nm spot radius exciting a 600 nm radius sphere, (x=0) plane.

(b) z-component of a subwavelength laser with 800 nm spot radius exciting a 600 nm radius sphere, (x=0) plane.

(c) z-component of a subwavelength laser with 600 nm spot radius exciting a 1000 nm radius sphere, (x=0) plane.

(d) z-component of a subwavelength laser with 800 nm spot radius exciting a 1000 nm radius sphere, (x=0) plane.

Figure 2. z-component of a subwavelength laser propagating in the -direction and exciting a micro-sphere suspended in air (symmetry exploited in the model). The outer region is a PML and the snapshot is taken at the plane. The laser electric field drops to exactly of its peak value at laser spot radius distance. The cases considered of the spheres radii and spot radii are (600 nm, 1000 nm) and (600 nm, 800 nm), respectively.

The far-field is automatically computed in the multiphysics solver using Stratton-Chu formula that utilizes the near-field information on the boundary between the air and the PML to compute the far-zone field.

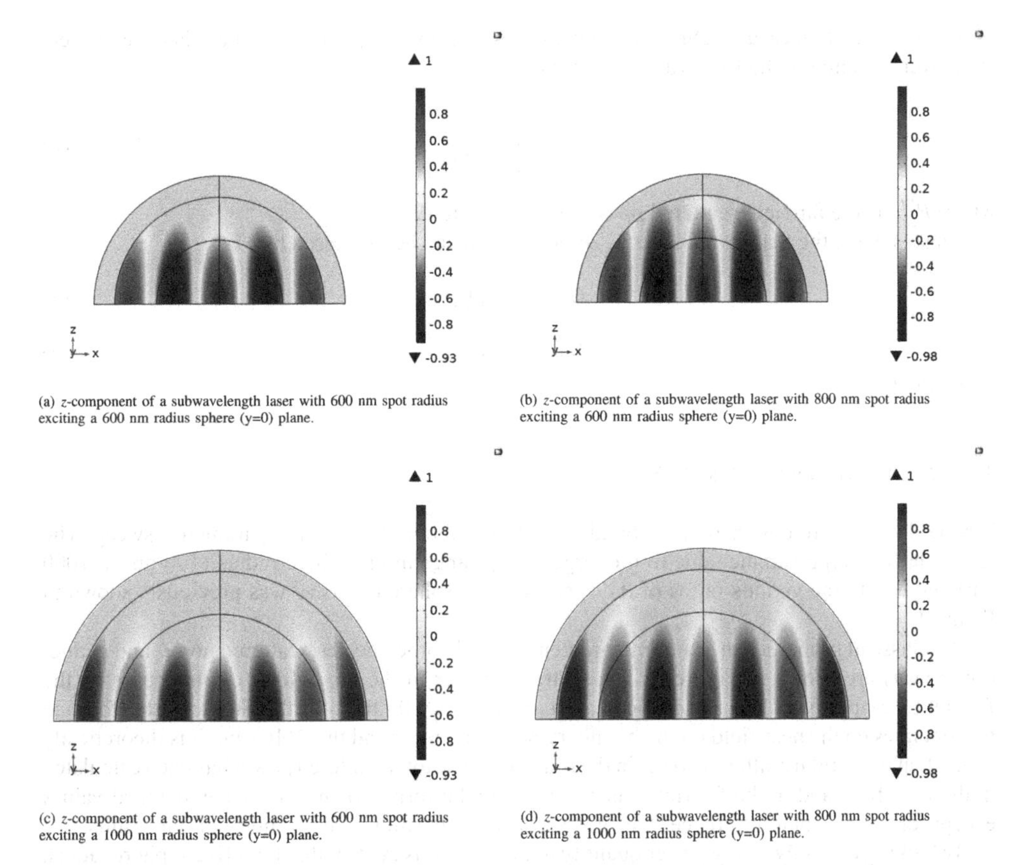

(a) z-component of a subwavelength laser with 600 nm spot radius exciting a 600 nm radius sphere (y=0) plane.

(b) z-component of a subwavelength laser with 800 nm spot radius exciting a 600 nm radius sphere (y=0) plane.

(c) z-component of a subwavelength laser with 600 nm spot radius exciting a 1000 nm radius sphere (y=0) plane.

(d) z-component of a subwavelength laser with 800 nm spot radius exciting a 1000 nm radius sphere (y=0) plane.

Figure 3. z-component of a subwavelength laser propagating in the -direction and exciting a micro-sphere suspended in air (symmetry exploited in the model). The outer region is a PML and the snapshot is taken at the plane. The laser electric field drops to exactly of its peak value at laser spot radius distance. The cases considered of the spheres radii and spot radii are (600 nm, 1000 nm) and (600 nm, 800 nm), respectively.

3 FAR-FIELD POSTPROCESSING EXPRESSIONS

The Far-Field postprocessing expressions for a transmitting nano-antenna excited by a subwavelength laser were extensively discussed in our previous work [11]. Here the main expressions to be evaluated in our computational results are restated.

Absorption, scattering and extinction powers are well known scattering properties that are conventionally calculated in scattering problems. Directivity calculations are needed to monitor the shape of the radiation pattern and to calculate the gain of the antenna. Forward Directivity D_F is mainly considered in this work since it is the maximum directivity in most of the cases.

The Standard Radiation Efficiency is the ratio of total scattered power to extinction power, i.e.:

$$e_{std} := \frac{P_{sc}}{P_{sc} + P_{abs}}, \tag{1}$$

where P_{sc} and P_{abs} are total scattered power in the far field and total absorbed power in the sphere, respectively.

A more suitable measure 'Directed Radiation Efficiency' [11] is needed such that it considers the power scattered in the forward direction, i.e.:

$$e_{dir} = \frac{P^{FF}_{sc,f}}{P_{sc} + P_{abs}}, \tag{2}$$

where $P^{FF}_{sc,f}$ is the far-field scattered power in the forward direction.

Consequently, the Standard Gain and Directed Gain can be expressed by

$$G_{std} = e_{std} D_f, \tag{3}$$

$$G_{dir} = e_{dir} D_f, \tag{4}$$

respectively.

4 COMPUTATIONAL RESULTS

Far-Field results are obtained corresponding to the different cases in the parametric sweep. The results have a large variance due to the large sweep range in the sphere radius (100 nm to 1000 nm) and due to the various ratios of the spot radius to sphere radius as was previously shown in Figure 1.

The classical properties in scattering problems (the absorbed power, scattered power and extinction power) are shown in Figure 4. The scattered power in the far field is calculated using the far-field variable (which is automatically obtained in COMSOL from the Stratton-Chu formula that operates on the near field on the boundary between the air and the PML) and it is theoretically equivalent to the total scattered power in the near field. The total scattered power has been calculated in the near field and in the far field (using Stratton-Chu formula) and they gave identical values except for a very small difference. This gives a strong indication of the validity of our results.

By looking carefully to any power quantity in Figure 4, it is evident that for a fixed sphere radius, the spot radius swept between 600 nm and 800 nm almost has no effect in the region where the spheres radii are below a specific critical sphere radius (around 240 nm radius in absorbed power plot and around 300 nm radius in scattered power plot). In this region, spheres are smaller than the laser spot radius and not comparable to it. In other words, for a fixed sphere radius below 240 nm (small spheres), the laser with 600 nm spot radius yields almost same absorbed and scattered power as that when laser has 700 nm or 800 nm spot radius. It is also noted that in this region, for any spot radius value, increasing sphere radius from 100 nm up to the critical sphere radius gradually increases absorbed power. However, generally speaking, for sphere radii above the critical sphere radius (radii where the spheres are comparable to or larger than the laser spot radius), the power is distributed in the plot such that it is almost proportional to both parameters (spot radius and sphere radius) which caused the peak power to be maximum (or almost maximum) in the case when both parameters are maximum (top right corner of the plot).

The Extinction Power shown in Figure 4(c) is the summation of absorption power and scattered power and is dominated by the scattered power which yielded a power plot very close to the scattered power plot.

Forward Directivity results are shown in Figure 5. Plot description and comments are similar to that of Figure 4. The main difference is that the critical sphere radius is 500 nm. The maximum directivity is at the top right corner of the plot (maximum sphere radius and maximum spot radius case).

Standard Radiation Efficiency, Directed Radiation Efficiency, Standard Gain and Directed Gain are plotted in Figure 6. Standard Radiation Efficiency shown in Figure 6(a) has very large values in most of the cases and the minimum values occur in the small spheres region in which, as expected, spot radius has no effect for a fixed sphere radius (see Equation (1) and Figure 4 above). By comparing Figure 6(a) with Figure 1, and by ignoring the small spheres in relative to spot radius,

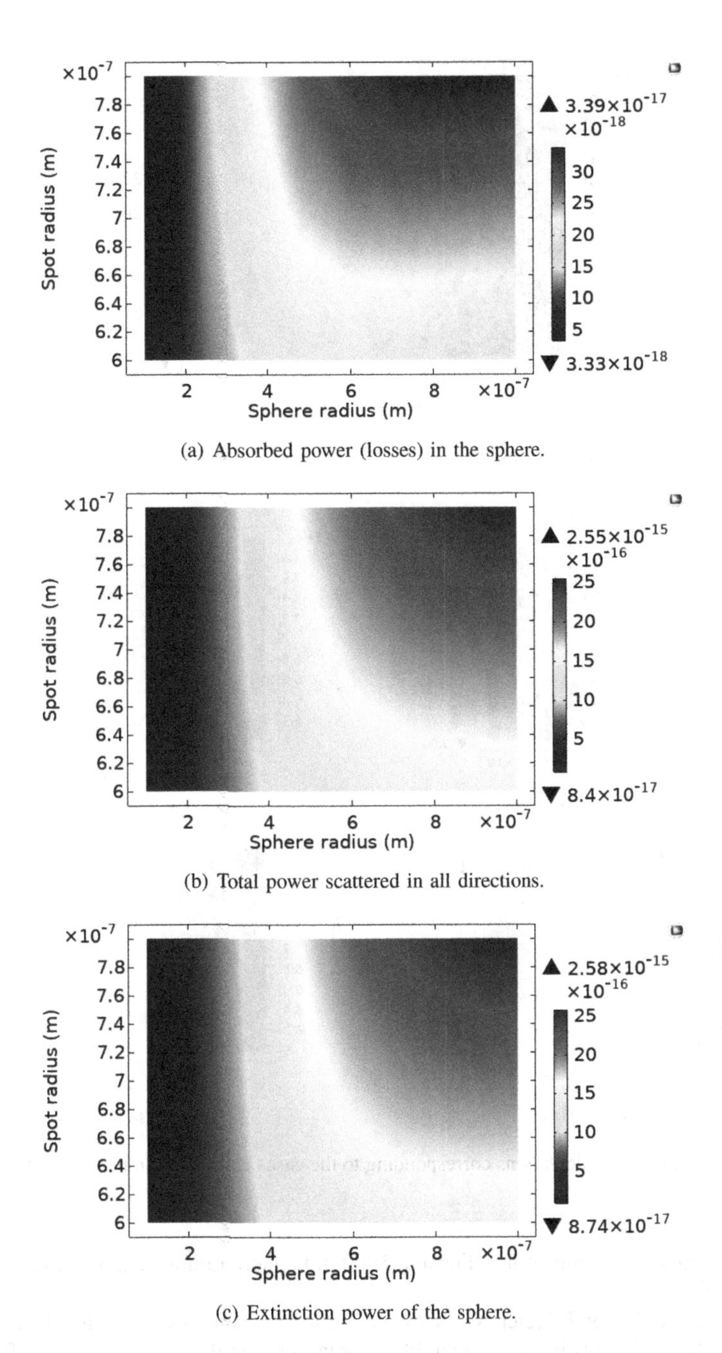

(a) Absorbed power (losses) in the sphere.

(b) Total power scattered in all directions.

(c) Extinction power of the sphere.

Figure 4. Absorption, scattering and extinction powers corresponding to the cases considered in the parametric sweep.

it can be deduced that the Standard Radiation Efficiency is almost constant for cases that have equivalent ratio of spot radius to sphere radius. This means that, in this region, Standard Radiation Efficiency is inversely proportional to the ratio of spot radius to sphere radius. By referring to Equation (1) it can be concluded that, for this region, the ratio of scattered power to absorbed power is determined by the ratio of spot radius to sphere radius. Maximum Standard Radiation

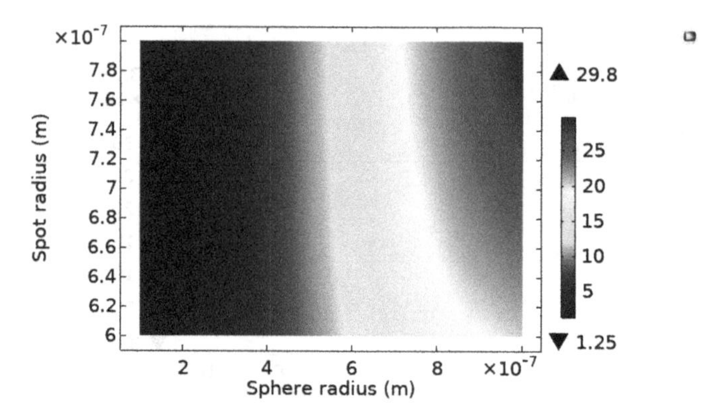

Figure 5. Forward directivity results.

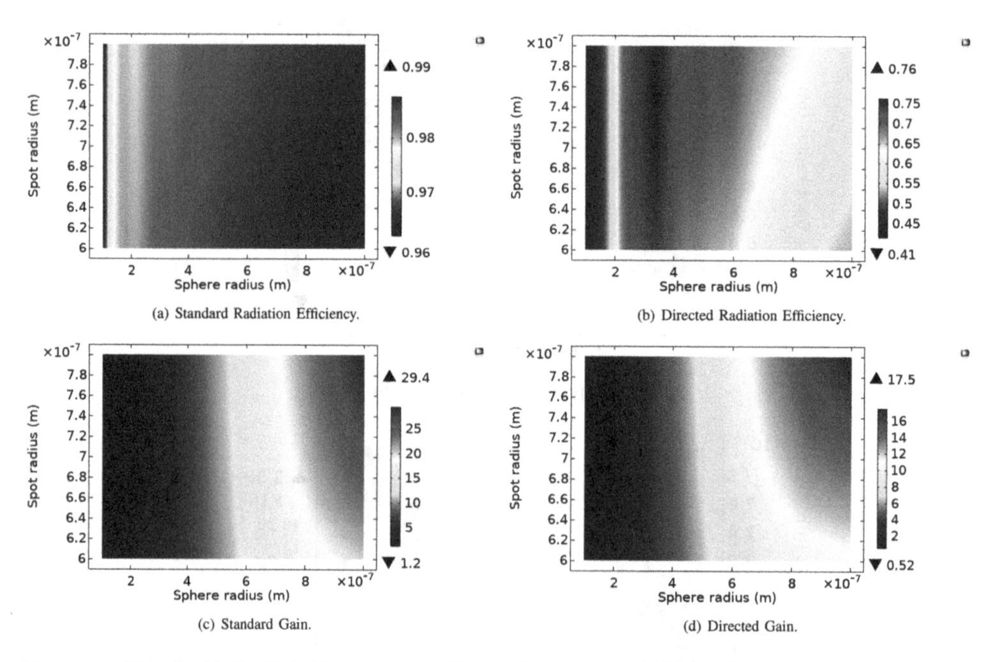

(a) Standard Radiation Efficiency.

(b) Directed Radiation Efficiency.

(c) Standard Gain.

(d) Directed Gain.

Figure 6. Main far-field calculations corresponding to the cases considered in the parametric sweep.

Efficiency occurs at 600 nm spot radius and 1000 nm sphere radius (smallest ratio of spot radius to sphere radius).

The Directed Radiation Efficiency is shown in Figure 6(b) and it shows large dependence on the spot radius when sphere radii are comparable to or larger than the laser spot radius. By comparing Figure 6(b) with Figure 1, and by ignoring the small spheres in relative to spot radius, it is clear that the Directed Radiation Efficiency is almost constant for cases that have equivalent ratio of spot radius to sphere radius. This means that, in this region, Directed Radiation Efficiency is proportional to the ratio of spot radius to sphere radius. By referring to Equation (2) it can be concluded that, for this region, the ratio of scattered power in the forward direction (far field) to extinction power is determined by the ratio of spot radius to sphere radius. To sum up, the ratio of spot radius to sphere radius determines the proportional relationship between absorbed power, scattered power and forward scattered power (in this region). In the large spheres region, minimum

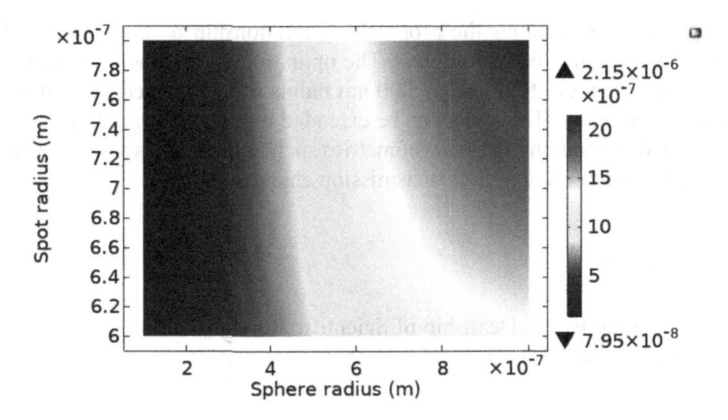

Figure 7. Absolute value of the z-component of the far-field electric field at 1 m in the forward direction.

Directed Radiation Efficiency occurs at 600 nm spot radius and 1000 nm sphere radius (smallest ratio of spot radius to sphere radius). It equals around 53%.

Standard Gain and Directed Gain are shown in Figure 6(c) and Figure 6(d), respectively. They result from multiplying Forward Directivity (Figure 5) with Standard and Directed Radiation Efficiencies, respectively. In the comparable and large spheres (compared to spot radius), both quantities are highly governed by the large forward directivities distribution which caused the maximum directivity case to coincide with the maximum gain case. The gain plots being similar to directivity plot also means that the description of the gain values in Figure 6(c) and Figure 6(d) are similar to that in the power quantities in Figure 4 (and hence also similar to Figure 5 description). The critical sphere radius is slightly shifted (different in the two gain measures) and the maximum is in the top right corner, i.e. at the maximum parameters (800 nm spot radius and 1000 nm sphere radius). As was previously discussed, Directed Gain is the main concern for the nano/micro-antenna operation.

The polarization diversity technique explained in [11] can also by used in this work by replacing the nano-dipole by our nano/micro-sphere. By following a similar technique, the results of the absolute value of the z-component of the far-field electric field at 1 m in the forward direction are calculated and shown in Figure 7. The plot description is also similar to that in Figure 4 (and hence also similar to Figure 5, Figure 6(c) and Figure 6(d)) with the critical sphere radius is around 400 nm. Maximum value is in the top right corner, i.e. at the maximum parameters (800 nm spot radius and 1000 nm sphere radius). This is also the optimum in the Directed Gain sense. The optimum case in our sweep is for a subwavelength laser having a spot radius equivalent to its wavelength.

5 CONCLUSION

A basic study was conducted by exciting a sphere with a subwavelength laser and calculating the far-field results to find the optimum nano/micro-spherical antenna for wireless communications. The subwavelength laser has a 800 nm wavelength and its spot radius ranged from 600 nm to 800 nm. The sphere radius ranged from 100 nm to 1000 nm. Far-field results showed a large variance due to the large sweep range and due to the various ratios of spot radius to sphere radius. Some spheres were totally overwhelmed by the laser spot (which caused them to be independent of the spot radius). In the comparable and large spheres region (in relative to spot radius), absorption power, scattered power, gain measures and the z-component of the far field variable (forward direction) are generally proportional to both spot radius and sphere radius. It was concluded that in the comparable and large spheres region (in relative to spot radius), the Standard and Directed Radiation Efficiencies are determined and are inversely proportional and proportional to the ratio of laser spot radius to sphere radius, respectively. Therefore, for the same region, the ratio of spot radius to sphere radius determines the proportional relationship between absorbed power, total scattered power and forward scattered power. The optimum micro-spherical antenna for wireless communications in the range studied is a 1000 nm radius sphere excited by a 800 nm spot radius laser (wavelength is 800 nm). This work can be extended by considering all cases of spot radius to wavelength ratio and by studying various geometries such as dipole, bow-tie, yagi-uda antennas in order to obtain a transmitter with better transmission characteristics.

ACKNOWLEDGMENT

This work was supported by the Deanship of Scientific Research at Jordan University of Science and Technology under Grant 2017/399.

REFERENCES

[1] C. Höppener and L. Novotny, "Imaging of membrane proteins using antenna-based optical microscopy," *Nanotechnology*, vol. 19, no. 38, p. 384012, 2008.

[2] H. A. Atwater and A. Polman, "Plasmonics for improved photovoltaic devices," in *Materials For Sustainable Energy: A Collection of Peer-Reviewed Research and Review Articles from Nature Publishing Group*. 1em plus 0.5em minus 0.4em World Scientific, 2011, pp. 1–11.

[3] E. Cubukcu, E. A. Kort, K. B. Crozier, and F. Capasso, "Plasmonic laser antenna," *Applied Physics Letters*, vol. 89, no. 9, p. 093120, 2006.

[4] L. Novotny, *Principles of Nano-Optics*. 1em plus 0.5em minus 0.4em Cambridge: Cambridge University Press, 2012.

[5] A. E. Miroshnichenko, I. S. Maksymov, A. R. Davoyan, C. Simovski, P. Belov, and Y. S. Kivshar, "An arrayed nanoantenna for broadband light emission and detection," *physica status solidi (RRL)–Rapid Research Letters*, vol. 5, no. 9, pp. 347–349, 2011.

[6] L. Novotny, "From near-field optics to optical antennas," *Phys. Today*, vol. 64, no. 7, pp. 47–52, 2011.

[7] K. Sendur and E. Baran, "Near-field optical power transmission of dipole nano-antennas," *Applied Physics B*, vol. 96, no. 2–3, p. 325, 2009.

[8] K.-M. See, F.-C. Lin, T.-Y. Chen, Y.-X. Huang, C.-H. Huang, A. M. YesÌ§ilyurt, and J.-S. Huang, "Photoluminescence-driven broadband transmitting directional optical nanoantennas," *Nano letters*, vol. 18, no. 9, pp. 6002–6008, 2018.

[9] P. Ghenuche, S. Cherukulappurath, T. H. Taminiau, N. F. van Hulst, and R. Quidant, "Spectroscopic mode mapping of resonant plasmon nanoantennas," *Physical review letters*, vol. 101, no. 11, p. 116805, 2008.

[10] A. A. Arisheh, S. Mikki, and N. Dib, "Design of transmitting nano-dipole antenna using a subwavelength laser excitation method," in *2019 IEEE International Symposium on Antennas and Propagation and USNC-URSI Radio Science Meeting*, July 2019, pp. 1313–1314.

[11] A. Abu Arisheh, S. Mikki, and N. Dib, "A subwavelength-laser-driven transmitting optical nanoantenna for wireless communications," *IEEE Journal on Multiscale and Multiphysics Computational Techniques*, vol. 5, pp. 144–154, 2020.

[12] =2 plus 4 3 minus 4 COMSOL AB, "Comsol multiphysicsÂ®5.5." [Online]. Available: https://comsol.com =0pt

[13] P. B. Johnson and R.-W. Christy, "Optical constants of the noble metals," *Physical review B*, vol. 6, no. 12, p. 4370, 1972.

Proceedings of the 1st International Congress on Engineering Technologies – Kiwan & Banat (Eds)

Practical performance analysis of Carrier Aggregation (CA) with 256 QAM in commercial LTE network

Eman S. Abushabab
College of Engineering. University of Dubai, Dubai, UAE

Mohamed Mahmoud
Technology Planning, EITC (du) Company, Dubai, UAE

Mohamed Saad
Department of Computer Engineering. University of Sharjah, Sharjah, UAE

Ahmed Alshal & Ayman Elnashar
Technology Planning, EITC (du) Company, Dubai, UAE

ABSTRACT: The main goal of this work was testing the performance of Carrier Aggregation (CA) in commercial LTE network and its comparison with the single carrier LTE network, and ensuring that the functionality of LTE-Advanced with CA meets the performance expectations for the commercial network, especially with 256QAM modulation. The aim of this article was to portray how CA technology is applied to the commercial LTE network and analyze its performance. We used CA in the downlink with two Carrier Components (CC) of 35 MHz aggregated bandwidth in the network with one CC of 20MHz at band 3 (i.e., 1800MHz) and the other of 10 MHz at band 20 (i.e., 800MHz). The field measurement results achieved the maximum downlink data rate of 215 Mbps in the Near-cell scenario, which is close to the theoretical peak throughput and 92 Mbps in the Far-cell scenario (worst case scenario).

Keywords: Carrier Aggregatioin; Long Term Evolution-Advanced, Inteenet of Things, Reference Signal Received Quality, Reference Signal Received Power, Channel Quality Indicator

1 INTRODUCTION

Carrier Aggregation (CA) is one of the main features and most important technology components in Long Term Evolution-Advanced (LTE-A). In particular, CA boosts average data rate, strengthens network capacity, extends the coverage range and makes traffic-management simpler [1].

Internet of Things (IoT), Cloud gaming and VR/AR applications have increased the demand for data rate. CA has been proposed to enable high data rates by overcoming the spectrum limitation problem and enabling the use of more than one carrier frequency (channel) at the same time to increase the overall bandwidth. The System bandwidth can be contiguous or consist of several non-contiguous bands [2]. CA enhances both peak data rates and average data rates while increasing downlink coverage and simplifying multi-band traffic management at the same time. The evolution of data rates through CA is shown in Figure.1 [3].

CA technology can be very effective as the aggregation of five carriers can provide higher data rates. Every aggregated carrier is referred to as a component carrier (CC). The bandwidth of the CC can be 1.4, 3, 5, 10, 15 or 20 MHz Since it is possible to aggregate a maximum of five CCs, the maximum aggregated bandwidth is 100 MHz. The recently launched smartphones support CAT11/12 and beyond with peak throughput of 1Gbps.

DOI 10.1201/9781003178255-15

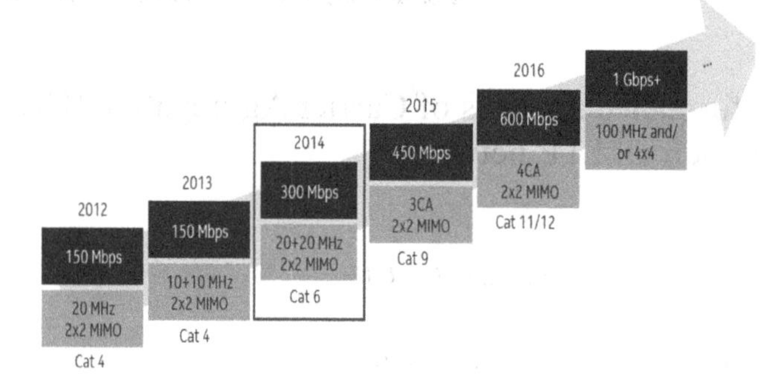

Figure 1. Data rate evolution in the downlink with carrier aggregation [3].

The easiest way to arrange aggregation is by using contiguous CCs within the same (as defined for LTE) operating frequency band, the so-called contiguous intra-band. This may not always be possible, thanks to operator frequency allocation scenarios. For non-contiguous allocation, it could either be intra-band, i.e. the CCs belong to the identical operating band, but have a niche, or gaps in between, or it could be inter-band, in which case the CCs belong to different operating frequency bands which were analyzed in this paper.

There are some related works regarding field measurement associated with CA. In [4], the results were provided by using five carrier components that work on 3.9GHz and 3.6GHz bands. In [5], the field measurement was done in a test network with one base station and three CCs at a 2.7GHz frequency band. However, a limitations of these studies was that the results have been obtained in interference-free conditions and in the absence of commercial User Equipment. In [1], the authors presented the performance of CA in real commercial networks using two CCs. In [1], the performance of CA has been discussed over 64QAM and the performance had been assessed using SINR and throughput.

The main goal of this work was assessing the performance of CA in real commercial networks with the highest modulation scheme (i.e., 256-QAM) with commercial User Equipment (smartphones) [6], [7].

The performance of CA in this work had been assessed using the following parameters: Reference Signal Received Quality (RSRQ), Reference Signal Received Power (RSRP), Block Error Ratio (BER), Channel Quality Indicator (CQI) and Throughput.

2 MEASUREMENT ENVIRONMENT

2.1 *Cells configuration*

The cell configuration that appears in Figure 2 was used in order to obtain our measurement results. The primary cell operates at 1.8GHz and the secondary cell operates at 800MHz According to the link budget, the radius of the primary cell (LTE 1800) was 870m (urban), while the radius for the secondary cell (LTE 800) is 1500 m. We assumed PC and SC denote the primary cell and secondary cell, respectively.

2.2 *Measurement environment parameters*

The parameters in Table 1 describe the test environment in Dubai (urban and suburban area). The downlink (DL) two-CC CA was measured in a driving test. The theoretical peak data rate of 20-MHz BW LTE with 2×2 MIMO was 150Mbps, and the theoretical peak data rate of 10-MHz BW LTE with the same conditions was 75Mbps. As a result, the theoretical peak data rate of the DL two-CC CA in this measurement is 225 Mbps without 256QAM.

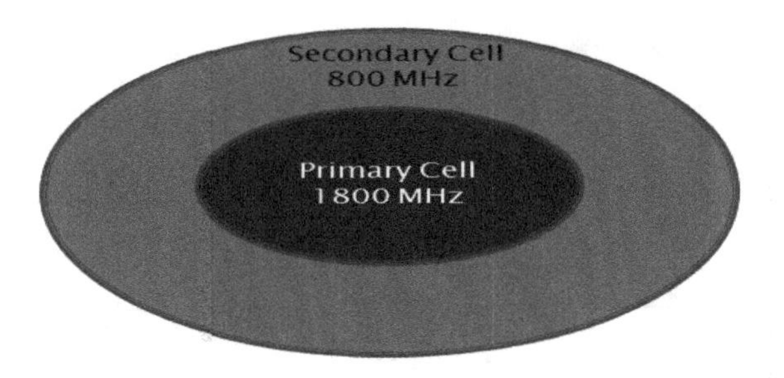

Figure 2. Cell configuration for the system.

Table 1. Measurement environment parameters

Number of the CCs	2
Carrier frequency of the 1st CC	1800 MHz (3GPP Band 3)
BW (number of RBs) of the 1st CC	20 MHz (100 RBs)
Carrier frequency of the 2nd CC	800 MHz (3GPP Band 20)
BW (number of RBs) of the 2nd CC	15 MHz (50 RBs)
Antenna gain	16 dBi
Antenna configuration	4X2
Antenna height	25 m
Cell transmit power	−43 dBm
Cell Radius	1.5 km
Drive test speed	Stationary
Measurement area size	Far and near Cell
Number of cells	1
Inter site distance	500 m

The scenario of CA deployment in the commercial netw-ork was close to that of the CA Scenario 2 in [4], which is the scenario of an interband non-contiguous CA in collocation. The antenna gain for both bands was 16 dBi Table 1 summarizes the values of all measurement environment parameters.

Our measurement area was split into two regions: the Near-Cell region and the Far-Cell region, in order to pinpoint the best and worst scenarios.

2.3 *Measurement scenario*

In this paper, the performance of CA has been assessed using collocated LTE 800 and LTE 1800 commercial networks. In other words, the same physical sites are used for both (primary and secondary) cells.

The tests were performed and the measurements were taken in the same site (stationary) with the same device supporting both LTE bands (CAT 12, 3GPP Rel-12 device). The KPIs are derived from the device aspect through post-processing scripts, as illustrated in Figure.3.

LTE eNB will get the information of device capability from the UE Capability information message sent (in the initial access/attach to the network) from device to network, which will contain the support of carrier aggregation for both bands LTE1800 and LTE800, that is the most important thing network pay attention in the UE Capability Information is CA Band combination as specified in the RRC (Radio Resource Control) message.

Once the network knows that the device supports CA, it will add the secondary cell (sCellToAddModList-r10) and (pucch-ConfigDedicated-v1020) for defining the uplink control channel for reporting the Ack/Nac for aggregated Carrier. The sequence can be seen here:

a) Network sends RRC asking for UE capability including band combination
b) Device replies with the supported band combination

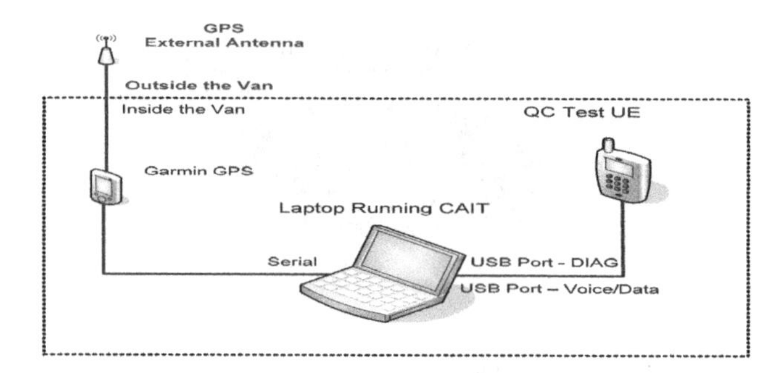

Figure 3. Measurement scenario.

c) If supported bands match with network configured carrier aggregation; then network will add the secondary cell to the device and will specify the uplink control channel
d) When the device start asking for data to be downloaded, SR (schedule request) will be sent to network
e) Then network will activate carrier aggregation on the MAC layer (through MAC CE activation)

We have obtained data from field measurements with a large sample size in order to get the average. In other words, for each iteration, we have obtained the results by averaging more than ten downloads of 20 GB files from a File Transfer Protocol (FTP) server connected to the LTE core network.

3 MEASUREMENT RESULT AND ANALYSIS

As mentioned previously, we have split our measurement environment into two regions: Near-cell region and Far-cell region in order to guarantee that the best and worst cases are taken into consideration.

3.1 *Near-cell region: The best-case scenario*

The Range for near cell testing was about 100–150 meters away from the site. We ran the download tests, where a laptop was tethered to the user equipment (UE), and ran our FTP downloads.

Various parameters have been used to assess the performance of CA, such as the Reference Signal Received Quality (RSRQ) and Reference Signal Received Power (RSRP) to test cell selection, reselection and handover. Figure. 4 depicts the values of RSRQ and RSRP for both primary and secondary cells. These two important metrics provide information about the quality of the channel. In particular, RSRP gives an indication of the average power received from a single reference signal and the typical range for it is around -44 dB (good) to -140 dB (bad). Moreover, RSRQ indicates the quality of the received signal, and its typical range varies from -19.5 dB (bad) to -3 dB (good). Based on our measurement results, the average value for RSRP and RSRQ was -55dB and -6dB for the PC and SC, respectively. This was quite acceptable.

Another performance parameter used to test the quality of the communication channels is the Channel Quality Indicator (CQI), which is an indicator that carries the information and shows how good or poor the communication channel is. We found that, for the near cell, the CQI was 14.9 for the PC and 11.4 for the SC which is within the acceptable range.

To measure the number of erroneous blocks received to the total number of blocks sent, the Block Error Ratio (BLER) was also used as a performance metric. The average value of the BLER was 5% most of the time.

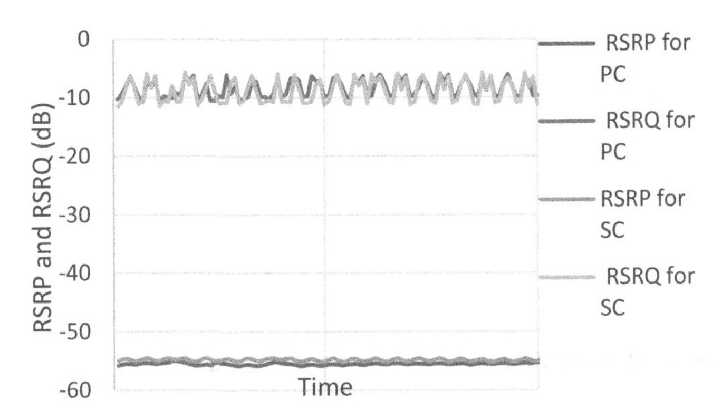

Figure 4. RSRP and RSRQ for PC and SC.

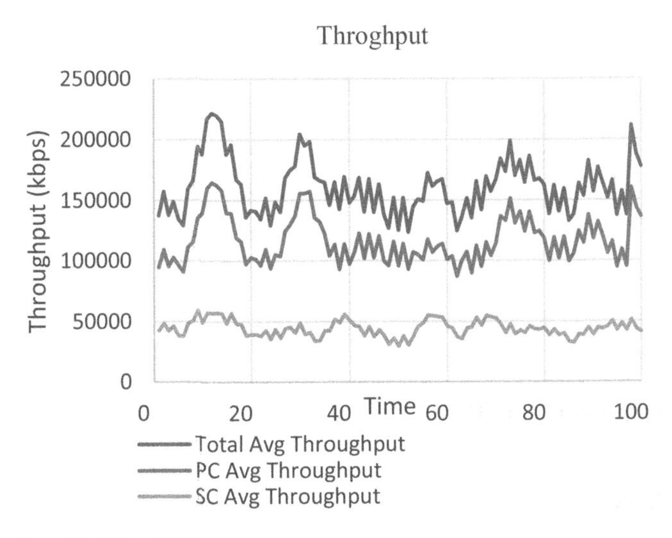

Figure 5. Throughput for PC and SC.

The Throughputs for the PC, SC and the total throughput for the whole configuration are shown in Figure. 5. The maximum throughput that has been measured while using the CA technique was 215Mbps, which is very close to the theoretical value that mentioned previously.

For the Near-cell region, the 256 QAM utilization was 99% for PC and 27% for SC. This gives us an indication that 256QAM is more applicable to a lower frequency.

3.2 *Far-cell region: The worst-case scenario*

The same performance parameters have been used to assess the Far-cell Region. Figure. 6 demonstrates the values of RSRQ and RSRP for both primary and SCs. Based on our measurement results, the average values for RSRP and RSRQ were -86.2dB and -0.7 dB, respectively, for the PC. The average values for RSRP and RSRQ were -5.2dB and -11.7dB, respectively, for the SC. Again, this is considered acceptable for the far-cell worst-case scenario.

The average value of BLER was 8.5% for the cell edge users. The Throughput for the PC, SC and the total throughput for the whole configuration is shown in Figure. 7. The maximum throughput that has been measured using the CA technique was 92 Mbps in the worst cases.

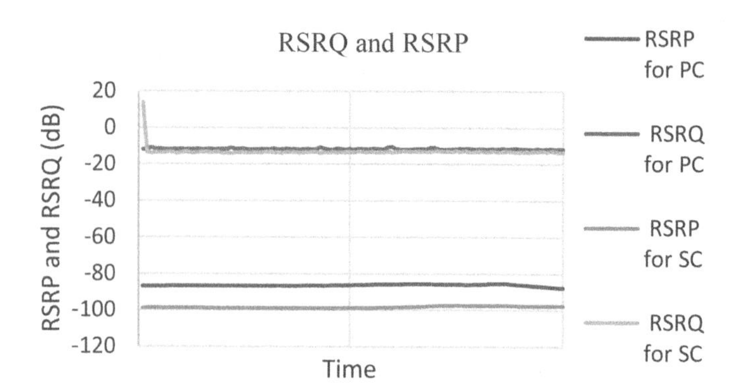

Figure 6. RSRP and RSRQ for PC and SC.

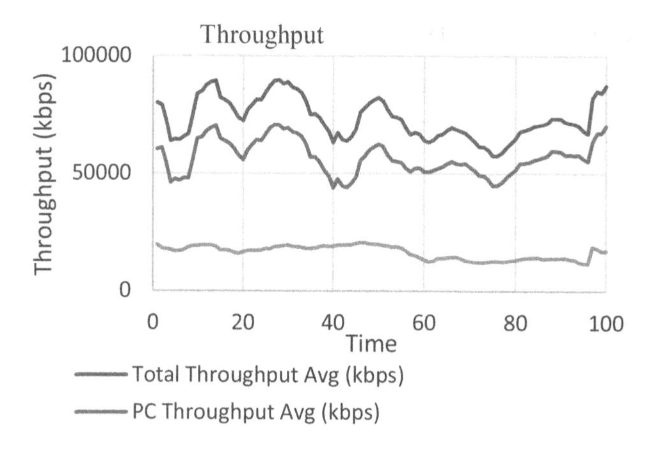

Figure 7. Throughput for PC and SC.

4 CONCLUSION

In this paper, the DL two-CC CA output was aggregated with 20MHz and 10MHz bands and was calculated by drive testing in both commercial urban and suburban areas. The performance evaluation shows that CA can provide 215 Mbps in the near-cell scenario and 100 Mb / s in the far-cell scenario in the real commercial networks with a very high peak user data rate. Based on the results of the measurements in this report, we are able to confirm that CA dramatically improves the user data rate when applied to commercial networks. The 256QAM modulation is more applicable to the lower frequency band.

REFERENCES

[1] K. I. Pedersen, F. Frederiksen, C. Rosa, H. Nguyen, L. G. U. Garcia, & Y. Wang, "Carrier aggregation for LTE-Advanced: functionality and performance aspects," IEEE Commun. Mag., vol. 49, no. 6, pp. 89–95, June 2011.

[2] S. Kundu, S. Gupta & D. Allstot, " Frequency- Channelized Mismatch-Shaped Quadrature Data Converters for Carrier Aggregation in MU-MIMO LTE-A"IEEE Trancactions on circuits and system., vol. 64., 2017.

[3] NOKIA, " LTE – Advanced Carrier Aggregation Optimization", Nokia Network whitepaper, 2017.

[4] K. Werner, H. Asplund, B. Halvarsson, A. K. Kathrein, N. Jalden, & D. V. P. Figueiredo, "LTE-A field measurements: 8×8 MIMO and carrier aggregation," in Proc. IEEE 77th Vehicular Technology Conf., 2013.
[5] K. Saito, Y. Kakishima, T. Kawamura, Y. Kishiyama, H. Taoka, & H. Andoh, "Field measurements on throughput performance of carrier aggregation with asymmetric bandwidth in LTE-Advanced," in Proc. FutureNetworkandMobileSummit, 2012.
[6] S. Lee, S. Hyeon, J. Kim, H. Roh & W. Lee, "The Useful Impact of Carrier Aggregation" , IEEE vehicular technology magazine, March 2017.
[7] Ayman Elnashar, M. El-Saidny, "Practical Guide to LTE-A, VoLTE and IoT: Paving the Way Towards 5G" Wiley, July 2018. http://eu.wiley.com/WileyCDA/WileyTitle/pro ductCd-1119063302.html

Proceedings of the 1st International Congress on Engineering Technologies – Kiwan & Banat (Eds)
 ISBN 978-0-367-77630-5

Round opportunistic fair downlink scheduling in wireless communications networks

Mohammad M. Banat & Razan F. Shatnawi
Department of Electrical Engineering, Jordan University of Science and Technology Irbid, Jordan

ABSTRACT: Round opportunistic fair (ROF) scheduling is proposed as a heuristic algorithm for improving fairness in wireless downlink transmission scheduling. ROF represents a trade-off between the achievable aggregate throughput on one side, and fairness on the other side. Even though the proposed algorithm is usable with other types of multiple access, a simple single-channel time-slotted system is assumed.

The proposed algorithm takes rounds over the users to be scheduled for downlink transmissions. However, it does not schedule users cyclically as in round robin scheduling because it prioritizes users with better channel conditions. At the same time, the proposed algorithm does not always schedule the user with the absolute best channel conditions to allow fairness in user access to system resources.

To keep the presentation of ROF as simple as possible, mathematical analysis that only demonstrates the operational mechanism of the algorithm is included. Fairness and other performance capabilities of ROF are evaluated using computer simulations. Simulation results show that ROF achieves good levels of fairness compared to other well-known algorithms such as proportional fairness scheduling.

Keywords: Downlink Transmission; Opportunistic Scheduling; Fairness

1 INTRODUCTION

Wireless spectrum efficiency is becoming more and more significant with the increasing demand on wideband wireless services [1] In a cellular network, channel conditions between the base station (BS) and different users have been generally different and statistically independent random time variation patterns [2] A signal that is transmitted over a wireless channel can suffer three types of (almost independent) superimposed fading effects: path loss, shadowing and multipath [2] Path loss is commonly modelled by an inverse power-law function of the received signal power on the distance between the transmitter and receiver. Shadowing takes the form of slow and random variations of the received signal power. Shadowing is mainly due to the presence of large obstacles in the signal transmission path, and is generally independent of the transmitter-receiver distance. Multipath fading takes the form of fast and random variations of the received signal strength. This effect is due to constructive and destructive interference among the received multipath components.

A scheduling policy is a rule, or set of rules, used to specify which user is scheduled to transmit/receive during a time slot. Opportunistic scheduling gives higher transmission priority to users with better channel conditions. Achievable throughput is an increasing logarithmic function of the signal to noise ratio (SNR) [3]. Sending to the user with the best channel conditions (or equivalently to the user with maximum achievable transmission rate) maximizes the downlink throughput [4] This approach is known as MaxRate scheduling. Several throughput-optimal scheduling algorithms have been presented in [5–7], and elsewhere. Opportunistic scheduling can cause severe unfairness

 DOI 10.1201/9781003178255-16

to users that are far from the BS. As a result of their large path losses, such users suffer low probabilities of having good channel conditions, and therefore, can be given only few and far-between transmission opportunities.

Fairness and spectral efficiency are very important issues in resource allocation in multiuser wireless networks. Spectral efficiency is measured by the normalized aggregate throughput in bits/s/Hz Fairness and spectral efficiency can involve contradicting network behaviours. A trade-off (that usually depends on the type of services provided by the network) between the two quantities should be sought.

Fair wireless scheduling, especially in time division multiple access (TDMA) systems has been extensively studied (e.g., in [8–10]) A major weakness in many previous works is that the channel is classified as either "good" or "bad" Such a coarse classification is too simple to characterize real wireless channels. Such a classification allows only a few degrees of freedom for the purpose of designing a scheduling algorithm.

The opportunistic framework in [11] takes into account three scheduling requirements: temporal fairness utilitarian fairness, and minimum performance requirement for each user. A scheduling scheme for the Qualcomm high data rate (HDR) system has been proposed in [12] This scheduling scheme exploits the time-varying channel conditions and is based on the proportional fairness (PF) concept defined in [13] More recently, [14] has provided a performance comparison between scheduling policies in the time and frequency domains for the LTE downlink.

The remainder of this paper is organized as follows. In section II, we introduce the channel model and outline our main assumptions. The proposed opportunistic scheduling algorithm is introduced in section III. Simulation results are presented in section IV, along with performance comparisons to other scheduling algorithms. The paper is concluded is section IV.

2 CHANNEL AND SYSTEM MODELS

We assume a downlink Rayleigh fading channel in a single cell system, with path loss, log normal shadowing and multipath fading The BS is assumed to be at the center of the cell, the radius of which is denoted as The BS is assumed to serve, in a time-slotted manner, fixed users, each equipped with one antenna The BS is assumed to always have packets to send to all users.

Assuming the transmitted signal power is P_t, the power received by user u is given by

$$P_u = |h_u|^2 P_t \tag{1}$$

where h_u is the channel gain expressed in the form [15]

$$h_u = \sqrt{c d_u^{-\alpha} s_u m_u} \tag{2}$$

where c is the mean path gain at a reference distance of 1 km, d_u is the distance in km between user u and the BS, α is the path loss exponent (PLE), s_u is the power scaling due to shadow fading and m_u is the phasor sum of the multipath components. α is typically between 2 and 4. The shadow fading power scaling factor s_u is assumed to follow the log-normal distribution. Let

$$s_{u,\mathrm{dB}} = 10 \log_{10} s_u \tag{3}$$

where $s_{u,\mathrm{dB}}$ follows a zero-mean Gaussian distribution with a variance σ_s^2. Typical values of σ_s are around 8 dB. As commonly accepted in the literature [16], we assume shadow fading is exponentially correlated. The multipath fading factor m_u is a zero-mean unit-variance complex Gaussian random variable.

The received SNR of user u is given by

$$Z_u = \frac{P_u}{P_n} \tag{4}$$

where P_n is the received zero-mean additive white Gaussian noise (AWGN) power. Substituting (1) and (2) into (4) yields the SNR of user u as follows:

$$Z_u = cd_u^{-\alpha} \frac{P_t}{P_n} s_u |m_u|^2 \tag{5}$$

Conditioned on s_u, the SNR is an exponentially distributed random variable, with a mean value

$$\overline{Z}_u = [Z_u | s_u] = cd_u^{-\alpha} \frac{P_t}{P_n} s_u \tag{6}$$

The received SNR by a cell edge user u^*, is given by

$$\eta = cD^{-\alpha} \frac{P_t}{P_n} s_u^* |m_u^*|^2 \tag{7}$$

Averaging η over shadow fading and multipath fading results in the mean cell edge SNR $\overline{\eta}$ as

$$\overline{\eta} = cD^{-\alpha} \frac{P_t}{P_n} \tag{8}$$

Like in [17], we use $\overline{\eta}$ to represent the acceptable noise level. Substituting (8) into (6) yields

$$\overline{Z}_u = \overline{\eta} \left(\frac{D}{d_u} \right)^{\alpha} s_u \tag{9}$$

Note that a new mean value of the SNR must be calculated every time s_u changes. The same is true when the distance between the user and the BS changes.

The modulation scheme that is used in the system under consideration is M - QAM. It is well-known that, at a given SNR, both the information bit rate and the bit error rate of M - QAM increase for larger values ofTherefore, when a BS transmission (BST) is assigned to the scheduled user, M is chosen such that it is higher when the user channel conditions are better. It was shown in [17] that the feasible transmission rate is a logarithmic function of the SNR. Following [15], we adopt the following expression for the achievable normalized throughput (in bits/s/Hz) as a function of the SNR:

$$\xi_u = \log_2 \left(1 + \frac{Z_u}{K} \right) \tag{10}$$

where K is a constant system-efficiency factor that depends on the system design and the target bit error rate.

3 OPPORTUNISTIC SCHEDULING ALGORITHM

In this section, we present the proposed round opportunistic fair (ROF) scheduling algorithm. A brief mathematical background is initially presented for the sake of clarifying the working principles of the algorithm. However, full mathematical analysis will not be attempted here.

ROF is a heuristic opportunistic scheduling algorithm that maximizes the network throughput under various fairness restrictions. The main idea of ROF is to limit assigning BST's to only a dynamic group of users that will be known in this paper as the candidate users. This restriction aims at improving chances of users with generally bad channel conditions to get BST assignments. Hence, good levels of fairness in BST assignment to users are expected to be achieved. Assigning a BST to the candidate user with the best channel conditions achieves maximum throughput under the user candidacy restriction. The number of candidate users controls the trade-off between fairness and network throughput. When there is only one candidate user, ROF performs round-robin scheduling, while when all users are candidates, it functions in a purely opportunistic manner.

The group of "candidate" users consists of users that have the smallest values of an ascendingly sorted "waiting figure". At each scheduling round, ROF grants channel access to the candidate user that has the best channel conditions. The waiting figure of the scheduled user is increased, and the group of candidate users is updated before the following scheduling round.

The role of the scheduling algorithm is to decide which user is to receive data from the BS in a given time slot, based on channel conditions of the users. Therefore, it is assumed that the BS knows the channel conditions of all users. How this information is made available to the BS is beyond the scope of this paper.

Even though the BS can be assumed to use multiple channels to transmit to several users in the same time slot, this work is limited to single-channel transmissions. Scheduling users on multiple channels in the same time slot can be the basis of using ROF in multicarrier wireless networks.

It is assumed that a BST consists of several symbols that are proportional to the achievable throughput. Quantization to integer numbers of bits is not considered in this paper; and hence whenever throughput is mentioned, what is meant is the achievable throughput. In the simulation results below, we measure the achievable throughput normalized to a unit bandwidth.

The duration of a BST is fixed, regardless of the number of transmitted symbols. It is assumed that channel conditions do not change during a BST. However, the channel is assumed to vary independently from one BST to another. As mentioned earlier, all users are assumed to always have data to receive from the BS.

At the beginning of each time slot, the scheduler determines the best candidate user that should receive a BST The objective is to optimize the network throughput Below, we illustrate how the proposed algorithm works.

- Downlink transmissions happen at the ith multiples of the channel coherence time (to make sure that channel conditions do not change during a BST), where $i = 1, 2, \ldots$.
- Each user has a wait figure $w_u(i)$, which controls the number of time slots the user has to wait before competing for a BST. Wait figures of all users are initialized with very small random values. In fact, initial wait figures should all be set to zero. However, this would not provide any means to select the initial set of users that are candidates to get the BST.
- The integer $r_u(i) \in [1, U]$ is used to indicate the rank of user u for the purpose of receiving a BST. Users are ranked in ascending order of their wait figures. Precisely, the rank of user u at time is i equal to

$$r_u(i) = \sum_{k=1}^{U} 1(w_u(i) - w_k(i)) \tag{11}$$

where $1(x)$ is the discrete unit step function, given by

$$1x = \begin{cases} 1, & x \geq 0 \\ 0, & x < 0 \end{cases} \tag{12}$$

- The number of users that are candidates to receive a BST at a given time instant is denoted as U_x, where $1 \leq U_x \leq U$. Users with ranks $r_u(i) \leq U_x$ are candidates to receive a BST at time i.
- If user v receives a BST at time i, a constant quantity ρ is added to its current wait figure $w_v(i-1)$ to form the new wait figure $w_v(i)$. Mathematically, if user v receives a BST at time i, then

$$w_v(i) = w_v(i-1) + \rho \tag{13}$$

- The channel gain from the BS to user u at time i is denoted as $h_u(i)$. The set of channel gains $\{h_u(i)\}_{u=1}^{U}$ will be assumed to be statistically independent for different values of u and for different values of i. The channel gain is assumed to take the form in (2).

- The signal to noise ratio $Z_u(i)$ of user u during time slot i is an exponential random variable, the mean of which takes the form in (9). In this paper we use $Z_u(i)$ to calculate the achievable throughput.
- The performance measure of user u is $\xi_u(i)$, as given by (10) Other forms of $\xi_u(i)$ dependence on $Z_u(i)$ can be used to represent a wide range of QoS requirements.
- The user that is scheduled to receive a BST at time i is the one with the largest performance measure from among all candidate users (users with $1 \leq r_u(i) \leq U_x$). As it will turn out to be the case, U_x is an important parameter in the trade-off between temporal fairness and performance optimization. Note that $U_x = U$ means purely opportunistic transmission (i.e., MaxRate), while $U_x = 1$ means absolutely fair transmission (i.e., RR).
- It is therefore useful to define the indicator function:

$$I_u(i) = 1(U_x - r_u(i)) \tag{14}$$

- Note that user u can be a candidate for receiving a BST at time i only if $I_u(i) = 1$.
- We define the weighted performance measure $\beta_u(i)$ as

$$\beta_u(i) = \xi_u(i) I_u(i) \tag{15}$$

- Note that $\beta_u(i)$ is equal to zero for all users with higher ranks than U_x. This means that such users are excluded from competition for the BST.
- If v is the identification number of the user that has received a BST at time $i \geq 1$, then the following conditions must be met

$$1 \leq r_v(i) \leq U_x \tag{16}$$

$$I_v(i) = 1 \tag{17}$$

$$\beta_v(i) = \max_u \{\beta_u(i)\}_{u=1}^{U_x} \tag{18}$$

4 SIMULATION RESULTS

To produce the simulation results that are presented below, we have assumed a single-cell cellular system with one BS at the cell center. The system has no co-channel interference, meaning that the system is noise-limited, and that throughput computations are based on the SNR. The cell radius is 1 km, and the PLE is 4. The number of users is 40. User u is assumed to be separated from the BST by a distance equal to

$$d_u = 0.1 + 0.8\left(1 - \frac{u}{U}\right) \tag{19}$$

In other words, user U is closest to the BS and user 1 is farthest, with equal distance increments for the users in between. In throughput calculations using (10), the system efficiency factor K is equal to 8. The number of users that compete for a BST is 10. Each simulation experiment includes 50 runs of 400,000 BSTs (or time slots) each. In each run, measurements like throughput and airtime share are taken every 1000 BSTs. This means that 400 readings are taken in every run. Readings are averaged over the 50 runs to produce the experiment results.

The shadow fading component of the channel gain is assumed to change every 100 time slots. Knowing that a random sequence with exponential autocorrelation can be generated by a first order autoregressive model [18], shadow fading updates are performed according to the recursion

$$s_{u,\mathrm{dB}}^{\mathrm{New}} = \eta s_{u,\mathrm{dB}}^{\mathrm{Old}} + (1 - \kappa)\varepsilon \tag{20}$$

where $s_{u,\mathrm{dB}}^{\mathrm{New}}$ is the updated shadow fading power scaling factor in dB, $s_{u,\mathrm{dB}}^{\mathrm{Old}}$ is the old shadow fading power scaling factor in dB, κ is the autocorrelation coefficient and ε is a zero-mean white Gaussian

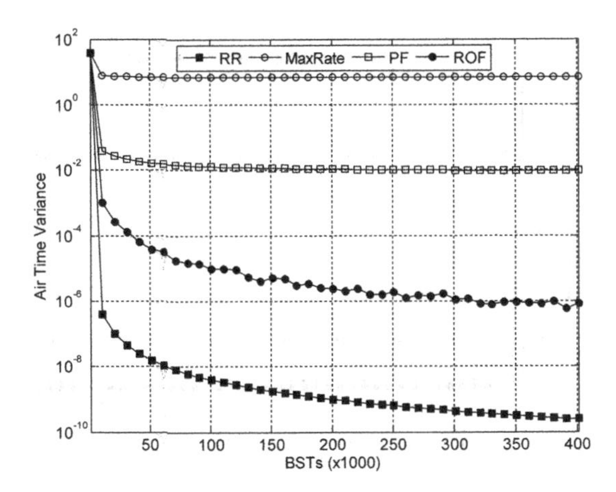

Figure 1. Airtime fairness comparisons.

sequence with variance σ_ε^2 that is statistically independent of $s_{u,\mathrm{dB}}^{\mathrm{Old}}$. Note that from (20) we should have

$$\sigma_s^2 = \kappa^2\sigma_s^2 + (1-\kappa)^2\sigma_\varepsilon^2 \tag{21}$$

Solving (21) for σ_ε^2, one obtains

$$\sigma_\varepsilon^2 = \frac{1-\kappa^2}{(1-\kappa)^2}\sigma_s^2 = \frac{1+\kappa}{1-\kappa}\sigma_s^2 \tag{22}$$

The shadowing autocorrelation coefficient is assumed to be 0.8, while its standard deviation is assumed to be 8 dB.

Although many fairness indicators have been proposed and used in the literature, we opt to use the variance of a set of measurements to quantify fairness in the measured quantity. This is motivated mainly because the variance measures the extent of variation of the measured quantity. When the variance of a measured quantity is lower, the measurements are closer to their average value, and hence the fairness is higher. Obviously, the converse is true as well.

Below we present comparisons between the proposed scheduling algorithm and RR scheduling, MaxRate scheduling and PF scheduling. We also study the effects of path loss and shadowing on the results.

In Figure 1 we have plotted the variance of the numbers of BSTs assigned to all users using ROF and a number of other scheduling schemes. ROF performs very well in this aspect, compared to MaxRate and PF. Note that in the long run, ROF achieves an airtime variance that is several orders of magnitude lower than those achieved by PF and MaxRate. Obviously, no scheduling scheme can achieve better air time fairness than the RR scheme, and this is indeed the case according to Figure 1.

In Figure 2 we have plotted the variance of the average normalized throughput achievable by users using ROF and a number of other scheduling schemes. ROF performs better than all three other scheduling schemes in this aspect. It outperforms PF and MaxRate by orders of magnitude in terms of throughput fairness among the users, while also slightly outperforming RR.

In Figure 3 we have plotted the ROF average normalized throughput versus the number of candidate users for PLE values 2 and 4, and shadowing standard deviation (SSD) values 6 and 10 dB In all cases, and as expected, the normalized throughput increases with the number of candidate users. As pointed out earlier, when the number of candidates is one (intersection of throughput curve with the vertical axis on the left), ROF performs exactly like an RR algorithm. RR achieves lowest throughput and highest fairness. On the other hand, when the number of candidates equals the number of users (intersection of throughput curve with the vertical axis on the right), ROF

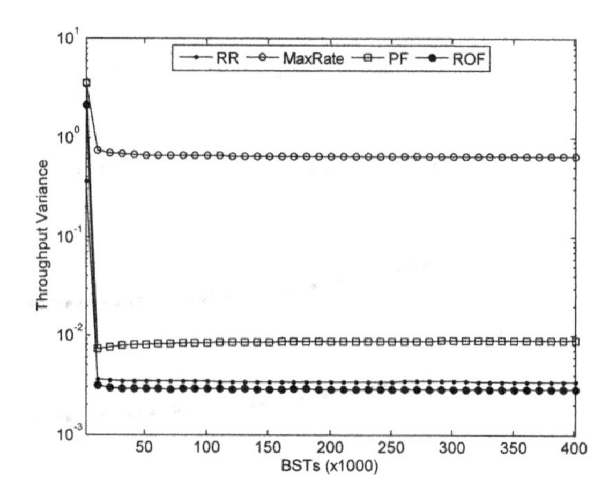

Figure 2. Throughput fairness comparisons.

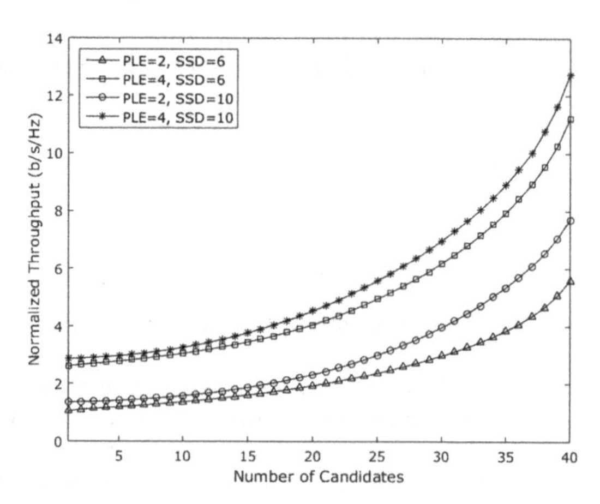

Figure 3. Effect of PLE and shadowing standard deviation on ROF average normalized throughput.

performs exactly like a MaxRate algorithm, which is purely opportunistic. MaxRate achieves highest throughput and lowest fairness.

As can be seen from the figure, a higher PLE leads to a higher achievable throughput. This is because a higher PLE causes larger variations in the received SNR, and hence in the achievable throughput. Given the opportunistic nature of the scheduling scheme, larger variations in the achievable throughput can be utilized to increase the overall network throughput.

Similarly, higher SSD deviations lead to higher achievable throughput. This is because a higher SSD causes larger variations in the received SNR, and hence in the achievable throughput. Given the opportunistic nature of the scheduling scheme, larger variations in the achievable throughput can be utilized to increase the overall network throughput.

5 CONCLUSION

We have presented ROF, a new heuristic algorithm for fair scheduling of opportunistic wireless downlink transmissions. Computer simulations have been used to study throughput and fairness performance of the new algorithm. Results have been compared to those of other well-known scheduling algorithms. Our results indicate that, compared to other algorithms, ROF achieves substantial fairness improvements at the cost of some throughput reduction This paper has opened more areas for research. Many ideas can be extended from this work. We can summarize them in the following points: 1. extend our work from single channel to multi-channel 2. assign different time shares to users. 3. study the delay statistics. 4. work with multiple services classes.

REFERENCES

[1] L. Zhang, M. Xiao, G. Wu, M. Alam, Y. C. Liang & S. Li, "A Survey of Advanced Techniques for Spectrum Sharing in 5G Networks," *IEEE Wireless Communications*, vol. 24, no. 5, pp. 44–51, October 2017.

[2] D. Tse & P. Viswanath, Fundamentals of Wireless Communication, New York: Cambridge University Press, 2005.

[3] A. Goldsmith, Wireless communications, Cambridge University Press, 2005.

[4] A. J. Goldsmith & P. P. Varaiya, "Capacity of fading channels with channel side information," *IEEE Transactions on Information Theory*, vol. 43, no. 6, pp. 1986–1992, Nov. 1997.

[5] M. Andrews & L. Zhang, "Scheduling algorithms for multi-carrier wireless data systems," in *MobiCom'07*, Montreal, Canada, September 9–14, 2007.

[6] S. Liu, L. Ying and R. Srikant, "Scheduling in multichannel wireless networks with flow-level dynamics," in *SIGMETRICS'10*, New York, NY, USA, 2010.

[7] Y. Chen, X. Wang & L. Cai, "On achieving fair and throughput-optimal scheduling for TCP flows in wireless networks," *IEEE Transactions on wireless communications*, vol. 15, no. 12, pp. 7996–8008, September 2016.

[8] G. Song & Y. G. Li, "Cross-layer optimization for OFDM wireless networks—Part I: theoretical framework," *IEEE Transactions on Wireless Communications*, vol. 4, no. 2, pp. 614–624, Mar. 2005.

[9] T. S. E. Ng, I. Stoica and H. Zhang, "Packet fair queueing algorithms for wireless networks with location-dependent errors," in *INFOCOM 1998*, San Francisco, CA, USA, March 29-April 2, 1998.

[10] V. Bharghavan, S. Lu & T. Nandagopal, "Fair queuing in wireless networks: issues and approaches," *IEEE Personal Communications*, vol. 6, no. 1, pp. 44–53, Feb. 1999.

[11] X. Liu, E. K. P. Chong & N. B. Shroff, "A framework for opportunistic scheduling in wireless networks," *Computer Networks*, vol. 41, no. 4, p. 451–474, 2003.

[12] P. Bender, P. Black, M. Grob, R. Padovani, N. Sindhushyana & A. Viterbi, "CDMA/HDR: a bandwidth efficient high speed wireless data service for nomadic users," *IEEE Communications Magazine*, vol. 38, no. 7, pp. 70–77, Jul. 2000.

[13] A. Jalali, R. Padovani & R. Pankaj, "Data throughput of CDMA-HDR a high efficiency-high data rate personal communication wireless system," in *VTC 2000-Spring*, Tokyo, Japan, May 15–18, 2000.

[14] O. Grøndalen, A. Zanella, K. Mahmood, M. Carpin, J. Rasool & O. N. Østerbø, "Scheduling Policies in Time and Frequency Domains for LTE Downlink Channel: A Performance Comparison," *IEEE Transactions on Vehicular Technology*, vol. 66, no. 4, pp. 3345–3360, April 2017.

[15] J.-G. Choi & S. Bahk, "Cell-throughput analysis of the proportional fair scheduler in the single-cell environment," *IEEE Transactions on Vehicular Technology*, vol. 56, no. 2, pp. 766–778, Mar. 2007.

[16] M. Gudmundson, "Correlation model for shadow fading in mobile radio systems," *Electronics Letters*, vol. 27, no. 23, pp. 2145–2146, 7 Nov. 1991.

[17] S. Catreux, P. F. Driessen & L. J. Greenstein, "Data throughputs using multiple-input multiple-output (MIMO) techniques in a noise-limited cellular environments," *IEEE Transactions on Wireless Communications*, vol. 1, no. 2, p. 226–234, Apr. 2002.

[18] W. Wei, Time Series Analysis: Univariate and Multivariate Methods, Addison Wesley, 1994.

Proceedings of the 1st International Congress on Engineering Technologies – Kiwan & Banat (Eds)
© 2021 Taylor & Francis Group, London, ISBN 978-0-367-77630-5

Simulation assisted leak detection in pressurized systems using machine learning

Ameer Mubaslat, Ahmad AlHaj & Saud A. Khashan
Jordan University of Science and Technology, Irbid, Jordan

ABSTRACT: In this paper, we utilize Computational Fluid Dynamics (CFD) generated data to train a Recurrent Neural Network (RNN) for detecting leaks in pressurized fluid distribution systems. The obtained results support the validity of implementing Machine Learning techniques in approximating active leak locations in a single pipe setup. This paper also discusses the validity of implementing these techniques for implementation on a fluid distribution network.

Results obtained utilizing the RNN model show adaptive behavior with the system's consistent response to different configurations of pipes and boundary conditions. The predictions for the leak localities are more accurate and more economically feasible than those obtained with currently used methods.

Keywords: Fluid; Machine Learning; ANSYS; MATLAB; Leak Detection

1 INTRODUCTION

The analysis of either long or short pressurized pipe networks, like those used in oil and gas pipeline and steam generation lines, is very important for maintaining efficient and safe operations. Such analysis is most needed to improve failure predictions, which can set the basis for risk analysis, integrity assessment, and managerial improvement used by operation and maintenance teams [1]. Furthermore, comparative assessment studies concerning pipeline failures addressing their frequencies, causes, consequences, similarities, and managerial differences proved important in ensuring effective, safe, environmentally clean, and economically feasible operation [10, 13]. Many of these assessments were conducted for energy facilities and oil shale pumping pipelines in the United States, Europe, and China [2].

Current leak detection methods rely on detecting leak propagation through constantly monitoring an array of pressure sensors distributed across the system, consequently reporting obtained sensory data to a proper Building Management System (BMS) for analysis [4]. This method, though functional, does not serve the goal of detecting the exact location of the leak and relies on the use of a large number of pressure sensors and classical measures such as pumping at high pressure with a visually detectable fluid to obtain the precise leak locale [6]. Other systems rely on the use of many separate valves and sensors at different points to physically detect regions of pressure drops in the system. The proximity to the leak can be reduced by increasing the number of sensors between elbows and joints. Modern and classical methods are considered expensive and time-consuming to operate, maintain, and install [3, 5].

Commercialized modern leak detection systems rely on monitoring pressure disturbances and drops throughout the network according to a predetermined profile, and then reporting such information to a relevant software to approximate a damaged segment of the network [3, 4].

The introduction of Machine Learning in the analysis of mathematical and physical models describing pressurized systems is projected to yield significant improvements [7], most notably on the accuracy of predicting the leak locales without relying on an excessive number of sensing locations.

DOI 10.1201/9781003178255-17

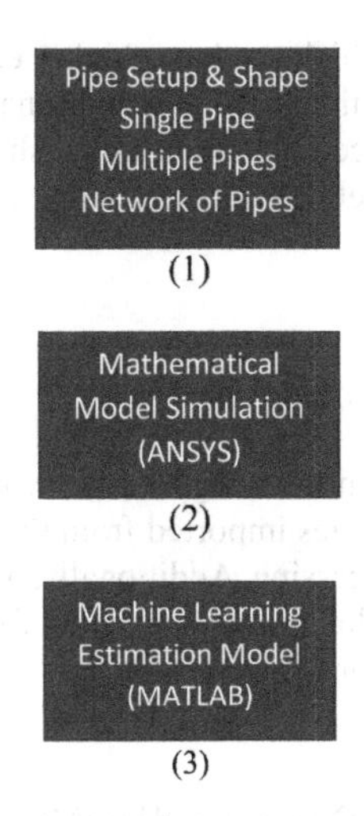

Figure 1. Areas of focus in the study.

In our work, prediction of leak localities is made through the utilization of a Recurrent Neural Network (RNN) that analyzes pressure readings (pressure predictions in our work), which associate with several relevant parameters. These parameters are diameter, pipe roughness, dynamic and kinematic viscosity, as well as fluid density. The pressure predations are acquired from simulations. Real-time readings and feedback from an operator can be expedited as well [9]. This approach allows for the introduction of adaptive behavior that allows the algorithm to function under different configurations of pipe setups, boundary conditions, and pressure profiles over time [8, 33].

2 MODELING FLUID FLOW AND LEAK SCENARIOS

In this paper, a simulation-based approach is used to collect pressure readings from pipe setups. This is accomplished through utilizing the Computational Fluid Dynamics CFD program, using the Fluent ANSYS toolkit. Through the CFD software different parameters of fluid flow are processed under a predetermined set of boundary conditions and program configurations [14].

Achieving a functioning RNN software capable of approximating leak locales requires obtaining a reliable dataset to serve as a benchmark for leak induced pressure-drop profiles. This approach allows for future development from simulating and processing single pipe setups into the more complex setup of a network of pipes. The different areas of focus that were included in the study are seen in Figure 1.

The different pipe setup configurations shown in Figure 1 are meant to be separately simulated and studied to produce a more complex and reliable leak detection system [11–13].

The following two sections describe the procedures and configurations revolving around the CFD modeling process.

2.1 *Meshing and simulation parameters adjustments for ensuring higher computational accuracy*

Several approaches were adopted in the CFD simulation and meshing setup to assist in producing outputs with higher accuracy which in turn are utilized as input to the RNN. We employed three dimensional (3D) simulation, instead of two dimensional (2D) rendering and processing [2, 4, 18]. This approach allows for improved fluid flow investigation around leak locales. It does so while increasing accuracy for debugging any design-based irregularity in the setup structure. Given the presence of turbulent flow, 2D simulations assume symmetry at one or more axis. Thus, it sidelines proper fluid diffusion in 3D space which produces higher error percentages. Furthermore, we use a refined mesh. Though it comes with a high computational cost, refined mesh allows for higher CFD accuracy. The cost can be reduced by clustering mesh on the boundary layer region adjacent to the walls. RANS Turbulence modeling was used with consideration for wall functions [15–17].

2.2 *Simulation setup parameters*

The geometry consists of a one-meter long pipe that is 35 cm in diameter and is constructed from PVC.

Through mesh analysis and after processing, it is shown that numerical error can be reduced to less than 3%, with material properties imported from the program's internal material dataset to ensure a fit between meshing and processing. Additionally, mesh refinement was introduced around the inlet, outlet, and leak surface. This process accounts for inconsistencies in the calculation of the numerical models in the processor [17].

3 APPLYING MACHINE LEARNING ALGORITHMS FOR LEARNING AND LEAK APPROXIMATION

Machine Learning allows for automatic and continuous training from input datasets without explicit programming. The training process begins with observations (datasets) of input and target data [7]. This allows the different algorithms to approximate patterns and behaviors in data. As such, these patterns are captured by functions which we can later use to anticipate the future behaviors of our system [9]. The main purpose is to allow the system's autonomous training to produce a method capable of producing reliable decisions. Nonetheless, the system requires additional training and datasets to properly build the model [19].

One of the main advantages of ML is that it allows a system to analyze massive quantities of data. While delivering faster, more accurate results to identify one or more output decisions, additional time and resources are required to properly train the system [8, 20]. Machine Learning Models are divided into four main categories, as listed below:

a) Supervised machine learning.
b) unsupervised machine learning.
c) Semi-supervised machine learning.
d) Reinforcement machine learning.

These categories conform to a set of uses and are limited to certain functions. The model which most properly fits the data and application of training a system on a complex set of input that is constantly updated and changing with one output parameter is the Supervised Machine Learning Algorithm [20, 21].

The following section describes the Models and methods used in conjunction with the RNN, as well as the modification made to the datasets used for the training process.

3.1 *Recurrent neural networks and machine learning*

Recurrent Neural Networks are structured to exhibit temporal dynamic behavior, as to be influenced by past inputs and targets outside of training, unlike basic feed-forward neural networks. This allows for the system's constant improvement while in operation. This gives them an advantage over Convolutional Neural Networks (CNNs) as it allows them to process unsegmented and connected time series [27].

The learning function utilized is based on gradient descent with backpropagation to update the weights with 4952 timesteps and 73 sets (iterations) and a target accuracy of 70% coupled with a learning rate of 0.001.

Table 1. Unmodified simulation output.

Readings		Targets
Pressure Reading (KPa)	Reading Position (meters)	Leak Position (meters)
4952 Value	4952 Value	Single Value

Table 2. Feature enhanced simulation output (L: leak location).

Readings					Targets
P KPa	X (m)	P1	P2	P3	L (m)
4952 Value	4952 Vlaue	Square value of P	Log Value of P	Square Root of P	Single Value

Provided that the simulation data acquired from the CFD program could prove insufficient, it is augmented by data from new inputs derived from the simulation data. This process is known as feature expansion [22–24].

Tables 1 and 2 demonstrate how the simulation output data in our problem is expanded using two expansion kernels to extract more features from the data [26]. These expansion kernels are based on:

a) nonlinear logarithmic kernels.
b) polynomial expansion to the powers of 2 and 0.5.

Feature expansion techniques on the input, seen in Tables 1 and 2, allow us to produce three additional sampling categories for each of the 73 iterations, generating a matrix of [4952×4×73] resembling the readings input stream. This aids in increasing the accuracy of the training process.

Targets are fed as Comma-Separated Value (CSV) data streams, initially run through cell2mat() function to generate a numerical array that can be processed in MATLAB, it is then run through the de2bi() and transpose() function to generate transposed functional binary targets [4×73] matrix the Neural Network's target input stream.

4 RESULTS

CFD simulations produced an averaged pressure reading over the cross-sectional area and across the pipe's length. Figure 2 shows four samples of a segment of the pressure profile highlighting the most significant disturbance in the pressure surrounding the leak location. Table 3 on the other hand, compares the values of four sets of data, belonging to four different leak scenarios (locations) in the first quarter of the pipe, with the inlet gauge pressure at 70 KPa. These two categories represent how the Neural Network will create approximation functions for the leak approximation process.

Due to current COVID-19 restrictions, and lack of access to physical prototyping tools, training datasets were obtained solely from the pressure reading outputs of the CFD. As such, this emphasized the importance of increasing dataset size and accuracy.

Results in Figure 3 provide a proof of concept for the utilization of an RNN for leakage detection in a pipe based on simulated pressure training data. This implies that obtaining similar results when analyzing leaks and their locations in a network of pipes setup, mentioned in Section III, is highly anticipated. Furthermore, the processes of data generation and model training can be automated to

Table 3. Pressure readings comparison for different scenarios.

Pressure KPa	66.9914	67.3025	67.4743	67.8149
Leak Location	2.9m	4.9m	6.9m	9.4m
Reading Location	1 m	1 m	1 m	1 m

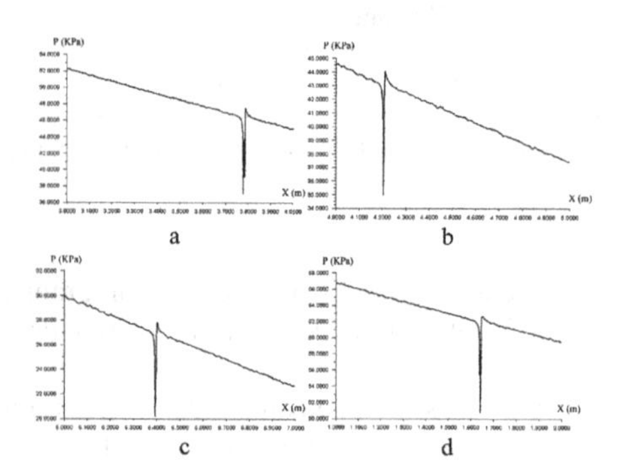

Figure 2. Pressure disturbances surrounding leak locations. x-axis: location on pipe's length X (meters), y-axis: pressure P (KPa). leak at 3.8 m, b. leak at 4.2 m.

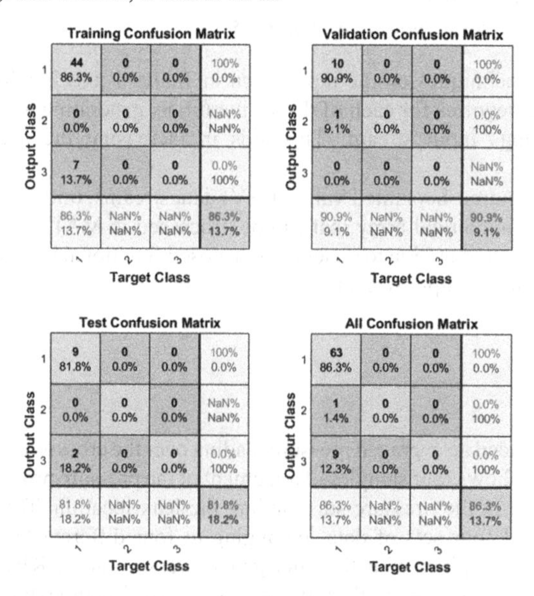

Figure 3. Confusion matrix.

continuously enhance prediction accuracy. This would prove most beneficial in the case of complex networks that require a series amount of time for dataset generation.

Results also show that the system can accept a constant feed of pressure readings from failed training iterations that are corrected by a designated operator. This can serve to enhance the accuracy and reliability of the system, especially when it is utilized on different pipe configurations or a more complex network of pipes.

5 CONCLUSIONS

A network of pipes is inherently a structure of multiple connections of pipes similar in nature to the one modeled in this paper. While an RNN is sufficient for representing a single pipe, it does not account for the diverging paths that are characteristic of a network. As such, it is necessary that the RNN implementation we utilized be adjusted to account for those diverging paths in a network. One possible way for such an implementation would be to represent each pipe segment enclosed between two junctions by its own separate RNN. The total network composed of multiple RNNs could then be trained over the data obtained from simulating the entire network.

REFERENCES

[1] C. a. S. K. a. C. X. Guo, "Experimental study on leakage monitoring of pressurized water pipeline based on fiber optic hydrophone," Water Supply, vol. 19, no. 116, 2019.

[2] M. H. H. a. A. M. a. A. A. M. a. A. A. a. L. W. K. a. O. R. a. O. C. Ishak, "Effects of Aspect Ratio in Moulded Packaging Considering Fluid/Structure Interaction: A CFD Modelling Approach," Journal of Applied Fluid Mechanics, pp. 1799–1811, 2017.

[3] R. a. P. V. a. P. J. a. P. A. a. L. E. a. J. L. Pérez, "Pressure sensor distribution for leak detection in Barcelona water distribution network," Water Science & Technology: Water Supply, vol. 9, 2009.

[4] C. S. C. X. Z. T. K. Chan, "Review of Current Technologies and Proposed Intelligent Methodologies for Water Distributed Network Leakage Detection" Access IEEE, vol. 6, pp. 78846–78867, 2018.

[5] A. a. M. N. a. C. D. a. A. C. Sadeghioon, "Water pipeline failure detection using distributed relative pressure and temperature measurements and anomaly detection algorithms" Urban Water Journal, vol. 10, no. 1080, pp. 1–9, 18 4 2018.

[6] X. L. T. Z. S. C. X. L. Yu Shao, "Time-Series-Based Leakage Detection Using Multiple Pressure Sensors in Water Distribution Systems," Sensors (Basel), vol. 19, no. 14, 2019.

[7] J. Schmidhuber, "Deep learning in neural networks: An overview", Neural Networks, vol. 61, pp. 85–117, 2015.

[8] K. B. Adedeji, Y. Hamam, B. T. Abe and A. M. Abu-Mahfouz, "Towards achieving a reliable leakage detection and localization algorithm for application in water piping networks: An overview", IEEE Access, vol. 5, pp. 20272–20285, Sep. 2017.

[9] A. Soldevila, J. Blesa, S. Tornil-Sin, E. Duviella, R. M. Fernandez-Canti and V. Puig, "Leak localization in water distribution networks using a mixed model-based/data-driven approach", Control Engineering Practice, vol. 55, pp. 162–173, 2016

[10] D. Chatzigeorgiou, K. Youcef-Toumi and R. Ben-Mansour, "MIT Leak Detector: Modeling and Analysis Toward Leak-Observability," in IEEE/ASME Transactions on Mechatronics, vol. 20, no. 5, pp. 2391–2402, Oct. 2015, doi: 10.1109/TMECH.2014.2380784.

[11] R. Pérez, V. Puig, J. Pascual, J. Quevedo, E. Landeros and A. Peralta, "Methodology for leakage isolation using pressure sensitivity analysis in water distribution networks", Control Engineering Practice, vol. 19, no. 10, pp. 1157–1167, 2011.

[12] U. Baroudi, A. A. Al-Roubaiey and A. Devendiran, "Pipeline Leak Detection Systems and Data Fusion: A Survey," in IEEE Access, vol. 7, pp. 97426–97439, 2019, doi: 10.1109/ACCESS.2019.2928487.

[13] J. Mashford, D. de Silva, D. Marney and S. Burn, "An approach to leak detection in pipe networks using analysis of monitored pressure values by support vector machine", Third International Conference on Network and System Security, pp. 534–539, 2009.

[14] Clark, R. J., & Bade Shrestha, S. (2015). A review of numerical simulation and modeling of combustion in scramjets. Proceedings of the Institution of Mechanical Engineers, Part G: Journal of Aerospace Engineering, 229(5), 958–980.

[15] H. a. J. P. V. Patil, "Mesh convergence study and estimation of discretization error of hub in clutch disc with integration of ANSYS," IOP Conference Series: Materials Science and Engineering, vol. 402, no. 10, pp. 12–65, 2018.

[16] M. Javadiha, J. Blesa, A. Soldevila & V. Puig, "Leak Localization in Water Distribution Networks using Deep Learning," 2019 6th International Conference on Control, Decision and Information Technologies (CoDIT), Paris, France, 2019, pp. 1426–1431, doi: 10.1109/CoDIT.2019.8820627.

[17] J. Blesa & R. Pérez, "Modelling uncertainty for leak localization in water networks", IFAC-PapersOnLine, vol. 51, no. 24, pp. 730–735, 2018.

[18] M. V. Casillas, V. Puig, L. E. Garza-Castañón & A. Rosich, "Optimal Sensor Placement for Leak Location in Water Distribution Networks Using Genetic Algorithms", Sensors, vol. 13, no. 11, pp. 14984–15005, 2013.
[19] L. Duan, M. Xie, J. Wang & T. Bai, "Deep learning enabled intelligent fault diagnosis: Overview and applications", Journal of Intelligent and Fuzzy Systems, vol. 35, no. 5, pp. 5771–5784, 2018.
[20] I. Santos-Ruiz, F. López-Estrada, V. Puig & J. Blesa, "Estimation of Node Pressures in Water Distribution Networks by Gaussian Process Regression," 2019 4th Conference on Control and Fault Tolerant Systems (SysTol), Casablanca, Morocco, 2019, pp. 50–55, doi: 10.1109/SYSTOL.2019.8864793.
[21] Ahmed, Nesreen & Atiya, Amir & Gayar, Neamat & El-Shishiny, Hisham. (2010). An Empirical Comparison of Machine Learning Models for Time Series Forecasting. Econometric Reviews. 29. 594–621. 10.1080/07474938.2010.481556.
[22] Nargesian, Fatemeh & Samulowitz, Horst & Khurana, Udayan & Khalil, Elias & Turaga, Deepak. (2017). Learning Feature Engineering for Classification. 2529–2535. 10.24963/ijcai.2017/352.
[23] Xie ZX., Hu QH., & Yu DR. (2006) Improved Feature Selection Algorithm Based on SVM and Correlation. In: Wang J., Yi Z., Zurada J.M., Lu BL., & Yin H. (eds) Advances in Neural Networks – ISNN 2006. ISNN 2006. Lecture Notes in Computer Science, vol 3971. Springer, Berlin, Heidelberg
[24] Dor, O., & Reich, Y. 2012. Strengthening Learning Algorithms by Feature Discovery. Information Sciences 189:176–190.
[25] Hofmann, Thomas & Sch, Bernhard & Smola, Alexander. (2006). A Review of Kernel Methods in Machine Learning.
[26] Sherstinsky, Alex. "Fundamentals of Recurrent Neural Network (RNN) and Long Short-Term Memory (LSTM) Network." Physica D: Nonlinear Phenomena 404 (2020): 132306. Crossref. Web.

Proceedings of the 1st International Congress on Engineering Technologies – Kiwan & Banat (Eds)
 ISBN 978-0-367-77630-5

A nonlinear regression-based machine learning model for predicting concrete bridge deck condition

Aqeed Mohsin Chyad
Department of Civil and Construction Engineering, Training and Energy Researches Office, Ministry of Electricity, Baghdad, Iraq

Osama Abudayyeh
Department of Civil and Construction Engineering, Western Michigan University, Kalamazoo MI, USA

Maha Reda Alkasisbeh
Department of Civil Engineering, Hashemite University, Al-Zarqa, Jordan

ABSTRACT: Understanding the process of concrete bridge deck deterioration and evaluating its condition are important for maintaining a healthy transportation infrastructure and for allocating the necessary funds for bridge maintenance, rehabilitation, or reconstruction actions. Therefore, it is important to investigate the factors impacting bridge condition to enable the development of predictive techniques. The main objective of this paper is to study the impact of average daily traffic (ADT), age, and deck area on the concrete bridge deck deterioration. Michigan concrete bridge deck condition data for the past 25 years was analyzed to determine the impact of these factors on concrete decks. An optimum machine learning algorithm that is based on nonlinear regression modeling has been developed to predict the deterioration rates of bridge decks under these impacting factors. This study has revealed that ADT, age, and deck area have a significant effect on the deterioration of concrete bridge decks.

1 INTRODUCTION

Knowing the condition of a concrete bridge deck at any time of its service life along with the factors affecting it can help in making significant decisions that serve transportation network officials in this field [1]. This study will contribute to the improvement of the condition assessment process for concrete bridge decks which is an important step towards achieving an accurate and reliable assessment of concrete bridge deck condition. Although some research attempts in the literature have considered the condition assessment of bridges, there is still a lack of investigations that explore the significant factors that can cause the deterioration either for the overall bridge condition or for a specific element in the bridge itself.

Concrete bridge decks are typically the first component of a bridge that needs repair after construction, more so than other parts, since they are exposed to severe conditions during their service life such as deicing salts and heavy traffic. Due to these factors, deterioration of concrete bridge decks and corrosion of the steel reinforcement can occur [2 & 3]. This is especially significant given the huge numbers of bridges the Federal Highway Administration (FHWA) has in its inventory to be monitored and maintained [4].

2 PROBLEM STATEMENT AND SIGNIFICANCE

According to ASCE (2017), there were approximately 614,387 bridges in the United States in 2016. Forty percent of these bridges were 50 years or older. Nine percent of these bridges were

DOI 10.1201/9781003178255-18

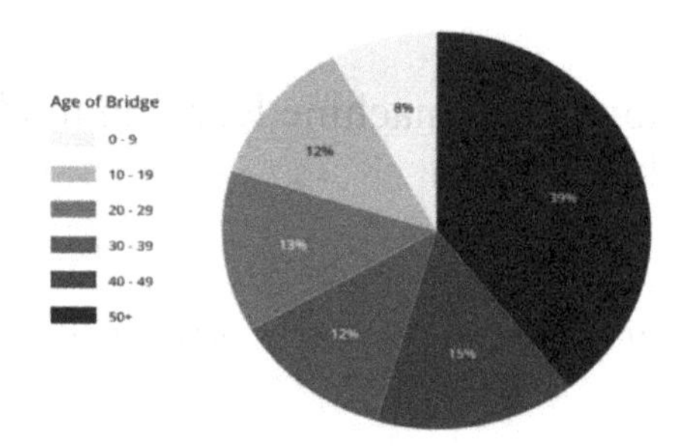

Figure 1. The number of the bridges in the U.S.A according to their ages.

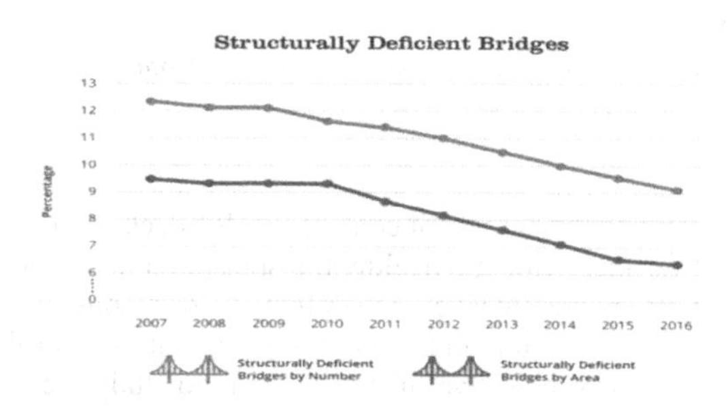

Figure 2. The deterioration rates of bridges according to their numbers and areas.

considered structurally deficient. The estimated allocated expenditures for the nation's backlog of bridge rehabilitation or repair actions were about $123 billion. As the service life of most bridges was designed at 50 years, major rehabilitation or repair will be required as more bridges approach their design life. The Figures 1 and 2 below show the number of bridges and their ages as of 2016, as well as the deterioration rates of bridges within that time.

The condition assessment of concrete bridge decks is complicated, as it is influenced by several factors, including ADT, main structure type, material type, and design loads. Some of these factors have significant effects on deck deterioration while others are less impactful. In recent years, many methods have been developed to predict the condition of concrete bridge decks and to better understand the bridge behavior. The most popular methods used to evaluate the condition of bridges, as well to measure deterioration rates over time are deterministic methods (regression models) and stochastic methods (Markov chain and probability distribution functions) [5].

Each of these methods has unique characteristics and limitations for evaluating the condition of bridge decks. For example, these models assume the presence of state independence that means past conditions have no effect on the predicted ones. These models are unable to consider the result of major maintenance actions on the deterioration process. Additionally, these models assume discrete transition time intervals.

3 METHODOLOGY

The data used to assess the condition of U.S. bridges is taken from the NBI, and represents information collected by the states during regular periodic inspection of their bridges [6]. The main considerations in determining bridge deficiencies are the bridge element condition ratings. These condition ratings present an overall description of the general condition of the bridge element being rated.

The NBI database includes condition ratings for each major element in U.S. bridges. Condition ratings vary from 0 to 9. Typically, a bridge deck is considered structurally and functionally deficient if the bridge deck is rated as 3 or less, while it is considered structurally and functionally efficient if it receives a condition rating of 7 or more [7]. Condition ratings of bridges can be used as an indicator for needed actions such as maintenance, replacement, and rehabilitation. Investigators do not involve condition ratings of 3 or less when developing bridge deterioration models because those condition ratings are a serious condition and instant conduct such as rehabilitation or reconstruction are necessary [8]. At these condition ratings, transportation agencies will basically be looking for the required resources to immediately correct the problems.

Since the condition prediction method is crucial in making bridge maintenance decisions, it is very important to choose an optimum model. Mathematical and statistical models have been developed to assess and predict the condition of bridge elements using the NBI database [9]. The NBI database includes condition ratings for the main bridge components for a period of 25 years. The most frequent probability distributions used for predictive modeling are the Exponential, Weibull, and Lognormal distributions. The Normal and Gamma distributions are also occasionally useful [10]. Some studies have used the Weibull method to develop predictive deterioration curves for bridges without evaluating whether it is the best model to use [8]. Therefore, investigating the best probability distribution model for the condition assessment of concrete bridge decks is a significant part for bridge management system.

Two main issues have been explored in this study: (1) the data needed to track bridge deterioration rates, and (2) the factors that affect bridge deterioration rates. NBI condition ratings are used to rank bridges and can be used to track concrete bridge deck deterioration rates [6]. Data from 1992 to 2016 have been used in this study for Michigan concrete bridge decks. Since inspections are performed biennially, NBI records of condition ratings are available for each of the structural parts of the bridge. These condition ratings are then transformed into consistent NBI condition codes, which are also identified by the FHWA.

To achieve an accurate evaluation of deterioration rates, inspection data must be treated to eliminate the effects of issues other than maintenance that may result in an increase or decrease in the condition ratings. These issues include repair and miscoding [3]. Therefore, the filtration of NBI data is a significant and essential step to remove the impact of inappropriate data.

4 MODEL DEVELOPMENT

While it has been observed from previous study that the lognormal and Markov chain bridge deck condition prediction models are comparable, the authors attempted to develop an optimized approach by developing a new nonlinear regression-based machine learning model. In this approach, the best fit nonlinear regression curve is created for the data generated from the lognormal and the Markov chain models. A combination between the best prediction of each model and for each condition rating is developed. For example, if the lognormal method is the best prediction in the condition rating of 9 and the Markov chain model is the best fit in the condition rating of 8, then the combination can be created between the results of lognormal and Markov chain methods at a condition rating of 9 and 8, respectively. Therefore, evaluating each model for predicting the condition rating of concrete bridge decks is necessary. This evaluation is accomplished by calculating the error rates for the results of each model. Under the new approach, the condition rating of concrete bridge decks can be expressed by the following nonlinear regression model:

$$CR(t) = \beta_1 + \beta_2 t + \beta_3 t^2 + \beta_4 t^3 + \beta_5 t^4 \quad (1)$$

Where:

- $CR(t)$ = The condition rating of the bridge deck at any time of service.
- t = Bridge deck age or number of years since last major reconstruction.
- $\beta 1$ to $\beta 5$ = Coefficients that can be determined based on the inspection data.

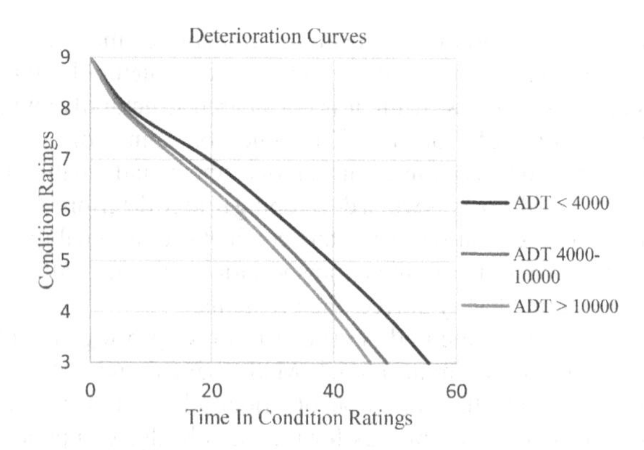

Figure 3. Deterioration curves of concrete bridge decks under impact of ADT.

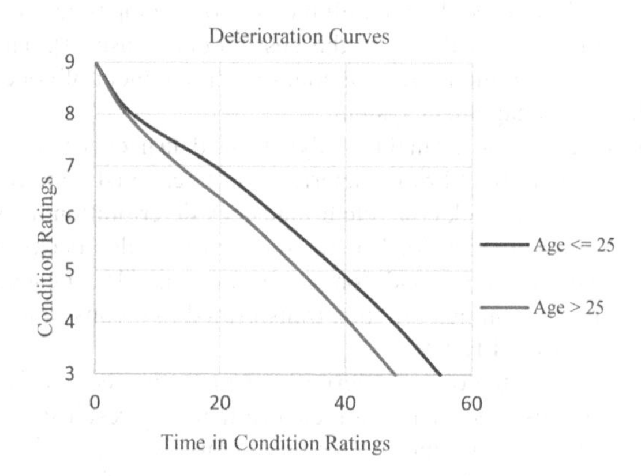

Figure 4. Deterioration curves of concrete bridge decks under impact of deck age.

5 RESULTS AND DISCUSSION

The new nonlinear regression approach demonstrated that the prediction error in the combined model was less than each of the two models (i.e. Markov and Lognormal). The model has been used to assess the condition of concrete bridge decks and predict their expected remaining service life.

There are several factors that may influence the deterioration rates of bridge decks. In this pilot study, the ADT, deck age, and deck area factors were selected to investigate their impacts on concrete bridge deck deterioration and to predict bridge deck performance. Figures 3 to 5 show the deterioration rates of concrete bridge decks under impact of the three factors.

The ADT factor has a major effect on the deterioration rates of concrete bridge decks, as shown in Figure 3. For example, the concrete bridge decks can stay 6.5, 5.5, and 5 years before dropping from condition ratings of 9 to 8 while they can take 48, 42, 40 years before dropping from condition ratings of 9 to 4 when the ADT less than 4000, between 4000 and 10,000 and more than 10,000 vehicles per day respectively.

The effect of age on the deterioration rates of bridge decks is similar to the effect of ADT. The progress of age showed a significant effect on the deterioration of the bridge decks as shown in fig. 4. It showed that concrete bridge decks can take 6 and 5 years to move from condition ratings of 9 to 8 while they can stay 40 and 48 years to drop form condition ratings of 9 to 4 when the deck ages less than or equal and more than 25 years respectively.

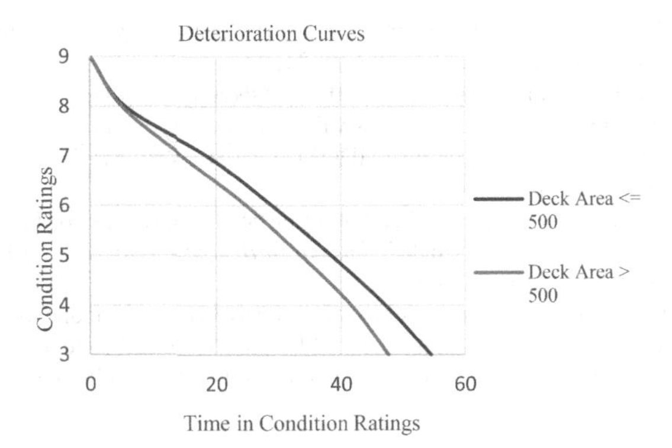

Figure 5. Deterioration curves of concrete bridge decks under impact of deck area.

Similar to ADT and age, the size of the bridge deck area can have a considerable effect on the deterioration rate for all condition ratings of concrete bridge decks (Figure 5). For example, Bridge decks may take 5.5 and 5 years to drop from condition ratings of 9 to 8 and stay 47 and 41 years to transfer from condition ratings of 9 to 4 when the deck area less than or equal and more than 500 m^2 respectively.

6 SUMMARY AND CONCLUSIONS

This study contributes to the improvement of the condition assessment process for concrete bridge decks which is an important step towards achieving an accurate and reliable performance prediction model. Although some research attempts in the literature have considered the condition assessment of bridges, there is still of exploration of the significant factors that can cause the deterioration either for the overall bridge condition or for specific element in the bridge itself.

The Michigan data from 1992 to 2016 were validated using nonlinear regression-based machine learning algorithm to study the impact of factors such as ADT, age, and deck area on the deterioration rate of concrete bridge decks. The results obtained from this study are generally consistent. This paper illustrated that three factors have significant impact on the deterioration rates of concrete bridge decks. Additionally, this study revealed that concrete bridge decks with small ADT and deck area values and in early deck ages deteriorate at a slower rate than those decks with large ADT and deck area values and at older deck ages. Thus, these factors must be considered when determining whether maintenance, repair, and rehabilitation actions are required.Moreover, ranges of ADT, age, and deck area can be established for each factor to provide a more detailed impact analysis for the deterioration rates of concrete bridge decks.

REFERENCES

[1] Tolliver, D., & Lu, P. (2011). "Analysis of bridge deterioration rates: A case study of the northern plains region." *Journal of the Transportation Research Forum*, 50(2), 87–100.

[2] Caner, A., Yanmaz, A. M., Yakut, A., Avsar, O., & Yilmaz, T. (2008). "Service life assessment of existing highway bridges with no planned regular inspections." *Journal of Performance of Constructed Facilities,* 10.1061/(ASCE) 0887-3828, 22, 108–114.

[3] Bolukbasi, M., Mohammadi, J., & Arditi, D. (2004). "Estimating the future condition of highway bridge components using national bridge inventory data." *Practice Periodical on Structural Design and Construction*, 10.1061/(ASCE)1084-0680, 9(1), 16–25.

[4] Goodwin B. T. (2014). *Bridge deck condition assessment using destructive and nondestructive methods* (Master's thesis). Missouri University, Rolla, MO.

[5] Jiang, Y., Saito, M., & Sinha, K. C. (1998). "Bridge performance prediction model using the Markov chain." *Transportation Research Record 1180*. 25–31.
[6] Federal Highway Administration (FWHA). (2016). *Tables of frequently requested NBI information.* Department of Transportation, U.S. https://www.fhwa.dot.gov/bridge/britab.cfm. (October 2015).
[7] Michigan Department of Transportation (MDOT). (2015). *National bridge inventory rating scale.* Michigan Department of Transportation, MI.
[8] Nasrollahi, M., & Washer, G. (2015). "Estimating inspection intervals for bridges based on statistical analysis of national bridge inventory data." *Journal of Bridge Engineering*, 10.1061/(ASCE)BE.1943-5592.0000710, 04014104-1-04014104-11.
[9] Moomen, M. (2016). Deterioration modeling of highway bridge components using deterministic and stochastic methods (Master's thesis). University of Purdue, West Lafayette, Indiana.
[10] Meeker, Q. W., & Escober, A. L. (1998). Statistical methods for reliability data, Wiley, USA.

Proceedings of the 1st International Congress on Engineering Technologies – Kiwan & Banat (Eds)

CFD simulation of ducted mechanical ventilation for an underground car park

Rafat F. Al-Waked*
Department of Mechanical and Maintenance Engineering, German Jordanian University, Amman, Jordan

ABSTRACT: A mechanical ventilation system of an underground car park consisting of four basement levels with a total parking capacity of 835 cars is examined. The purpose of the current study is to undertake a performance-based assessment of the proposed alternative solution and to validate the performance of the proposed ventilation system of the car park by the use of computation fluid dynamics (CFD) at design conditions. Generally, the proposed car park ventilation system design is considered not acceptable and needs to be modified. This is because the average Carbon Monoxide (CO) concentration levels at 1.5 m above floor level throughout most of the car park areas is around 60 ppm or more. Moreover, significant areas of Basement 1 and Basement 4 are higher than the maximum allowed peak CO concentration of 100 ppm. The results have indicated that the current amount of the supplied air system is not sufficient and needs to be increased in order to ventilate the car park and meet the performance requirements.

Keywords: CFD, Ventilation, Car park, Performance based, CO concentration.

1 INTRODUCTION

Motor vehicles account for 70% of oxides of nitrogen (NO_X) emissions, 52% of volatile organic compounds (VOC) emissions and 23% of fine particulate (PM) emissions. Exposure to high levels of these pollutants could cause health effects such as respiratory disease and heart disease [1, 2]. Carpark ventilation systems are used to dilute vehicle emissions to an acceptable level. In areas where cars use gasoline as a fuel, the dilution of Carbon Monoxide (CO) is seen as the defining parameter for carpark air quality because it is emitted at levels generally over twice that of the other contaminants [3].

Commercial computational fluid dynamics (CFD) models have been used as an effective tool to predict the indoor air quality inside [4]. The validation of such CFD codes was achieved by comparing simulation results with an already published data from a given carpark [5]. Khalil et. al. [6, 7] investigated the performance of the ventilation system of an underground carpark using a steady state CFD simulation. Under Egyptian regulations, the adopted limits for CO concentration shouldn't exceed 50 ppm in the normal operating conditions and 100 ppm in the extreme peak conditions. They have reported that ASHRAE Standard 62-2010 [8] recommends a ventilation rate of 7.6 L/s.m^2 or 6 air changes per hour. Furthermore, they conclude that most carparks in the U.S. use lower flowrates while maintaining a healthy indoor environment.

The current study involves a simulation of the underground car park ventilation system proposed for a commercial facility by using CFD-ACE+ software. The performance based CFD simulation is intended to demonstrate the effectiveness of the proposed car park ventilation strategy in diluting CO emissions and to make recommendations for providing ventilation solutions in areas where the design does not perform adequately.

*http://www.gju.edu.jo/content/dr-rafat-al-waked-6240

DOI 10.1201/9781003178255-19

Table 1. Mechanically ventilated air quantities.

Car Park Level	Supplied air (m^3/s)	Exhausted air (m^3/s)
Basement 1	17.80	0.00
Basement 2	19.92	26.62
Basement 3	27.31	39.63
Basement 4	14.88	19.88

2 CAR PARK DETAILS

The underground car park provides parking for 835 cars distributed across four basement levels as shown in Figure 1. The car park is co-located beneath two buildings. Due to the size of the car park, the car park design requires a performance based assessment to ensure that the ventilation system meets the requirements of the Building Code of Australia and Australian Standard [9, 10]. Furthermore, the internationally recognized Permanent International Association of Road Congress Technical Committee on Road Tunnels (PIARC) motor vehicle emission data has been used to determine the ventilation requirements [11]. Consequently, it is intended to provide an amount of exhaust ventilation based on the values listed in Table 1.

The overall maximum dimensions for each Basement floor are approximately 166 m long, 44 m wide and 2.8 m floor-to-floor height. Figure 1 shows that Basement 1 is opened to ambient air from the Eastern side through two open areas of 4 m^2 and 24 m^2 covered by louvres. The main entrance and exit into/out of the entire car park are located at Basement 1. The supply air grills are located at the western wall of the basement. Basement 1 is connected to Basement 2 via two ramps (ramp 5 and ramp 7). The total car park capacity of Basement 1 is 132 cars. Basement 2 differs from Basement 1 in that it does not have an opening to the ambient air. Therefore, a mechanical ventilation system has been designed to reduce the concentration of the CO inside the space. In addition to ramp 5 and ramp 7 that connects Basement 2 and Basement 1, ramp 4 connects Basement 2 to Basement 3. The total car park capacity of Basement 2 is 214 cars with an average travel distance of 160 m to exit Basement 2 and an additional 45 m to exit the car park at Basement 1. Basement 3 is connected to Basement 2 via ramp 4 and to Basement 4 via ramp 2. The total number of car spaces on Basement 3 is 246 cars. Each car travels 170 m to exit Basement 3 and an additional 95 m to exit the car park. Basement 4 is connected to Basement 3 via ramp 2. The total number of car spaces at Basement 4 is 243 cars. Each car travels 170 m to exit Basement 4 and an additional 265 m to exit the car park. The exhaust grills for Basement 3 and Basement 4 are located at the eastern sidewalls of the basement. Whereas the supply air grilles are located at the western sidewalls of the basement.

Basement 4 is considered the most critical level in the car park due to its location and the nature of air flowing into it as shown in Figure 1. Further details on this issue will be discussed in the results section.

The amount of CO in car exhaust depends on, among other things, the age of the car, whether a catalytic converter has been fitted, the engine running temperature (i.e. whether the car is started from cold conditions or not) and the speed of the vehicle. The emission rates according the data produced by PIARC are listed in Table 2. These CO emission levels are used in the current CFD analysis.

2.1 *CFD governing equations*

A three dimensional model was constructed using the commercial CFD software CFD-ACE+ [12]. A steady-state conditions utilizing the pressure-based solver and the Semi-Implicit Method for Pressure Linked Equations-Consistent (SIMPLEC) algorithm was adopted for this study. The current CFD model consisted of 2.75 million cells of unstructured, polyhedral elements. The

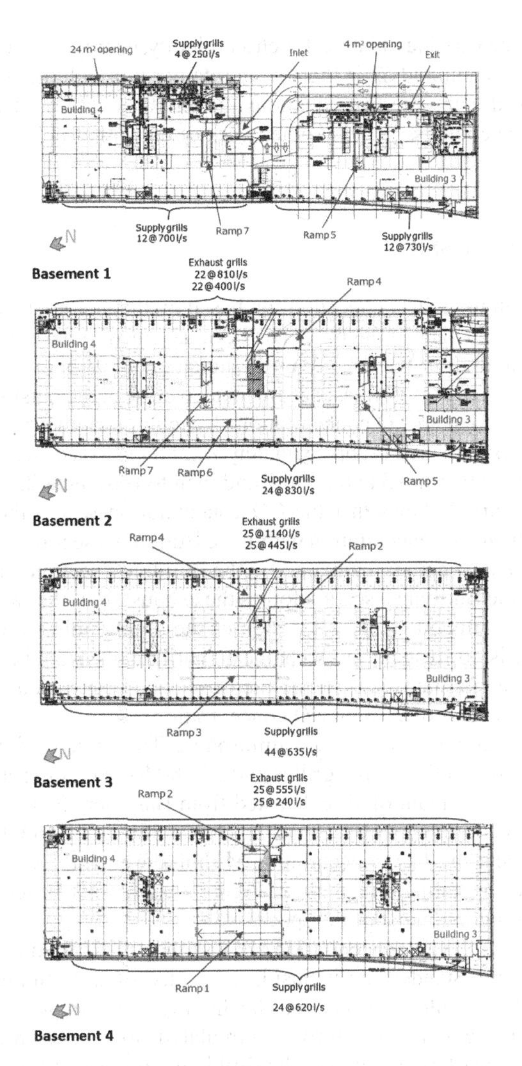

Figure 1. Floor plans of the underground car park.

Table 2. CO emission rate (g/min) from cars, PIARC [11].

Age of car (years)	Starting Period	Idling	Driving at 5km/hr
< 3	1.7	0.33	1.08
3 – 5	2.76	0.56	1.80
5 – 10	8.43	1.68	2.53
10 – 50	24.2	4.83	6.90

momentum, energy, species and pressure solving equations were discretized using the second order upwind discretization scheme.

The adopted general form of the utilized equations for continuity, momentum, energy and species could be written as:

$$\nabla \cdot (\rho \phi \vec{V}) = \nabla \cdot (\Gamma_\rho \nabla \phi) + S_\phi \tag{1}$$

where ϕ represents the variable of interest such as velocity, energy and/or species, ρ is the density, Γ is the diffusion coefficient and S_ϕ is the source rate per unit volume. Furthermore, turbulence modelling was achieved using the two equations model (k-ε) with its default constants. Further details of the CFD code could be found at other work of the author [13].

3 RESULTS AND DISCUSSION

Figure 2 shows the contours of CO concentration based on the one-hour exposure limit of 60 ppm and peak limit of 100 ppm. The ambient air has been assumed to have 0 ppm. Therefore, results presented in all the related figures are based on an increment of 91 ppm. Most of Basement 1 has a 51 ppm CO concentration or more. Areas that are located close to the outside fresh air opening have a significantly less than 51 ppm CO concentration. Although Basement 1 is supplied with 17,160 l/s, the extracted air to Basement 2 is approximately 23,380 l/s. Consequently, all the mechanically supplied fresh air flows into ramp 5 and ramp 7 and then to Basement 2.

A closer look at Figure 2 shows that the CO concentration within the Building 4 section of Basement 1 is higher than the concentration within the Building 3 section. The maximum allowable peak CO concentration of 91 ppm is maintained in the Building 3 section. The circulation of airflow at the northern part of the Building 4 section forces the exhaust from the cars in that area to disperse rapidly. The result of continuous rapid dispersion of the highly polluted car emissions is severe, especially where there is no fresh air or mechanical ventilation system in the vicinity. As a result, there are not only highly polluted areas within the Basement 1 car park areas, but those areas also exceed the allowed peak limit of 91 ppm CO concentration.

Basement 2 has no opening to the ambient atmosphere. The amount of air supplied to Basement 2 is 19,920 l/s of fresh air via the supply grilles and 23,380 l/s via ramp 5 and ramp 7 of a polluted air from Basement 1. The amount of air exhausted from Basement 2 is 26,620 l/s via the exhaust grilles and 16,680 l/s via ramp 4 to Basement 3. The 23,380 l/s of polluted air that comes from Basement 1 itself affects the CO concentration within Basement 2 even before the additional pollution from vehicles operating on this floor. Figure 2 shows that Basement 2 has unacceptable CO concentrations throughout most of the space.

Although areas with high CO concentration are less than those found in Basement 1, it is still higher than the acceptable Standard limits inside car parks. Areas with high circulation still exist at the northern side of the Building 4 section. This highlights the importance of installing mixing fans in the ceiling at these areas to disturb the circulated flow, hence mixing fresh air with the polluted air. The existence of ramp 7 within the Building 4 section of Basement 2 has contributed to the increase in CO concentrations. Air coming from Basement 1 through ramp 7 has an average CO concentration of 48 ppm, which is very close to the limiting value of 51 ppm. Similarly, air coming from ramp 5 to Building 3 section of Basement 2 has an average concentration of 43 ppm. As a result, the flow structure and CO concentrations within the basement are affected.

The amount of air supplied to Basement 3 is 27,940 l/s of fresh air via the supply air grilles and 16,680 l/s via ramp 4 of a polluted air from Basement 2. The amount of air exhausted from Basement 3 is equal to 39,625 l/s via the exhaust grilles and 4,995 l/s via ramp 2 to Basement 4. The 16,680 l/s of polluted air flowing from Basement 2 into Basement 3 is significant and affects the CO concentrations within Basement 3. Figure 2 shows that the area near ramp 4 is dominated by the high CO concentration of 33 ppm flowing from Basement 2. Although 33 ppm of CO is below the acceptable limit of 51 ppm, it affects the distribution of the CO across the whole basement level. Areas with high amounts of re-circulation exist at the northern and the southern sides of the basement, similar to Basement 2, and increases the potential for the accumulation of CO in these areas. The area close to the store at the center of the Building 4 section has high CO concentrations from the adjacent cars. Due to the shadowing effect of the store, the area located between the store and the exhaust has high CO concentrations as shown in Figure 2. This is because of the high airflow from ramp 4 to the exhaust grilles and from ramp 2 to Basement 4. Therefore, emissions

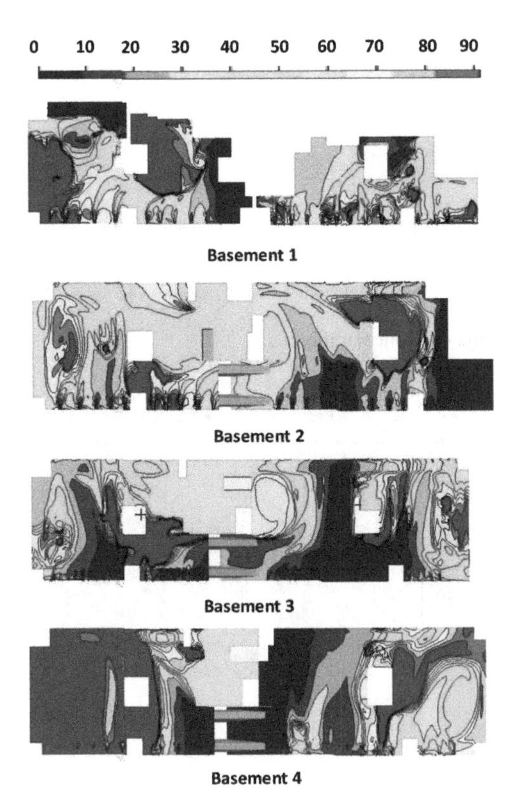

Figure 2. CO concentration (ppm).

from the car tend to flow away from ramp 4 toward the northern side of the Building 4 section and toward ramp 2 in the Building 3 section.

Basement 4 is located at the bottom of the car park with a single ramp connecting it with Basement 3. The supply air grilles are located along the western walls while the exhaust air grills are located along the eastern walls. The amount of polluted air flowing from Basement 3 to Basement 4 through ramp 2 is equal to 4,995 l/s with an average CO concentration of 59 ppm. Fresh air is supplied at a rate of 14,880 l/s and polluted air is exhausted at a rate of 19,875 l/s. Ramp 2 is located within the Building 4 section of Basement 4. The high CO concentration of 59 ppm makes it impossible to clean that area from pollutants as shown by Figure 2. Therefore, the northern part of Basement 4 is a highly polluted zone and the ventilation system within that part needs to be reassessed. The modelling shows that the amount of fresh air supplied to the basement is insufficient and should be increased. The amount of fresh air required to ventilate Basement 4 is almost 1,000 l/ s higher than the minimum recommended design flow rate. This minimum amount is based on the emission of the cars only and does not include the effect of polluted air flowing from Basement 3. As a result, the ventilation system of Basement 4 needs to be redesigned by taking into consideration the effect of the polluted air flowing from Basement 3 via ramp 2.

3.1 *Comparison*

The average CO concentration at 1.5 m above floor level needs to be lower than 51 ppm for the ventilation systems to be acceptable and valid.

Figure 3 shows the volumetric average of CO throughout each basement of the car park. The values presented here give a general idea of how well each basement compares to the others. Although Basement 4 has the least number of cars, the results show that it is has the highest

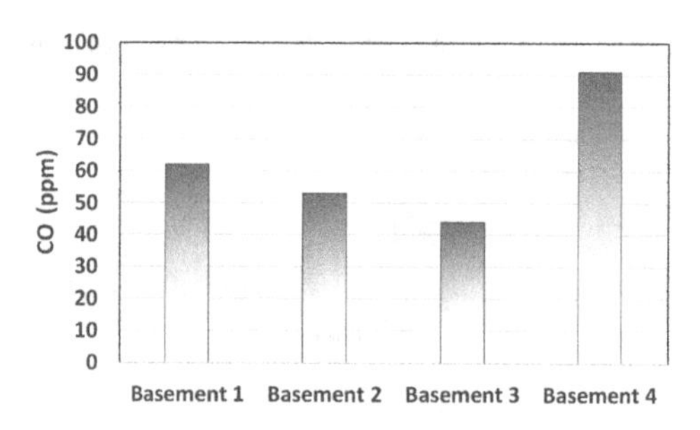

Figure 3. Volumetric average CO concentration.

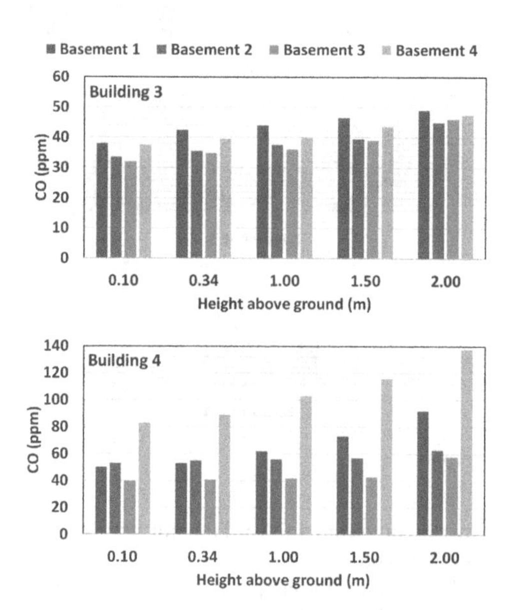

Figure 4. Average CO concentration.

concentration in the whole car park. In contrast, Basement 3 has the highest car emissions and still provides the lowest volumetric average CO concentrations among the rest of the car park sections. However, the average CO concentration is still considered high and reassessment of the car park ventilation design is required to improve the air quality and environment inside the car park.

The area-weighted average at different heights within each basement level is shown in Figure 4. It is shown that the average CO concentration within Building 3 sections of all basements satisfy the ventilation system design requirements.

This is because the polluted air flowing from Basement 2 to Basement 3 then to Basement 4 is transferred from Building 3 sections and is supplied into Building 4 sections of the car park via the connecting ramp. Therefore, Building 3's sections of the car park have a better ventilation arrangement than Building 4's sections. The best preforming areas of Building 4 sections of the car park are within Basement 2 and Basement 3 with CO concentrations less than 63 ppm and 59 ppm, respectively. This is because these two sections have the highest amount of fresh supply air. Basement 2 has an additional 4,500 l/s and Basement 3 has an additional 7,200 l/s more than the minimum fresh air requirements. These results confirm the necessity of providing more fresh air to the car park to cover any effect of polluted air flowing between the basement levels and to maintain the car park under acceptable CO concentrations.

4 CONCLUSIONS

A performance-based assessment of the proposed ventilation system was undertaken using the CFD technique at design conditions. Results have indicated that the current amount of the supplied air system is not sufficient and needs to be increased in order to ventilate the car park and meet the performance requirements. This is because the average CO concentration levels, at 1.5 m above floor level, throughout most of the car park areas was around 60 ppm or more. Moreover, significant areas of Basement 1 and Basement 4 were higher than the maximum allowed peak CO concentration of 100 ppm. On the other hand, the strategy of allowing air to flow towards lower basement levels through ramps has proven to be invalid due to the polluted air involved. Finally, integrating CO concentration sensors with mixing/jet fans control strategy could lower the peak CO concentration with the car park

ACKNOWLEDGMENTS

The support provided by the Deanship of Scientific Research of the German Jordanian University (GJU) Jordan are gratefully acknowledged.

REFERENCES

[1] Hodas N., Loh M., Shin H.M., Li D., Bennett D., McKone T.E., Jolliet O., Weschler C.J., Jantunen M., Lioy P., & Fantke P., Indoor inhalation intake fractions of fine particulate matter: review of influencing factors, Indoor Air, 26 (2016) 836–856.

[2] Yan Y., He Q., Song Q., Guo L., He Q., & Wang X., Exposure to hazardous air pollutants in underground car parks in Guangzhou, China, Air Quality, Atmosphere & Health, 10 (2017) 555–563.

[3] Cui J & Nelson J.D., Underground transport: An overview, Tunnelling and Underground Space Technology, 87 (2019) 122–126.

[4] Eshack A., Leo Samuel D.G., Nagendra S.M., & Prakash M., Monitoring and simulation of mechanically ventilated underground car parks, Journal of Thermal Engineering, 1 (2015) 295–302.

[5] Ayari A., Grot R.A., & Krarti M., Field evaluation of ventilation system performance in enclosed parking garages, ASHRAE Transactions, 106 (2000).

[6] Essam E. Khalil, Yasser M.M. Shoukry, Hesham A. Osman, & Hatem Harridy, Investigation of Ventilation System Performance of Tahreer Car Park Using CFD, Journal of Energy and Power Sources, 2 (2015) 81–89.

[7] Sherif M. Gomaa, Essam E. Khalil, Mahmoud Fouad, & Ahmed Medhat, Ventilation System Design for Underground Car Park, Open Journal of Technology & Engineering Disciplines, 1 (2015) 30–41.

[8] ANSI/ASHRAE, ANSI/ASHRAE Standard 62.1-2013, Ventilation for Acceptable Indoor Air Quality, in, ASHRAE, Atlanta, U.S., 2013.

[9] Standards Australia, The use of mechanically ventilation and air-conditioning in buildings. Part 2: Mechanical ventilation for acceptable indoor air quality, in: AS1668.2 – 2002, Standards Australia, Sydney, Australia, 2002.

[10] Standards Australia, The use of mechanically ventilation and air-conditioning in buildings. Part 2: Mechanical ventilation for acceptable indoor air quality – commentary in: AS1668.2 Suppl1 – 2002, Standards Australia,, Sydney, Australia, 2002.

[11] Permanent International Association of Road Congress (PIARC), Road Tunnels, Emissions, Ventilation, Environment', in, 1995.

[12] CFD Research Corporation CFD ACE User Manual, CFD Research Corporation Cummings Research Park, Huntsville, Alabama, 2002.

[13] Al-Waked R., Groenhout N., Partridege L., & Nasif M., Indoor air environment of a shopping centre carpark: CFD ventilation study, Universal Journal of Mechanical Engineering, 5 (2017) 113–123.

Proceedings of the 1st International Congress on Engineering Technologies – Kiwan & Banat (Eds)
© 2021 Taylor & Francis Group, London, ISBN 978-0-367-77630-5

Heat removal analysis for a fin which has longitudinal elliptical perforations using one dimensional finite element method

Abdullah H. M. AlEssa*
Department of Mechanical Engineering, Al-Balqa Applied University, AlHuson University College, Jordan

ABSTRACT: Heat removal from a straight fin with longitudinal elliptical holes is studied. 1D numerical simulation of the fin thermal dissipation performance is carried out by using the finite element technique. The perforations make full use of boundary-layer effect to improve heat transfer efficiency. The perforated fin is segregated into several elements using automatic mesh generation procedure. The number of finite elements can be changed as necessary according to the solution stability and convergence. The heat dissipation is calculated and matched up to that of the solid fin of similar measurements and physical properties. The assessment passes on satisfactory outcomes.

Keywords: Fin with Holes; Elliptical Holes; Finite Volume Element; Heat removal.

NOMENCLATURE

A	the traverse section surface area of the fin without elliptical holes [m^2]
A_{ep}	the traverse section surface area of the elliptical hole [m^2]
A_e	the traverse section surface area of the finite volume element [m^2]
h	the surface heat removal coefficient of the fin without elliptical holes [W/m^2.°C]
h_{ps}	the surface heat removal coefficient of the fin with elliptical holes [W/m^2.°C]
h_{pc}	the surface heat removal coefficient of the internal lining of the elliptical hole [W/m^2.°C]
h_s	the surface heat removal coefficient of the fin sides [W/m^2.°C]
h_t	the heat removal coefficient of the fin free end tip [W/m^2.°C]
k	the thermal conductivity of the fin substance [W/m.°C]
L	fin or finite element span [m]
l	the unity vector [m]
m	major elliptical perforation axis or longest elliptical perforation diameter [m]
n	minor elliptical perforation axis or smallest elliptical perforation diameter [m]
N	the elliptical holes number.
P	the outside circumferential of the fin [m]
P_{ep}	the elliptical hole perimeter [m]
Q	the heat removal of the fin [W]
Q1	- The heat removal from the fin with elliptical holes which depends on the sum of all the finite elements of the fin faces [W].
Q2	- The heat removal from the fin with elliptical holes which depends upon the heat conduction (Fourier's) Law of heat removal [W]
Q3	- The heat removal from the fin with elliptical holes which depends upon the first algebraic equation of the finite element equations system [W]
S	the distance between the adjacent perforations [m]
t	the fin depth [m]
W	the fin breadth [m]

*https://orcid.org/0000-0002-9592-4905

DOI 10.1201/9781003178255-20

SUBSCRIPTS

b	fin base
e	finite volume element
ep	elliptical hole
max	the greatest value
pf	the fin with elliptical holes
pc	internal surface of the elliptical hole lining
ps	the remaining surface of the fin with elliptical holes
s	the side faces of the fin with elliptical holes
sf	the fin without elliptical holes
t	the fin end free tip
x	coordinate x or at the track of the x axis
y	coordinate y or at the track of the y axis
z	coordinate z or at the track of the z axis
∞	the surroundings environment

1 INTRODUCTION

The development technology of heat exchangers required them to be in patterns that have a huge heat removal area in a unit of its volume. This large area can be achieved by using extended surfaces [1]. The performance of the extended surfaces (fins) can be improved by cutting them into pieces to obtain [2, 3]. The interruption of the fins increases the surface heat removal coefficient and occasionally boosts the heat removal surface area [4]. In passive thermal systems, the extended surfaces are designed to work by natural convection. Natural convection in enclosed spaces has numerous applications in thermal engineering. The heat transfer by natural convection in vertical channels like perforations have been studied and they are fully established by many researchers [5]. The best profile of the fin (perforated, wavy, slotted, triangular, serrated, rectangular and pin) is still being investigated [6]. The fin profile modification can be done by removing some matter from the fin to make openings, slits or punctures in the fin body [7, 8]. The fin's previous modification by cutting made it light, compact and economical. This means that the modified fin can attain the greatest heat removal with the least material spending [9, 10]. One in style of heat removal intensification procedure implies the use of perforated fins in which there is an increase in convection heat removal coefficient [11, 12]. The heat removal intensification in fins with holes becomes significant when the fin is manufactured from substances having big thermal conductivity [13, 14]. In this study the question of heat removal from a rectangular fin with elliptical holes made through the fin thickness is analyzed. The problem is numerically solved for natural convection heat transfer mode. The solution assumes heat transfer by conduction through the fin body in one dimension. This is satisfactory when the Biot numbers in the tracks of y and z axis is very small (less than 0.01). If the Biot numbers are more than 0.01 then a two- or three-dimensional heat transfer solution must be taken into account.

2 ASSUMPTIONS FOR ANALYSIS

The investigation and consequences in this paper have the following assumptions:

1. Steady-state heat transfer by natural convection.
2. The fin has a uniform material.
3. The two fins (with and without holes) have the same constant thermal conductivity.
4. The two fins have no heat generation inside the fin material.
4. The base temperature of the two fins is invariable.
5. The ambient temperature is constant.

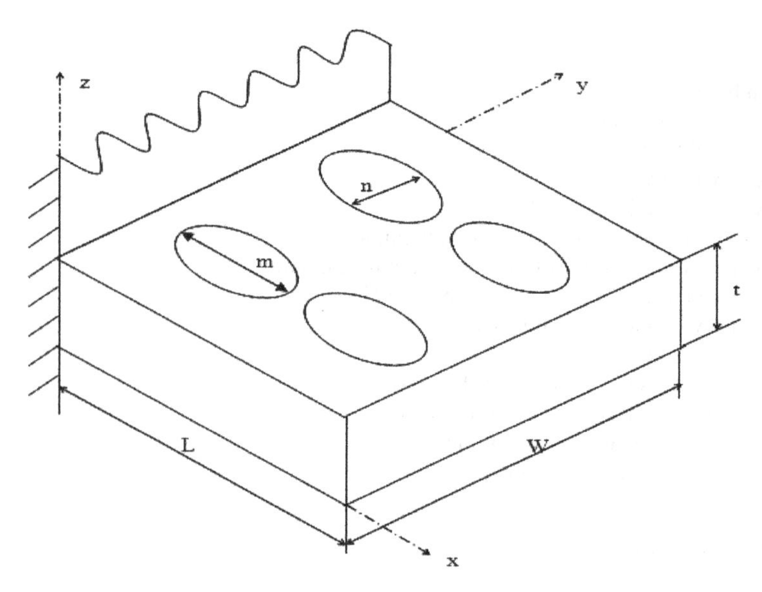

Figure 1. Perforated fin with embedded longitudinal elliptical perforations.

3 HEAT REMOVEAL INVESTIGATION OF THE PERFORATED FIN

The fin with elliptical holes that is investigated in this research is illustrated in the following figure (Figure 1).

In Figure 2, the symmetry portion is shown for the heat transfer problem formulation. This portion is shown shaded. For this shaded portion, the lateral Biot number in the direction of the (z) axis (Bi_z) is determined by the following formula:

$$Bi_z = h.t / 2k \tag{1}$$

Also, for the same hatched portion the lateral Biot number in the direction of the (y) axis (Bi_y) is determined by the following formula:

$$Bi_y = h.(S_y + n / 2) / k(2) \tag{2}$$

In this study, while the Biot number values in the direction of the (z) axis (Bi_z) and that of the Biot number in the direction of the (y) axis (Bi_y) is smaller than 0.01 then the heat removal in the (z) and (y) tracks are considered lumped and 1D heat removal can be accepted. For the Biot numbers (Bi_z) and (Bi_y) larger than 0.01, the solution of the heat removal should be in 2D or 3D. In this article the factors of the fin with elliptical holes are picked as they direct to values of (Bi_z) and (Bi_y) smaller than 0.01. According to the Biot number values along the (z and y) directions which are less than (0.01), the differential heat removal equation of the fin is stated as below [15].

$$k\frac{d^2T}{dx^2} = 0 \tag{3}$$

This differential equation has the following two boundary conditions:

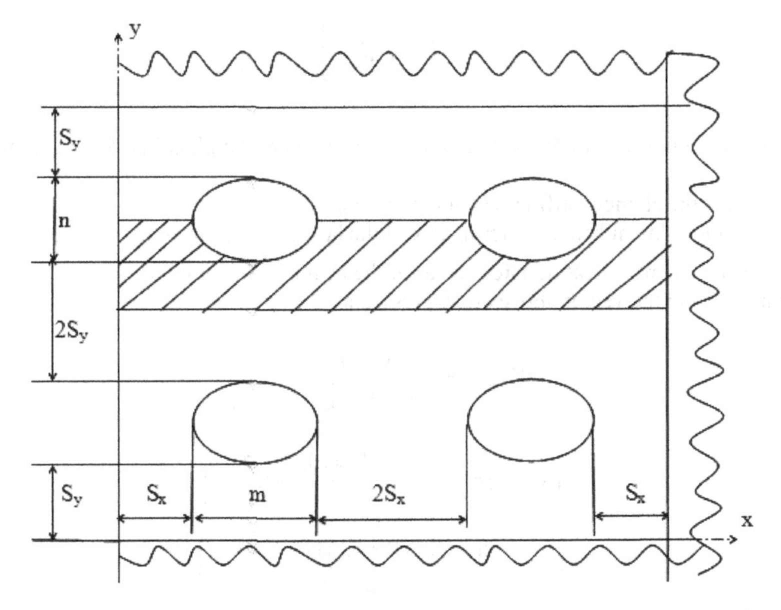

Figure 2. The symmetrical portion (hatched) of the perforated fin assumed for heat removal formulation.

1- The first boundary condition is at the fin fixed surface (x = 0)

$$T = T_b \tag{4}$$

2- The second boundary condition is at the left-over surface of the fin after perforation, the perforation internal surfaces and at the free end of the fin. This condition can be written according to Figures 1 and 2 and as described in [14] in the following form:

$$\begin{aligned} &k.A_e \tfrac{dT}{dx} l_x + h_{ps}.\,A_{ps}(T - T_\infty) + h_{pc}.\,A_{pc}(T - T_\infty) + \\ &h_t.A_t(T_t - T_\infty) = 0 \end{aligned} \tag{5}$$

In this current article, the differential heat removal equation of the fin shown in (3) is resolved by using the computational technique of 1D finite volume element. The matching variational approach equation as it is explained in [14] has the next notations:

$$\begin{aligned} I_n = &\tfrac{1}{2} \iiint_V k(\tfrac{dT}{dx})^2 dV + \tfrac{1}{2} \iint_{A_{ps}} h_{ps}(T - T_\infty)^2 dA_{ps} + \tfrac{1}{2} \iint_{A_{pc}} h_{pc}(T - T_\infty)^2 dA_{pc} \\ &+ \iint_{A_t} h_t(T_t - T_\infty) T\, dA_t \end{aligned} \tag{6}$$

The above variational approach equation is applied over each finite element. Then the equations of the finite elements are written in matrix notation to formulate the algebraic equations of the problem of the fin with elliptical holes as depicted in [14]. The formulas in matrix notations for the finite elements mentioned in eq. 6 are as follows.

$$[K_{cond}] + [K_{conv}]\ \vec{T} = \vec{P}_{conv} \tag{7}$$

$$\vec{T} = \begin{bmatrix} T_I \\ T_{I+1} \end{bmatrix} \quad (8)$$

where:

$\vec{T}$: is the temperature column vector, and T_I and T_{I+1} are the global node temperatures of the element.

$[K_{cond}]$: is the element stiffness conductivity matrix.

$[K_{conv}]$: is the element heat transfer matrix of the heat removal from the surface area of the fin.

$\vec{P}_{conv}$: is the element heat transfer vector of the heat removal from the surface area of the fin.

The matrices and the vector above are expressed as

$$[K_{cond}] = \frac{k * AA(I)}{LE(I)} \begin{bmatrix} 1 & -1 \\ -1 & 1 \end{bmatrix} \quad (9)$$

$$[K_{convs}] = \frac{h_{ps} * PP(I) * LE(I)}{6} \begin{bmatrix} 2 & 1 \\ 1 & 2 \end{bmatrix} \quad (10)$$

$$\vec{P}_{convs} = \frac{h_{ps} * PP(I) * LE(I) * T_{\infty}}{2} \begin{bmatrix} 1 \\ 1 \end{bmatrix} \quad (11)$$

The finite volume elements and the body discretization of the similar portion of the fin with elliptical holes are shown in Figures 3 and 4. As given away in figures 3 and 4, adjacent to each semi- elliptical holes there are four regions marked I, II, III and IV. These four regions replicate each other in the direction of the (x) axis around each hole. All replicated regions along the (x) axis are used in the formation the discretization network and the finite volume element equations.

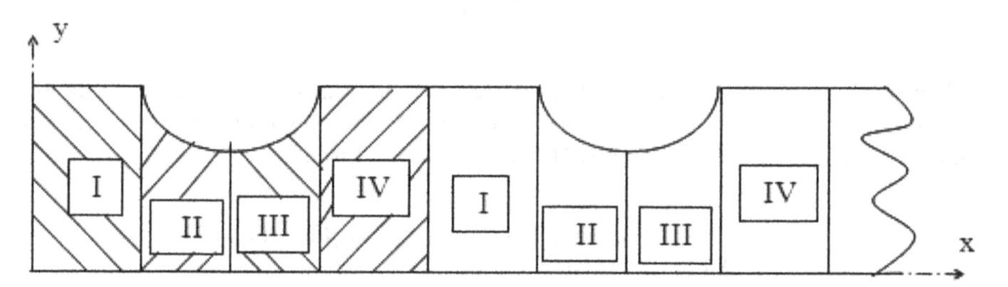

Figure 3. The symmetrical portion with the four regions around each perforation shown as in the computational formation of the fin with elliptical holes.

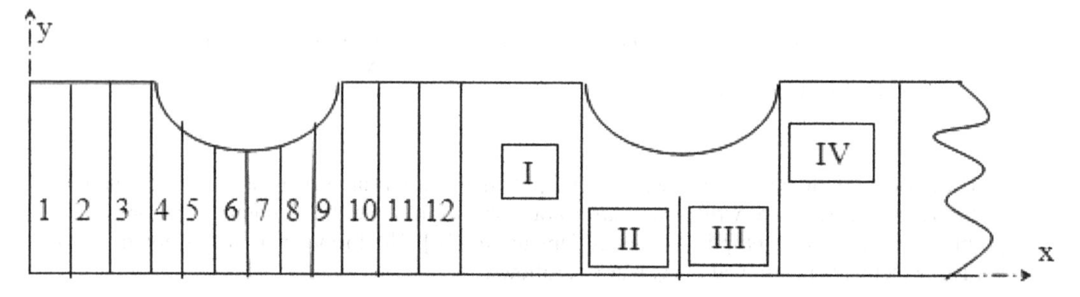

Figure 4. Symmetrical part with the four regions I, II, III, IV assumed in the mathematical derivation for the fin with longitudinal elliptical holes (1, 2, 3, ... serial numbers of the linear finite volume elements).

The number of holes (N_x, N_y) and their spacings (S_x, S_y) are related with the fin span and its breadth. They can be expressed and calculated by the below formulas:

$$L = N_x(2S_x + m) \quad (12)$$

$$W = N_y(2S_y + n) \quad (13)$$

To match up the fin with longitudinal elliptical holes with the fin without holes, their sizes (span, breadth, and height) are chosen to be similar. The heat removal surface areas of the two fins can be expressed according to the next below formulas:

$$A_{sf} = P.\,L + W.t \quad (14)$$

$$A_{pf} = A_f + N_x.\,N_y(P_{ep}\,.\,t - 2\,.\,A_{ep}) \quad (15)$$

$$P_{ep} = \pi\left(3.\,(m + n) - \sqrt{(3m + n)}\sqrt{(m + 3n)}\right) \quad (16)$$

$$A_{ep} = \pi\,.\,m\,.\,n \quad (17)$$

Regions I and IV are split into (N_f) finite volume elements each, while regions II and III are split into (N_t) finite volume elements each. The numbers of the finite volume elements (N_f and N_t) have been randomly selected in agreement with the automatic computer generation requirements of the solution mesh. The whole numbers of the finite volume elements (N_e) and the overall numbers of nodes (N_n) are articulated as:

$$N_e = N_x(2N_f + 2N_t) \quad (18)$$

$$N_n = N_e + 1 \quad (19)$$

Based on the finite volume element approach as described in [14], the finite element equations in the matrices notation form are deduced in condensed form according to the concept of the symmetric banded matrix. This means that the global characteristic matrix or the assembly matrix [GK] of the problem is stored according to the semi-banded width of (NB = 2) for a one Dim. scheme. The algebraic system obtained is analyzed by using the Choleski decomposition routine [14]. The result of this analysis is the temperature variation profile of the fin with holes throughout the x axis. Once the temperature variation profile throughout the x axis is obtained, then the heat removal rate from the fin with (Q_{pf}) can be intended by one of the next below terms:

1- The equation of heat removal that depends on the combination of arithmetic summation over all the fin surfaces of all finite volume elements as shown in the below formula:

$$Q1 = 2N_y.\sum_{I=1}^{N_e}\left(\frac{T_I + T_{I+1}}{2} - T_\infty\right)\left(h_{ps}\left(\frac{P(I) + P(I+1)}{2}\right)Le(I) + h_{pc}.L_{pc}(I).t\right) + Q_t + Q_s \quad (20)$$

Where (Q_t and Q_s) are the heat removal from the free end and the free sides of the punctured fin. They can be found below:

$$Q_t = A_t.\,h_t.\,(T_t - T_\infty) \quad (21)$$

$$Q_s = 2\sum_{I=1}^{N_e}\left(\frac{T_I + T_{I+1}}{2} - T_\infty\right)(h_s.Le(I) * t) \quad (22)$$

2- The formula for heat removal from the fin by using Fourier's conduction rule

$$Q_2 = -k.A.\left(\frac{dT}{dx}\right)_{at\ x=0} \quad (23)$$

The term (dT/dx) at surface (x = 0) can be estimated by using the two temperature values of the two surfaces of the first finite volume element of the fin as shown below:

$$Q_2 = k.t.W \frac{T_1 - T_2}{Le} \tag{24}$$

3- The formula of heat removal from the fin that is deduced from the algebraic equation system of the finite volume elements has the following form:

$$Q_3 = 2N_y \left(CK(1,1).T_1 + CK(1,2)T_2\right) \tag{25}$$

The constants CK(1,1) and CK(1,2) are the first and the second constants of the algebraic global assembly matrix. The theoretical highest heat removal of the fin is found when the temperature of the entire fin equals the fin attached surface temperature, which is expressed as:

$$Q_{pf,\,max} = (A_{ps}.h_{ps} + N_x.N_y.A_{pc}.h_{pc} + A_t.h_t + A_s.h_s)(T_b - T_\infty) \tag{26}$$

The fin effectiveness can be articulated as:

$$\eta_{pf} = Q_{pf} \;/\; Q_{pf,max} \tag{27}$$

To judge against the two fins the following formulas are used [15]:

$$\frac{T_x - T_\infty}{T_b - T_\infty} = \frac{\text{Cosh}\,(m.\,(L - x)) + (h_t\,/\,(m.\,k))\,\text{Sinh}\,(m.\,(L - x))}{\text{Cosh}\,(m.\,L) + (h_t\,/\,(m.\,k))\,\text{Sinh}\,(m.\,L))} \tag{28}$$

$$T_t = T_\infty + \frac{T_b - T_\infty}{\text{Cosh}\,(m.\,L) + (h_t\,/\,(m.\,k))\,\text{Sinh}\,(m.\,L)} \tag{29}$$

4 RESULTS AND DISCUSSION

The improvement of heat transfer produced by using the perforations is evaluated by comparing the perforated fin with its non-perforated equivalent. In the following comparison between the two fins (perforated and equivalent non-perforated), it is taken into account that the two fins have the same values of dimensions, thermal conductivity, heat removal coefficient, base and ambient temperatures.

4.1 *The temperature variation of the fin with holes*

In extended surfaces analysis, the first important information that should be known is the temperature variation throughout the fin body. They are calculated by using the finite volume element described above. To investigate the perforated fin thermal behavior, the temperature variation of the fin body throughout the x axis (T_{pf}) is drawn in Figure 5.

As pointed out in Figure 5, it can be said that the temperature distribution profiles show non-regular curves. The non-regularity of these curves is due to the fin perforations. The holes in the fin body make variations in sectional areas in the x axis direction (throughout the fin distance end to end) which lead to a difference in fin heat removal slowing. Figure 5 shows that the heat removal slowing of the fin with holes declines as the thermal conductivity rises so the curve lines become more regular. To make a side-by-side comparison between the temperature distributions of the fin with holes and the fin without holes, the temperature differences of the previous two fins (T_{sf} - T_{pf}) are plotted in Figure 6.

From Figure 6 it is apparent that the temperatures of the fin without holes are higher than those of the fin with holes in all distribution profiles. This is because the conduction heat removal slowing of the fin with holes is always greater than that of the equivalent fin without holes. Also,

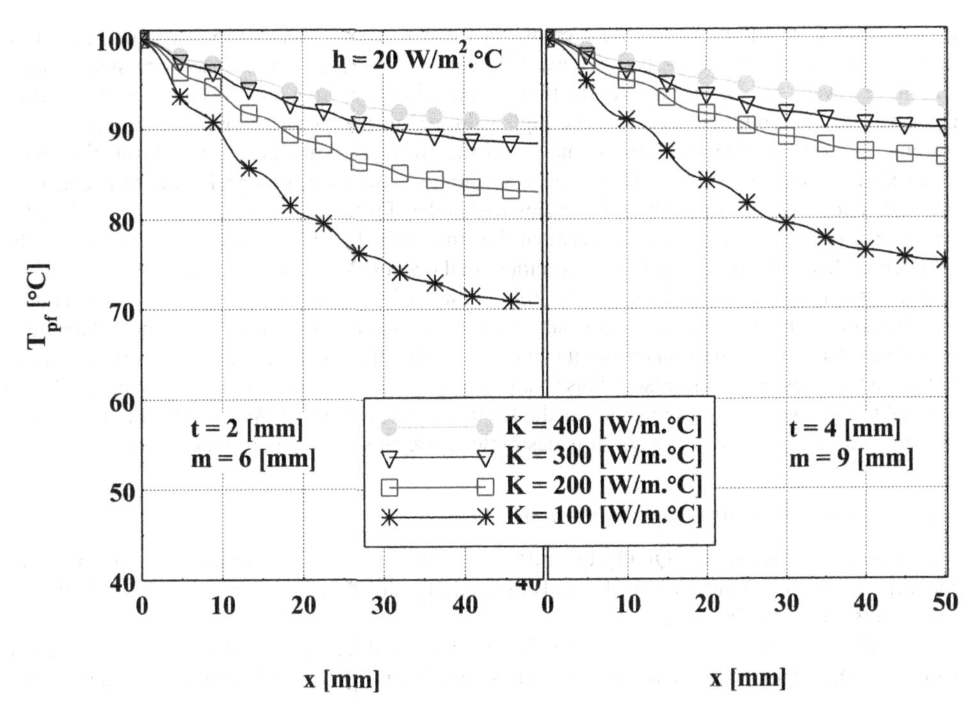

Figure 5. The temperature variation profiles of the perforated fin body related to different values of elliptical perforation diameters, fin thicknesses and fin thermal conductivities.

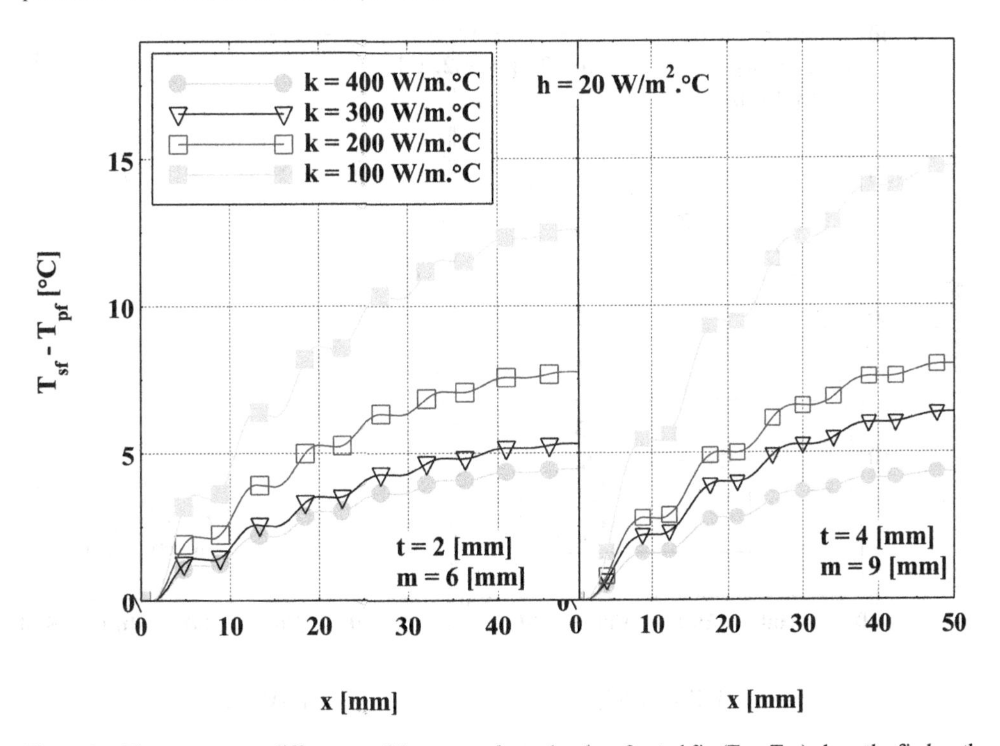

Figure 6. The temperature differences of the non perforated and perforated fin (T_{sf} - T_{pf}) along the fin length for various elliptical perforation diameters, fin thicknesses and the fin thermal conductivities.

it is obvious from Figure 6 that the increase in thermal conductivity of the two fins leads to a decrease in differences (T_{sf} - T_{pf}). These differences (T_{sf} - T_{pf}) diminish to very small values and move toward zero as thermal conductivity approaches very high values. This is because when the fins' thermal conductivity has very high values then the two fins (with or without holes) become the same temperature and come up to the fin base temperature (T_b). From the above discussion and according to the temperature variations standpoint, it is preferable to use as high thermal conductivities as possible for the fins with holes. Furthermore, from figures (5 and 6) it is recognized that the temperature fall between the fin attached surface and its free end rises as the elliptical hole diameter is raised. This is understood because the heat removal slowing of the fin with elliptical holes rises as the elliptical hole diameter is raised. Therefore, from a temperature variation profile point of view, it is advisable to use the smallest possible elliptical hole diameters. From the temperature variation curves it can be found that that the fin temperatures are increasing as the fin thicknesses are increased. This is simply explained as the heat removal slowing of the fin with elliptical holes becomes lessened as the fin thickness is enlarged. As a result of this and from the temperature variation judgment, it is desirable to use fins with higher thicknesses.

4.2 *Removal of heat from the fin with elliptical holes*

The rates of heat driven away (Q_1, Q_2, Q_3 and $Q_{pf,max}$) from the fin with elliptical holes are inspected for various variables of the fin as its thermal conductivity, elliptical hole diameter (m) and thickness. The results are shown in Figure 7.

From Figure 7 it is recognized that the values of (Q_1) are different from those of (Q_2 or Q_3). The values of (Q_1) appear more reasonable than those of (Q_2 or Q_3). This is because the values of (Q_1)

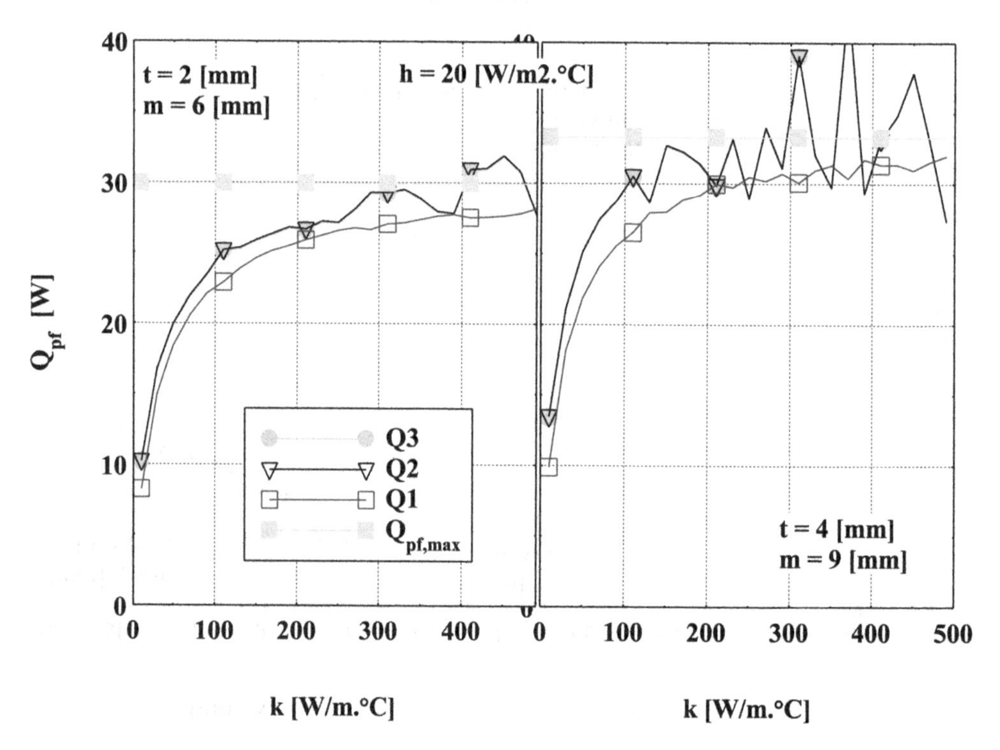

Figure 7. The removal heat from the fin with elliptical holes in terms of its thermal conductivity, hole diameter and its thickness.

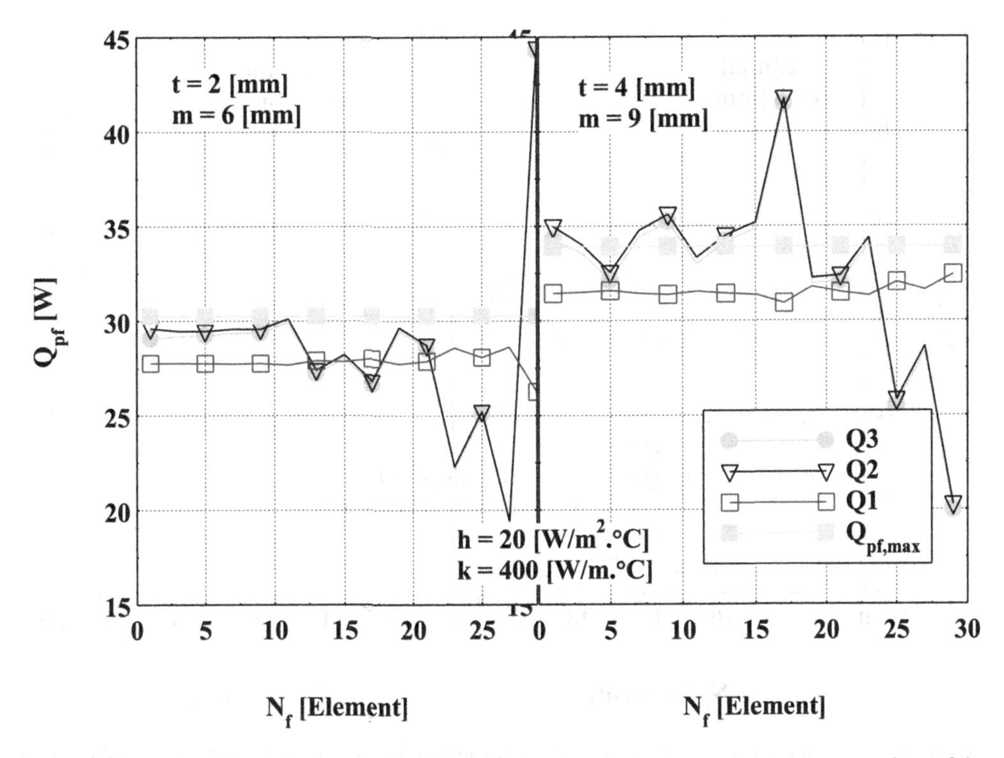

Figure 8. The removal heat rate of the fin with holes in terms of the finite volume element numbers of the uniform part of the fin (region of type I or IV).

do not go above the highest value ($Q_{pf,max}$) while the values of (Q_2 or Q_3) rise and fall around the highest value (Q_{pf},max). The explanation for these incorrect values of (Q2 nor Q3) is that these values are calculated by using the formulas (19, 20). These formulas depend upon the calculated temperatures (T_1, T_2) which belong to the two edges of the first finite volume element.

In Q_2 and Q_3 calculations, it is obvious that the errors in T_1 and in T_2 will significantly influence the values of Q_2 and Q_3. On the other side, the calculated values of Q_1 are dependent upon every one of the calculated temperatures (T_1,.........T_n) for all finite volume elements. This indicates that the very small mistakes in the finite volume element temperatures will overlook each other. To test the constancy and steadiness of the heat removal of Q_1, $Q_2$and Q_3, they are drawn in terms of the finite element volume number (N_f and N_t). This is done in Figures 8 and 9.

Once more, and from figures 8 and 9 it can be noticed that Q_1 seems has good uniformity values while Q_2 and Q_3 rise and fall in bad trends less or more than the (Q_{pf},max). From the preceding outcomes the heat removal from the fin with holes which was computed by using the summation technique over the fin heat removal surface area Q_1 is acceptable and it has been followed in this research calculation.

4.3 *Confirmation of heat removal rate from the fin with holes (Q_1)*

For more confirmation of the fin heat removal computed by using the summation technique Q_1, the difference ($Q_{pf,max}$ - Q_1) is computed in arbitrarily selected deferent values of fin thermal conductivity, fin thickness and elliptical hole diameter. The results are plotted in Figure 10.

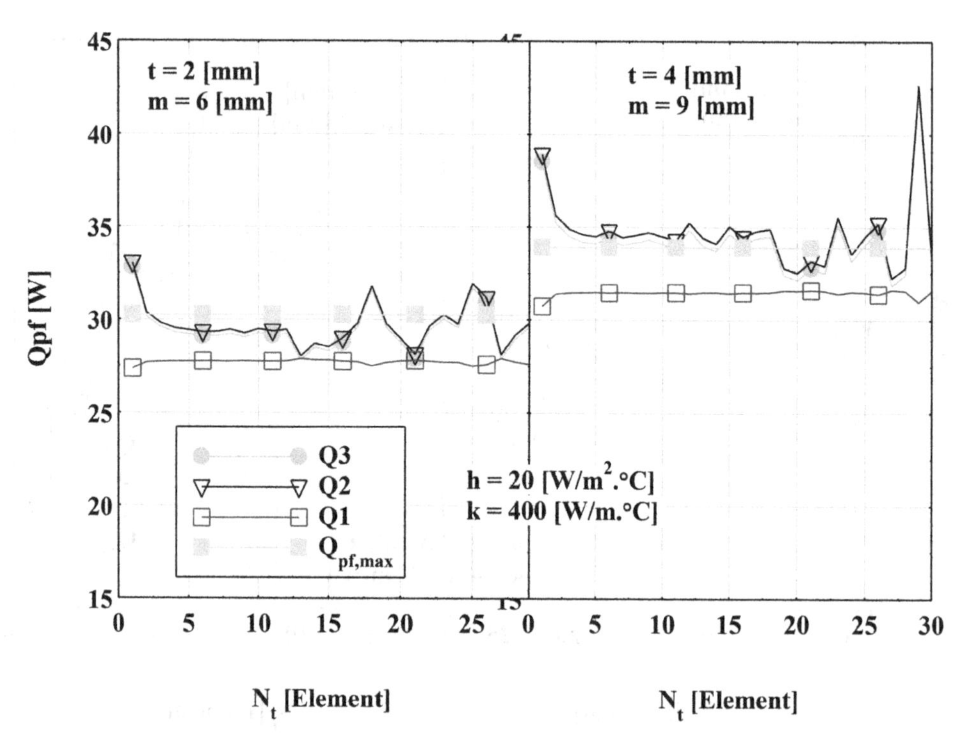

Figure 9. The removal heat rate of the fin with holes variation in the finite volume element serial numbers of the tapered part of the fin with holes (the elements belonging to the regions II or III).

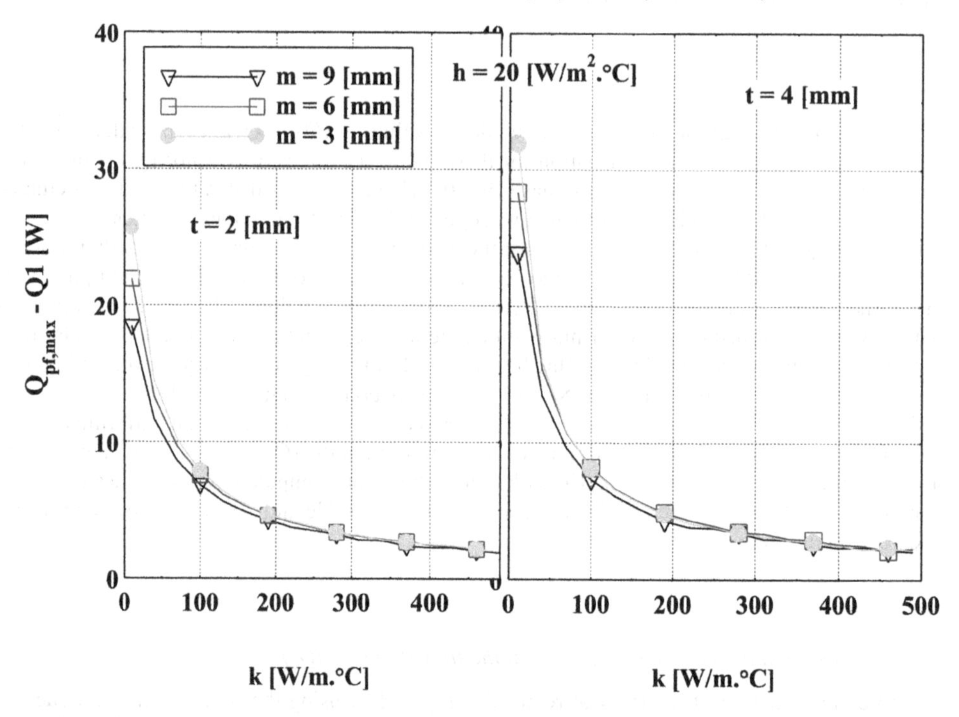

Figure 10. The holed fin heat removal difference between the greatest value ($Q_{pf,max}$) and that computed by using the summation technique Q_1, of the holed fin in terms of the thermal conductivity and the fin thickness.

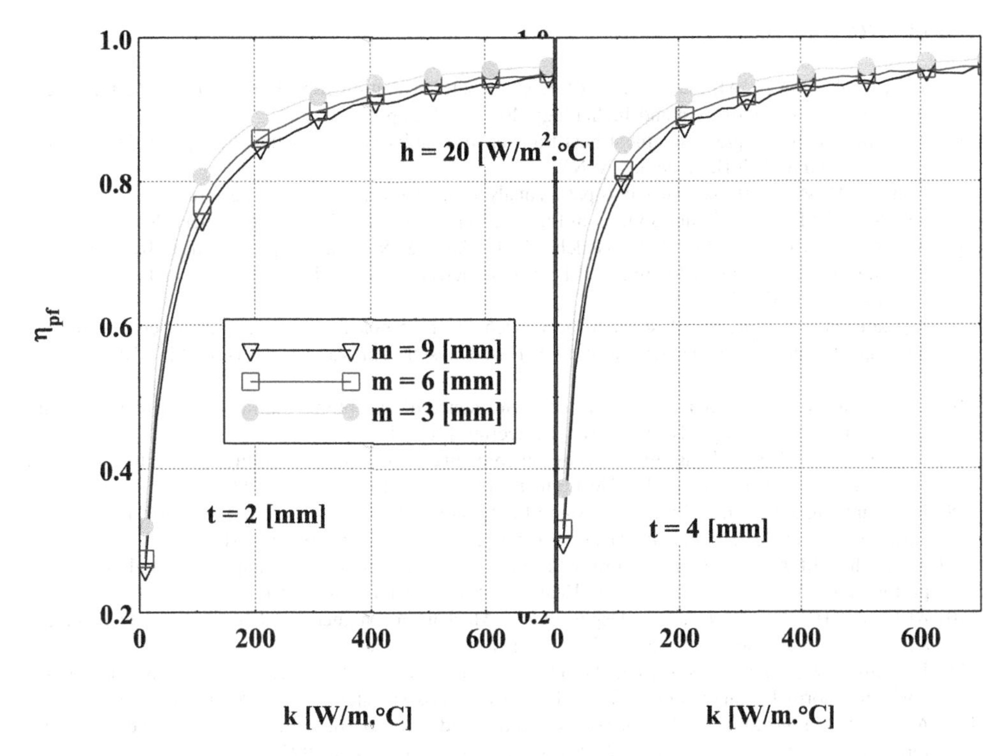

Figure 11. The efficiency of the fin with holes in related to its thermal conductivity for different fin thicknesses and elliptical hole diameters.

In the literature of the extended surfaces heat transfer, it is known that the difference (Q_{pf},max – Q1) always has a positive value and it comes to about zero as the fin thermal conductivity comes to very large values. The outcomes in figure 10 identify with this literature.

4.4 *Confirmation heat removal by using the efficiency (η_{pf}) of the fin with holes*

The efficiency of the fin with holes has been calculated with the value of Q_1 and it is drawn in figure 11. This efficiency has a regular line with very tiny jiggles and they unite as the fin thermal conductivity comes to very large values. This outcome appears satisfactory.

5 CONCLUSION

1- The answer of the single dimension solution of the fin with elliptical holes lead to satisfactory outcomes as the Biot Numbers are less than 0.01.
2- The fin heat transfer slowing of the fin with holes decreases as its thermal conductivity increases. So, it is advisable to use as high thermal conductivities as possible for the fin with holes.
3- The heat removal slowing of the fin with holes raises as the elliptical hole diameter is raised. Therefore, it is suggested to use small hole diameters.
4- The fin heat transfer slowing of the fin with holes declines as the fin thickness is raised. So, it is favorable to use as large fin thickness as probable.
5- The solution in this paper can be used to investigate the enhancement of heat removal from the perforated fin.

REFERENCES

[1] Tariq Azab & A.H. AlEssa, "Effect of Rectangular Perforation aspect ratio on fin performance". International Journal of Heat and Technology. 40(1), 2010, pp. 53–60.

[2] A. E. Bergls, Technique to augment heat transfer. In Handbook of heat transfer Applications, Second Edition, McGraw-Hill Book company, NY.

[3] Abdullah H. M. A. AlEssa. "Heat dissipation analysis of a fin with hexagonal perforations of its one side parallel to the fin base". Yanbu Journal of Engineering and Science (YJES). 7(october), 2013, pp. 21–30.

[4] Abdullah H. AlEssa, K. AlB. Alrawashdeh, M. H. Okour, & N. Talat. "Improvement of free convection heat transfer from a fin by longitudinal hexagonal perforations" JP Journal of heat and mass transfer. 15(2), 2018. pp. 457–481.

[5] R. Mullisen & R. Loehrke, A study of flow mechanisms responsible for heat transfer enhancement in interrupted-plate heat exchangers, Journal of Heat Transfer (Transactions of the ASME) 108, 377–385 (1986).

[6] B.V. Prasad & A. V. Gupta, " Note on the performance of an optimal straight rectangular fin with a semicircular cut at the tip", Heat transfer engineering 14(1), 1998.

[7] C.F. Kutscher, "Heat exchange effectiveness and pressure drop for air flow through perforated plates with and without crosswind". Journal of Heat transfer 116((May), 1994. pp. 391–399.

[8] B.T. Chung & J.R. Iyer,. "Optimum design of longitudinal rectangular fins and cylindrical spines with variable heat transfer coefficient". Heat transfer engineering. 14(1), 1993, pp. 31–42.

[9] Abdullah H. M. AlEssa. "One dimensional finite element solution of the rectangular fin with rectangular perforations". WSEAS Transactions on Heat and Mass Transfer. 1(10), 2006. pp. 762–768.

[10] Al-Essa, A.H., 2000, Ph.D. thesis, Department of Mechanical Engineering, University of Baghdad, Iraq and University of Science and Technology. Jordan.

[11] E.M. Sparrow & M. Carranco Oritz, "Heat transfer coefficient for the upstream face of a perforated plate positioned normal to an oncoming flow", Int. J. Heat Mass Transfer. 25(1), 1982, pp. 127–135.

[12] A. Aziz & V. Lunadini, "Multidimensional steady conduction in convicting, radiating, and convicting-radiating fins and fin assemblies". *Heat Transfer Engineering* 16(3), 1995, pp. 32–64.

[13] Abdullah H. M. AlEssa, & Nabeel S. Gharaibeh. "Effect of triangular perforation orientation on the heat transfer augmentation from a fin subjected to natural convection". Advances in Applied Science Research. 5(3), 2014, pp. 179–188.

[14] Singiresu Rao. "The Finite Element Method in Engineering", 5th Edition, Elsevier, 2010.

[15] Frank P. Incropera & David P. Dewitt, *Fundamentals of Heat and Mass Transfer* (Forth Edition), John Wiley and sons, New York (1996).

Proceedings of the 1st International Congress on Engineering Technologies – Kiwan & Banat (Eds)
 ISBN 978-0-367-77630-5

Enhancing the drying process of coated abrasives

Shuruq Shawish & Rafat Alwaked
Mechanical and Maintenance Engineering Department, German Jordanian University, Jordan

ABSTRACT: An Abrasives facility manufactures abrasive paper with a process that involves blending the paper product with various grades of abrasive ceramic using liquid adhesives. The purpose of this investigation is to investigate and develop a method of increasing the production rate of the adhesive paper through an increase in efficiency of the dryer. The project will incorporate a staged methodology to ensure that any proposed modifications to the plant provide real and tangible benefits. This paper details the findings of the first two stages of the project. While a number of changes can be implemented to improve the process, the effect of the airflow within the duct system is considered to be a core issue affecting performance and a particular focus of the next stage of the investigation.

Keywords: coated abrasives, drying, oven, heat recovery, air velocity.

1 INTRODUCTION

An Abrasives facility manufactures abrasive paper with a process that involves blending the paper product with various grades of abrasive ceramic using liquid adhesives. Once the wet abrasive and adhesive mix is applied to the paper, the continuous roll of paper is fed through a dryer as shown in Figure 1.

The dryer consists of a direct gas-fired burner which heats the air inside the oven enclosure to a maximum of 145°C through which the paper passes. To improve the effectiveness of the dryer, a fan draws air from within the factory through the dryer enclosure and discharges the hot moisture laden air directly outdoors. The production speed of the adhesive paper is dependent on the time it takes for the adhesive to dry as it passes through the dryer.

There are different methods of drying adhesive coatings, one of which involves the water in the coating being removed from the surface of the adhesive coating via evaporation. The water below the surface migrates to the surface at a speed based upon the rate of diffusion of the coating. This process continues until all of the water is raised to the surface and is evaporated. There are different technologies used in the drying process of adhesive paper. Among these technologies, there are

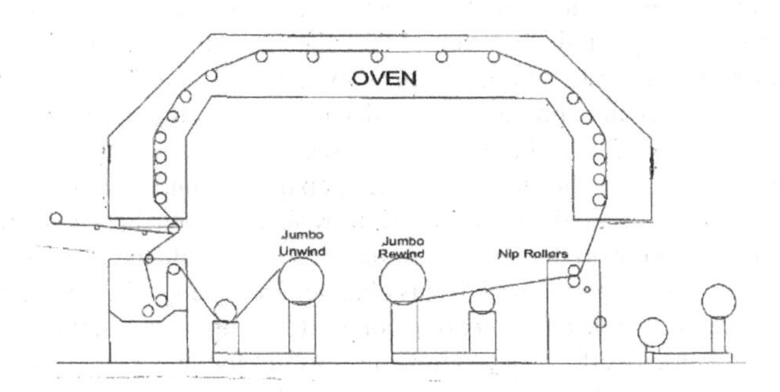

Figure 1. General view of the drying oven.

DOI 10.1201/9781003178255-21

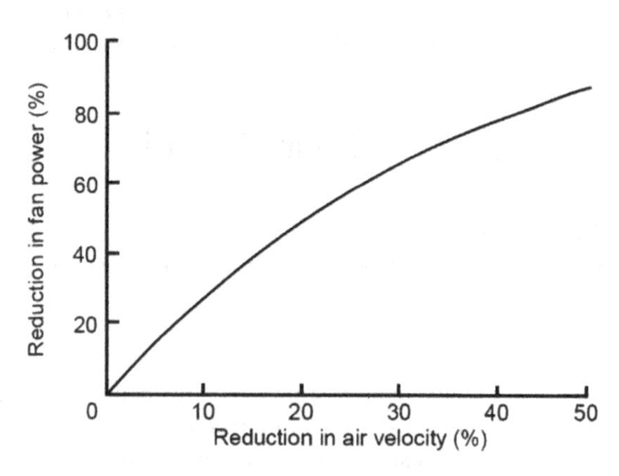

Figure 2. The relationship between reduction in air velocity and reduction in fans electrical power [7].

major parameters affecting the drying process, such as vapour pressure and relative humidity, temperature, air movement, diffusion of Moisture within the adhesive paper, and supply of heat. Energy is a major element in the running costs of a conventional drying oven which vents warm, moist air to the outside and therefore recovers none of the drying energy input. As the relative cost of energy rises, efficient heat utilisation has become a more significant factor and much attention has been given to techniques which might reduce expenditure on energy. These range from the simple upgrading of oven insulation to the development of heat pump ovens.

A new method is developed by adding another heat exchanger of waste gas and a heat exchanger of thermal conducting oil, thus incorporate the heat conducting oil system into the waste gas burning system [1]. The oil would absorb the remaining heat to cure the coated abrasives. The modelling of the moisture movement in the coating process is a very important issue when coating paper. The studying of a novel laboratory method to determine the liquid movement in the paper coating process was presented. The experimentally determined physical properties yield information relevant to the mass and energy balances of a real paper coating process. The water loss from the coating colour into a hygroscopic material can be done by applying vapour diffusion into the base paper as the governing mass transfer mechanism [2]. A modelling and simulation study regarding the heat transfer in hot pressing as well as in the impulse drying of paper has been performed. A simple heat transfer model could be used to simulate the heat transfer in cases in which the temperature was moderate and the applied pressure did not exceed 0.5 MPa [3]. The two major air drying techniques, where the sheet is in direct contact with the drying air, are through air drying and impingement air drying. An advantage intrinsic to through air drying is exceptionally high drying rates because the distance for heat and mass transport processes is reduced from the thickness of the sheet to the dimensions of pores and fibres. Use of through air drying combined with traditional cylinder drying into multiple technique dryer sections is constrained by two basic characteristics: the cost of the pressure drop for air flow through the sheet, and the tendency for through air drying to produce undesirable local moisture nonuniformity during drying [4]. For printing and heavier grades, combining cylinder and impingement air drying into a multiple technique dryer section can enable higher productivity through higher machine speed. The large differences in local moisture content and temperature across the sheet which develop quickly under high intensity impingement drying provide the potential for reducing drying time by sheet reversal between impingement drying cylinders [5]. Experimental work had been conducted to investigate the moisture movement in the paper coating process. The experimental data is obtained by coating a base paper with a laboratory coater and by scraping off some of the coating color after a certain amount of time [6].

Knowledge of air velocity effects on drying rate is useful for determining optimum air velocity. In fact, if the optimum air velocity were known for each step in the oven schedule, air velocity specifications could be added to dry- and wet-bulb temperatures for each step.

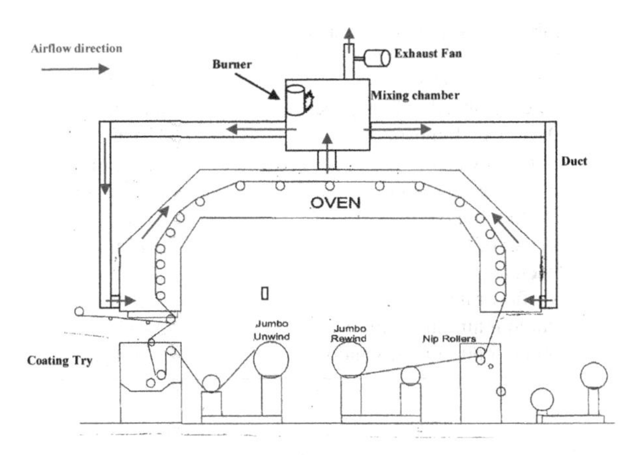

Figure 3. General view of the drying oven.

It is worth mentioning that the basic drying rate varies with air velocity at different temperatures and relative humidities. The drying rate increases with air velocity for moisture contents above approximately 40% to 50%. The rate of increase in the drying rate with air velocity gradually decreases and tends to level off with air velocities above approximately 3.05 to 3.550 m/s and moisture contents below approximately 80% to 90%. The main objective of this work is to investigate the operating conditions in the convection ovens. The drying process in convective ovens (Figure 3) starts by supplying hot and relatively dry air at the bottom ends of the oven. The hot dry air contacts the adhesive paper which dries the paper. As a result, the air gets warm and moist. Warm and moist air is extracted at the top of the oven duct system. A certain amount of the air is extracted out of the system and replaced by relatively cold and dry air. The fresh air and warm moist air are mixed at the mixing chamber while the gas burner supplies the heat required for the drying process.

2 PROBLEM FORMULATION

The drying process starts by supplying hot and relatively dry air at the bottom ends of the oven. The hot dry air contacts the adhesive paper which dries the paper. As a result, the air gets warm and moist. Warm and moist air is extracted at the top of the oven duct system. A certain amount of the air is extracted out of the system and replaced by relatively cold and dry air. The fresh air and warm moist air is mixed at the mixing chamber while the gas burner supplies the heat required for the drying process.

Several areas require further investigation and consideration in order to improve oven performance. These are: makeup air, gas burner performance, the air flow within the duct system, and the recirculated air. The state of the fresh air could be better controlled by providing sensible heating to the fresh air by using an indirect heat exchanger with exhausted hot air; or by active solar air heating.

3 PROBLEM SOLUTION

3.1 *Simulation parameters*

The current simulation is based on a mix of parameters of the abrasives and assumptions made by the authors. The air is assumed to behave as an ideal gas so the ideal gas laws are assumed to be valid during the current simulations. Moreover, these conditions are going to be referred to as the base case conditions, these parameters are summarised as follows (Table 1):

Table 1. Simulation parameters.

Burner Capacity	160 kW
Feeding Speed of Sandpaper	0.27 m/s
Drying air Temperature	135°C
Wet coat weight	120 g/m^2
Water concentration	0.0624 g/m^2
Width of Oven	1.6 m
Height of Oven	0.45 m
Oven length	17 m
Air flow rate	5.82 m/s
Outdoor air temperature	24°C
Outdoor air humidity ratio	12 g/kg$_{da}$

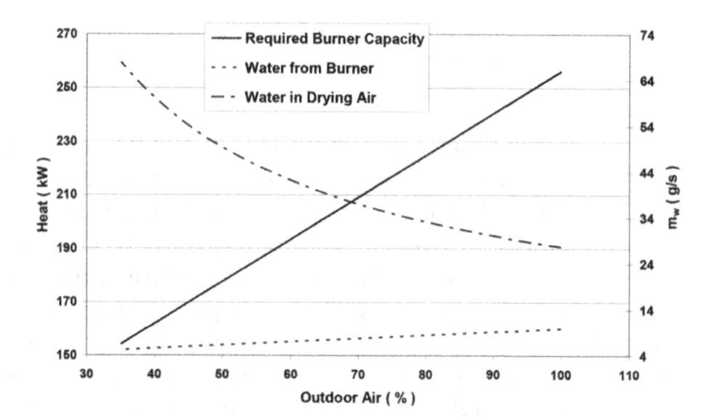

Figure 4. Heat requirements and the corresponding water content in the air.

3.2 *Outdoor air with no conditioning*

Using 100% unconditioned outdoor air is a cheap option to have. However, it is an extremely weather dependant option with no control on the air quality at the inlet of the oven.

In winter, the outdoor air tends to be cold and dry. Therefore, the air has a higher potential to contain evaporated water. However, the cold air temperature slows the total drying process due to reverse convection. Furthermore, colder air requires more heat input from the burner. As a result of the gas combustion, more water is provided to the supply air.

For an outdoor air of 297 K temperature and humidity ratio of 12 g/kg$_{da}$, the amount of heat required to dry the sandpaper fed at a speed of 0.26 m/s is shown by Figure 4. For a fresh air percentage of higher than 40%, the amount of heat required is higher than the current burner capacity. Therefore, a lower percentage is required to maintain the drying process flowing at an efficient speed. As the fresh air percentage decreases, the amount of circulated air increases, however, the amount of water vapour in the air increases, hence, the water transfer from the sandpaper into the air decreases.

The related cost of the current operation is shown in Figure 5 where the significance of recirculation on reducing the operating cost is clear. The operating cost for the base case consists mainly of the gas for the burner and the electricity for the supply and exhausts fans. As the percentage of outdoor air decreases, the required capacity of the exhaust fan decreases, hence the cost. Furthermore, there is less need for heat from the burner due to the high temperature of the recirculated air.

Based on these conditions, it is assumed that the Base case has 35% outdoor air supplied at 145°C, a velocity of 5.86 m/s and a water content of 68 g/s. The resulted monthly cost is estimated to be around $685.

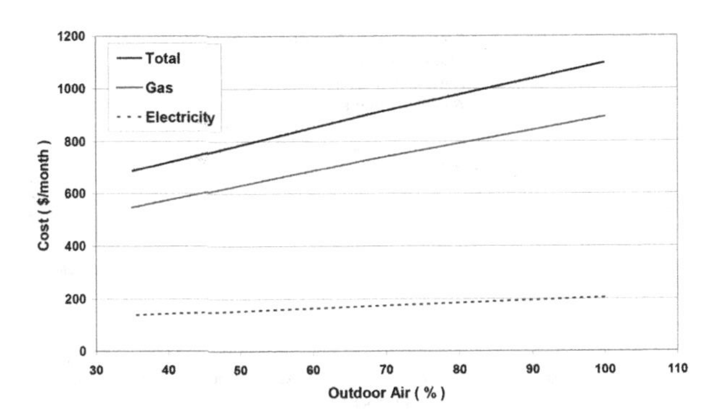

Figure 5. Related operating cost of the outdoor air without air conditioning system.

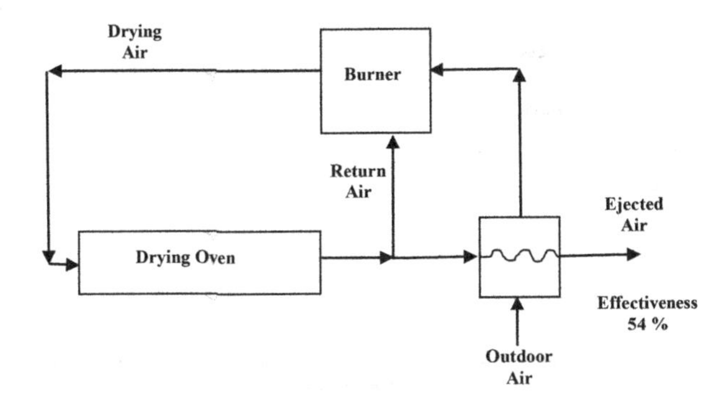

Figure 6. Systematic description of the heat recovery system.

3.3 *Outdoor air with heat recovery (option 1)*

Outcomes from analysing the base case have shown the need to increase the heating capacity of the burner and reduce the amount of water in the drying air. Due to the usage of the gas burner, the increase of burner capacity is translated to an increase in the water content of the drying air. One possibility of increasing the heating capacity of the drying air without increasing the capacity of the burner is the usage of a heat recovery system.

Installing a recovery heat exchanger enhances the control process of the temperature of the air entering the drying oven. However, it comes with the increased energy price used in the initial heating process. A general schematic of the heat recovery system is shown in Figure 6. The main advantage of the heat recovery system is the use of heat available in the ejected air to heat the outdoor air before it enters the burner. Therefore, it reduces the amount of heat required to heat the air to the 145°C temperature.

Figure 7 shows the effect of installing a heat recovery system with an effectiveness of 54% on the required capacity of the burner. Moreover, it shows that the amount of heat recovered from the ejected air is a function of outdoor air percentage. The amount of heat required to dry the sand paper is remained the same. However, the actual heat provided by the burner has dropped from 154 kW for 35% outdoor air for the base case to 126 kW.

More significantly, the amount of outdoor air that can be utilised in the drying process can be doubled after installing the heat recovery system. The importance of using more outdoor air than is used currently is the low water content. As mentioned earlier, the lower the water concentration in the drying air, the better and faster the water transfers from the sandpaper to the air.

The operating cost of the drying process behaves similarly to the base case operating cost and it relates proportionally to the percentage of outdoor air (Figure 8). The running cost of the recovered

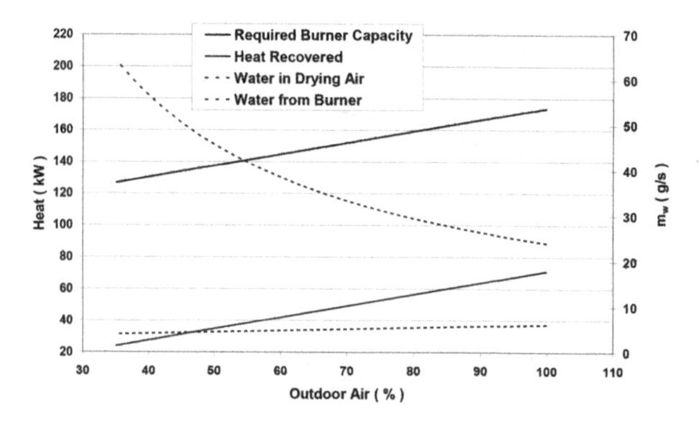

Figure 7. Heat requirements and the corresponding water content in the air (option 1).

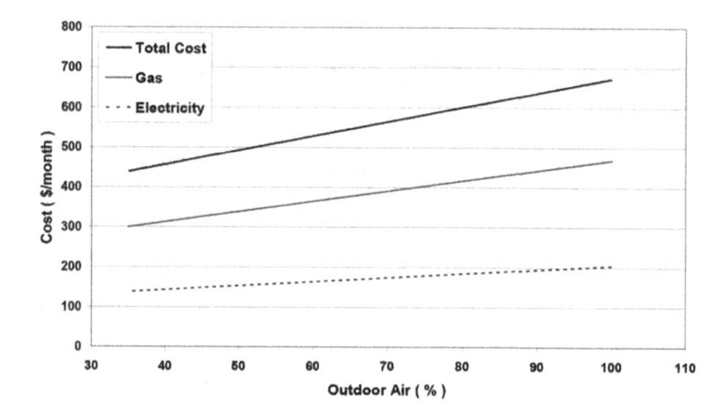

Figure 8. Related operating cost of the outdoor air with heat recovery system.

heat is zero dollars. This is due to the location of the exhaust fan at the outlet of the ejected air duct. Therefore, the exhaust fan drives the air through the heat exchange and out of the drying system at the same time.

In comparison with the base case at the operating conditions of 35% outdoor air supplied at 145°C, a velocity of 5.86 m/s, the water content has dropped from 68 g/s to 64 g/s. This is due to the lower heat generated by the burner. The operating cost has also dropped from $685 to $438.

3.4 *Outdoor air with heat recovery and heat pump (option 2)*

It is demonstrated that the usage of a heat recovery system with an effectiveness of 54% is an effective mean of reducing the thermal requirements of the drying process. However, it failed in decreasing the amount of water in the drying air. This is due to the need for recirculating air in order to achieve the drying requirements within the capacity of the existing Burner. One way to overcome this limitation is the use of heat pump at the outlet of the system.

A general description of the suggested system is shown in Figure 9. The outdoor air is extracted via a duct from outdoors to the dry heat exchanger. This process increases the temperature of the outdoor air without affecting its water content.

The ejected air is extracted to the evaporator of the heat pump after losing heat to the outdoor air which reduced its own temperature. At the heat pump evaporator, the ejected air loses temperature and water. The condensation of water gives heat to the working refrigerant. The refrigerant steam is compressed to a higher pressure and temperature as listed in Table 2.

The use of a heat pump reduces the amount of heat required from the gas burner. Therefore, the amount of water resulting from the combustion products is reduced. Recirculating air throughout

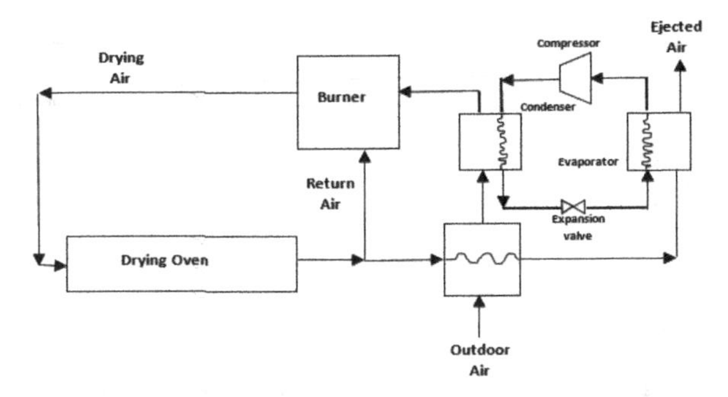

Figure 9. Systematic description of the heat recovery and heat pump at the exit.

Table 2. Heat pump specifications.

Refrigerant	R-134a
COP	1.75
Evaporator Temperature	20°C
Evaporator Pressure	0.57171 MPa
Condenser Temperature	90°C
Condenser Pressure	3.2442 MPa
Heat Exchanger Effectiveness	50%

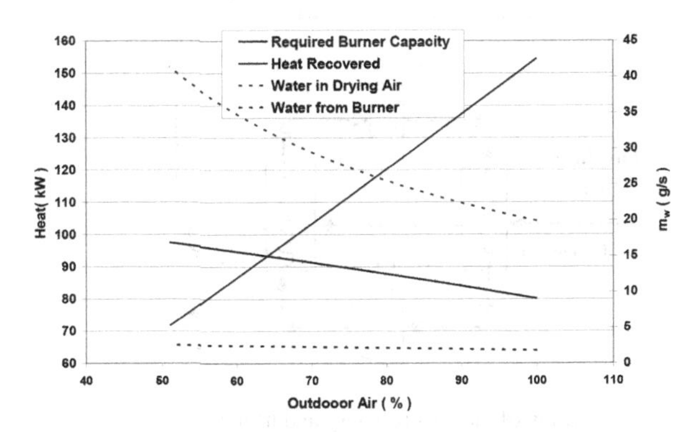

Figure 10. Heat requirements and the corresponding water content in the air with heat recovery and heat pump.

the system increases the amount of water in the drying air. Therefore, the use of any of the options already analysed depends on the amount of energy saved rather than the amount of water in the drying air. The most efficient way of reducing the amount of water in the drying air is via the use of 100% outdoor air (Figure 10).

Figure 11 shows the significant reduction in the amount of heat required from the burner. The main outcome of using the heat pump at the outlet of the system is increase in the required heat as the amount of recirculating increases. The amount of heat required from the Burner is 80 kW at 100% outdoor air and 97 kW at 50% outdoor air. These two cases are much lower than the results for the base case. The operating cost of the current system is inversely proportional to the percentage of the outdoor air. The lowest operating cost corresponds to the usage of 100% outdoor air due to the lowest gas cost. This cost starts to increase as the percentage of outdoor air get less and the need for more heat from the burner increase. Therefore, the best operating cost of the system (option 2) is around $688. Although this figure is similar to the base case and higher than the heat recovery

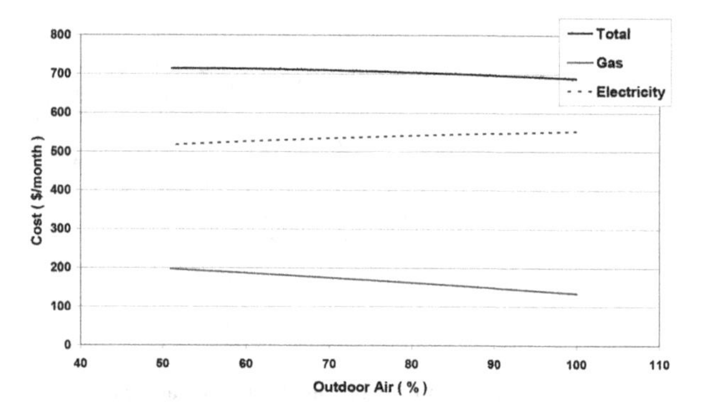

Figure 11. Related operating cost of the outdoor air with heat recovery and heat pump system.

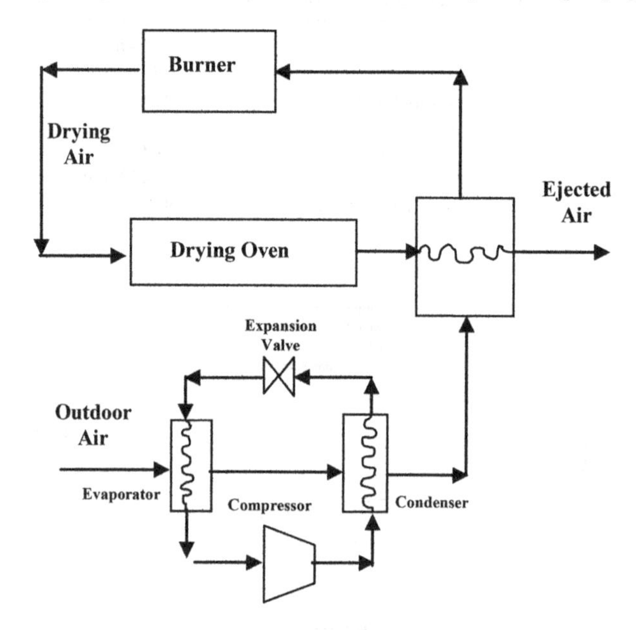

Figure 12. Systematic description of the heat recovery and heat pump at the inlet.

system alone, option 2 has the lowest water concentration in the drying air. As a result option 2 has the greater potential to increase the speed of the drying process.

3.5 *Outdoor air with heat pump and heat recovery (option 3)*

Another way of using a heat pump is to install the heat pump at the inlet rather than at the outlet of the drying oven as in option 2. Installing a refrigerated dehumidifier at the inlet increases the control over the air humidity. However, it increases the energy consumption of the drying air during the heating process as well.

The outdoor air flows into the drying oven through the evaporator as shown in Figure 12. By coming in contact with the cold surface of the evaporator, humidification of the wet air takes place. Consequently, the outdoor air becomes colder and dryer. By forcing the outdoor air to flow through the condenser of the refrigeration cycle, the air regains its temperature while maintaining its new low water content.

It is worth mentioning that the heat exchanger properties of the condenser and the evaporator are similar (effectiveness of 54%). The warm outdoor air flow through the heat exchanger increases

Table 3. Heat pump specifications for option 3.

Refrigerant	R-134a
COP	3.85
Evaporator Temperature	0°C
Evaporator Pressure	0.2928 MPa
Condenser Temperature	30°C
Condenser Pressure	0.7702 MPa
Heat Exchanger Effectiveness	50%

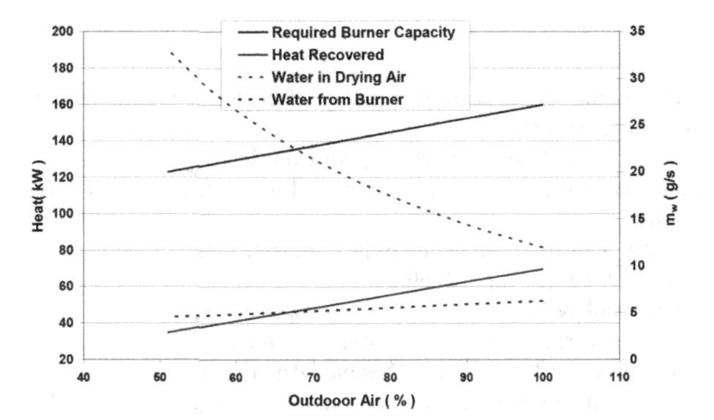

Figure 13. Heat requirements and the corresponding water content in the air (option 3).

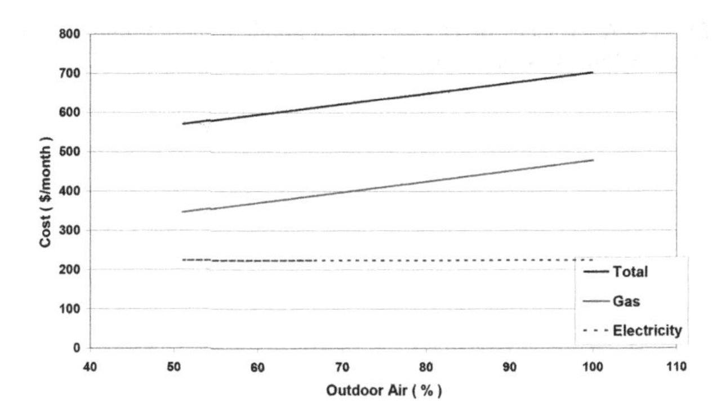

Figure 14. Related operating cost of the outdoor air with heat recovery and heat pump system (option 3).

its temperature to a relatively high level. The last stage of heating the outdoor air to 145°C is by forcing it to flow through the burner (Table 3).

Figure 13 shows the enhancement in the outdoor conditions both in increasing temperature and reducing water content. Compared to option 1 where only a heat exchanger is used, the heat required for 100% of outdoor air was around 173 kW. This heat requirement is reduced to around 160 kW. This is due to the heat recovered from the condenser of the refrigeration cycle. However, this is still higher than the heat requirement for option 2 which was around 80 kW.

The main reason behind using the heat pump at the inlet of the drying process is to reduce the water content in the drying air. The lowest value achieved thus far is 12 g/s which is obtained from option 3. However, the operating cost of option 3 is not the lowest among the investigated options. Figure 14. shows that the operating cost of option 3 at 100% outdoor air is around $700 per month which higher than the operating cost for option 2.

4 CONCLUSION

A review of the current operation of the drying oven at an abrasives plant has been undertaken to assess the potential for improvements to the drying process and improve production efficiency. The initial review of the existing drying process has indicated a number of possible areas of improvements. These are: makeup air, gas burner, air flow within the duct system, and recirculated air. While a number of changes can be implemented to improve the process, the effect of the airflow within the duct system is considered to be a core issue affecting performance and a particular focus of the next stage of the investigation.

REFERENCES

[1] SI Wen-yuan; Discussion on waste heat recovery of coated abrasives manufacture [J]; Diamond & Abrasives Engineering; 2013-02.

[2] C. G. Berg, J. A kerholm, & M. A. Karlsson, An experimental evaluation of the governing moisture movement phenomena in the paper coating process. I. theoretical aspects, Drying Technology, 19, 10, 2001, pp. 2389–2406.

[3] Johan Nilsson & Stig Stenstrom, Modeling of heat transfer in hot pressing and impulse drying of paper, Drying Technology, 19, 10, 2001, pp. 2469-2485.

[4] S. J. Hashemi, S. Sidwall, & W. J. Murray Douglas, Paper drying: a strategy for higher machine speed. I. through air drying for hybrid dryer sections, Drying Technology, 19, 10, 2001, pp. 2509-2530.

[5] S. J. Hashemi, S. Sidwall, & W. J. Murray Douglas, Paper drying: a strategy for higher machine speed. II. Impingement air drying for hybrid dryer sections, Drying Technology, 19, 10, 2001, pp. 2509-2530.

[6] C. G. Berg, J. A kerholm, & M. A. Karlsson, An experimental evaluation of the governing moisture movement phenomena in the paper coating process. II. Experimental, Drying Technology, 19, 10, 2001, pp. 2407-2419.

[7] William T., "Effect of Air Velocity on Drying Rate of Single Eastern White Pine Boards," United States Department of Agriculture, Research Note, 1997 FPL-RN-266.

Proceedings of the 1st International Congress on Engineering Technologies – Kiwan & Banat (Eds)

Streamline development of propellent-free topical pharmaceutical foam via quality by design approach for potential sunscreen application

Esra'a Albarahmieh, Razan Alziadat, Munib Saket, Mohammad Khanfar & Walid Al-Zyoud
German Jordanian University, Amman, Jordan

ABSTRACT: Catching up with recent interest in developing green pharmaceutical products, a key challenge is encountering formulation scientists to find proper ingredients, approach or dispense tools in a reasonable time due to manufacturing variability. Therefore, this work aims to demonstrate green, propellent-free, topical pharmaceutical preparation through utilization of the concept of quality by design (QbD) to eliminate or reduce the variability. A foam dosage form with ingredients suitable for potential sunscreen application was developed. Formulation ingredients were based on local Jordanian virgin olive oil. Twenty formulae were screened under the theme of QbD. One formulation successfully combined in relatively stable water in oil emulsion, whereby foam generation was achieved through a specialized dispensing tool. Textural characteristics were evaluated by a human panel (30) and found favourably acceptable for this formulation. Physicochemical characteristics of the investigated and optimized formulae were confirmed through light microscopy, viscosity and spectrophotochemical measurements. Sun protection factor (SPF) was determined as 12.1±0.6 (n=3); indicating potential sunscreen applications for almost two hours in the daytime. Critical quality attributes were determined throughout the development process and risk assessment measurements were estimated thereof. Such an outcome that will help to create preliminary predictive model for scalable industrial screening production of this green pharmaceutical topical preparation with local sourcing of generally regarded safe materials (GRAS).

Keywords: Foam; Propellent-free; Quality by Design; Sunscreen; Olive oil

1 INTRODUCTION

Historically, the pharmaceutical industry has responded efficiently to demands of eco-friendly initiatives with positive regulatory body enforcement. Among other examples, propellent-free aerosol inhalers with soft mist ejection for the pulmonary system was a breakthrough and an ideal example. On the other hand, foam formulation may have an added advantage [1]; especially on external body surfaces (i.e. topical). Therefore, the present work is attempting to develop a foam-able yet propellent-free formulation. For this investigation, sunscreen with topical (i.e. skin) application was chosen due to the interest in skin cancer protection [2].

Sunscreens based on emulsion formulation and UV light protection, such as titanium dioxide, are trending for efficient protection from harmful radiation with great consumer compliance and health outcomes [3]. Herbal sunscreen based on the sustainable Jordanian resource of olive oil and titanium dioxide or talc were principally used in this work and examined for potential water in oil emulsion formulation that could remain longer on the skin for maximum benefits. The evaluation type was based on a multidisciplinary approach of analytical and human assessment. We also studied the application of a holistic strategy of quality by design that realizes the critical issues relevant to modifying and or optimizing similar products with an focus on a risk assessment plan [4]. This

DOI 10.1201/9781003178255-22

includes a framework of defining critical quality attributes and assigning a quantitative risk priority number.

2 MATERIALS AND METHODOLOGY

2.1 *Materials*

Olive oil was sourced locally from Jordan (Alziadat Farm, Jordan) and obtained by cold extraction from the olive fruits. Sodium lauryl sulfate (SLS) and cetyl alcohol were purchased from AZ chem for chemicals (India). Tween 80 (Polysorbate 80) was purchased from Chem Cruz chemicals (USA). Amaranth dye (85%) was obtained from Difco Laboratories (UK) with a color Index of 16185. Titanium Dioxide (<100 nm particle size) and Talc were purchased from Sigma Aldrich. The rest of the reagents were of analytical grade. Distilled water was used entirely in this work.

2.2 *Methodology*

2.2.1 *Emulsion preparation*

Olive oil was evaluated using its acidity value, which is commonly used to evaluate its quality, as described in different editions of the British pharmacopeia and found to have a value of circa 2.9 g% (w/w) that falls in the range of virgin oil. Emulsions were prepared by factorial design of mixed ingredients and single stage mixing using a laboratory mixer (Heidolph RZR 2020, Heidolph Elektro GmbH & Co., Germany) at 6000 rpm for five minutes to obtain a homogenous mixture of milky or translucent product and avoid air entrapment. Talc or titanium dioxide was first dissolved in distilled water alone and then mixed with the olive oil with gradual addition of the rest of ingredients.

2.2.2 *Physicochemical characterization*

Emulsion viscosity was studied on a Brookfield viscometer (type-DV-1PR, Anton Paar GmbH, Germany) using a CPA 42Z spindle with a cone radius of 2.4 cm. All emulsions were prepared at 22±1°C. A light microscope (MSZ 5000 Stereo Zoom Microscope; Krüss, Hamburg, Germany) fitted with a digital camera (Krüss Optronics) was used to assess component homogeneity. Amaranth (water-soluble) dye was used to assess the type of the emulsion. The sun protection factor was calculated based on the traditionally used method [5] that relies on the input obtained from the assessment of the samples using the UV-Vis Shimadzu UV-1800 spectrophotometer (Shimadzu Corporation, Kyoto, Japan). Foam generation of reproducible amounts of the prepared emulsions were assessed using kindly-gifted commercial Airspary® foam dispensers (Layan company, Jordan) with a concomitant spread-ability test based on the diameter of spreading on Whatman® filter-cellulosic membranes. Textural characteristics on the skin were evaluated by recruiting 30 people from the 15-61 year age range and evaluated qualitatively based on a five-point scale that ranges between 1-5 that combined sensory attributes with spread-ability evaluation. Scores 1, 2, 3, 4 and 5 correspond to: too fluid, fluid but all right, agreeable foam, stiff but alright and too stiff respectively.

3 RESULTS AND DISCUSSION

3.1 *Setting quality target product profile for formulation screening*

The following criteria were set for evaluating the tested formulations (twenty) with factorial design of the added ingredients and with the variation of the added ingredients as shown in Table 1. The quality target product profile (QTPP) was set thereof as shown in Table 2.

Table 1. Design of experiment for the components utilized to produce fomable water in oil emulsion.

Variation element	Type	Range (w/w%)
Surfactant/solubilizer	SLS	1-20
	Tween 80	1-9
	Cetyl alcohol	1-9
Foam enhancer	SLS	1-20
UV sun protection material	Talc	2-10
	Titanium dioxide	2-6

Table 2. Quality Target Product Profile (QTPP) of potential fomable water in oil emulsion with potential sunscreen application.

QTPP	Minimally Acceptable Profile
Intended use	Sunscreen foam that provides protection from exposure to sunlight with SPF value of no less than 12.
Dosage form	Propellant-free foam with a light texture; suitable for hair-bearing skin.
Delivery system	Water-in-oil (W/O) emulsion.
Route of administration	Topical (Skin).
Labelling	Applied twenty minutes before sun exposure.
Application frequency	Re-applied every two hours.
Particle size distribution	Uniform and monodispersed.
Deliverable amounts	Consistent.
Spreadability	Easily spreadable.
Viscosity	Between the range of (2.5-3) cP.
Stability	Stable emulsion of not less than one month at R.T.

Table 3. Critical Quality attributes (CQA) established after screening of potential fomable water in oil (W/O) emulsion with potential sunscreen application.

Emulsion Stability	**Foam Production**	**Delivery system**	**Consistency of deliverable amount**	**Frequency of application**	**Target Parameter**	
Significant	Significant	Moderate	Moderate	No impact	SLS	Surfactants
Significant	Significant	Moderate	Moderate	No impact	Tween 80	
Significant	Significant	Significant	Moderate	No impact	Cetyl Alcohol	
Little impact	Moderate	No impact	Moderate	Significant	Titanium dioxide (TiO_2)	Sun protection material
Little impact	Moderate	No impact	Moderate	Significant	Talc	
No impact	Significant	No impact	Significant	No impact	Viscosity	
No impact	Moderate	No impact	Significant	No impact	Filling Volume	

3.2 *Most promising formulation based on the set QTPP coupled with Critical Quality Attributes (CQA)*

After physicochemical evaluation of the examined twenty formulations based on the set QTPP, which aided in the probing of the CQA shown in Table 3. The formula with a 6:4 oil to water ratio

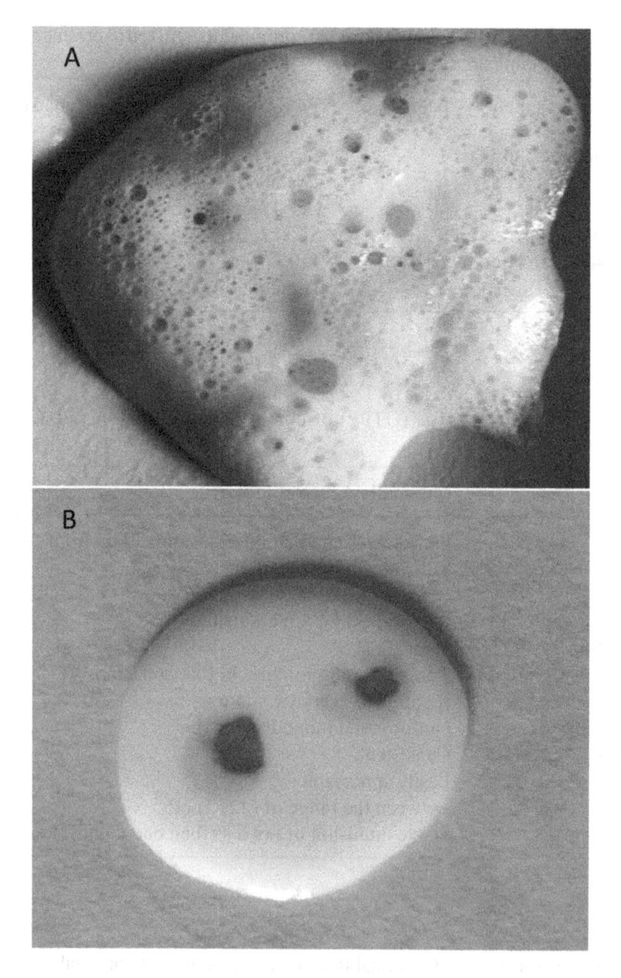

Figure 1. Representative photographs showing foam collected from the developed formable emulsion (Image A) that has been classified as w/o type using Amaranth water-soluble dye (Image B).

and 9% SLS that was used with 2% Titanium dioxide was superior in all the tests and achieved the best scores. Figure 1 shows the appearance of this formulation. This formula has shown stability over four months without phase separation and particle size that lie within colloidal range as confirmed by microscopic evaluation which entails further enhancement of the estimated shelf life. A risk assessment plan was then derived to create a platform for future workers to have a preliminary predictive model for scalable industrial screening production of this green pharmaceutical topical preparation with local sourcing of generally regarded safe materials (GRAS). The criticality of successful preparation of this foam-able emulsion, assessed using risk priority number (RPN), was found highly related to the quantity added of SLS above 2% followed by the added quantity of titanium dioxide or talc in the range of 6 or 2 respectively (Table 4).

4 CONCLUSIONS

This work was based on two main ideas. To fabricate an emulsion with foam-able properties in a green approach that utilizes propellant-free ingredients, and then to assess the success of the

Table 4. Risk assessment through assigned risk priority number (RPN) of the target parameters with significant CQA (See Table 3). RPN calculation was performed through multiplication of severity, occurrence, and detectability numbers of their used rating scales. The scale rating of severity was of 1-3: no critical impact, 3-5: moderate impact, 5-9: significant impact, whereas the occurrence rating scale was 1-3: now frequency, 3-5: moderate, 5-9: high frequency and the detectability rating scale corresponded to 1-3; immediately detected, 3-5: moderately detected, 5-9; not detected.

Emulsion Stability*	Foam Production	Delivery system	Consistency of deliverable amount	Frequency of application	Target Parameter	Levels		
15,192	**21,192**	**N/A**	**N/A**	**N/A**	**RPN**	**Low (<2%), High (≥2%)**	SLS	Surfactants
64,20	**30,15**	**N/A**	**N/A**	**N/A**			Tween 80	
18,112	**84,18**	**15,40**	**N/A**	**N/A**			Cetyl Alcohol	
N/A	**40,126**	**N/A**	**N/A**	**84,120**		**Low (≤2%), High (≥6%)**	Titanium Dioxide	Sun protection material
N/A	**126,20**	**N/A**	**N/A**	**120,18**			Talc	
N/A	**40,120**	**N/A**	**10,112**	**N/A**		**Low (≤2cP), High (>2cP)**	Viscosity	
N/A	**N/A**	**N/A**	**70, 108**	**N/A**		**Low (≤20%), High (>20%)**	FillingVolume	

*At Laboratory room conditions of 22±1°C and 60-65% RH for no less than one month.

preparation using the QbD approach. A successful water in oil emulsion was formulated based on olive oil and titanium dioxide. Sodium lauryl sulfate was used on both as a surfactant and foam aid. The QbD approach identified the most critical parameters for this product and shortlist recommendations for the type and range of ingredients that will ensure production of proper preparation of the investigated emulsion type with potential robustness. We conclude that, for design intent of similar product, the use of this study could be a reliable model. However, further quantitative work on the packaging process parameters and microbiological testing for stability and shelf-life estimation must be verified with the controls for acceptable product performance according to regulatory bodies such as the FDA.

5 DECLARATIONS

The authors declare that we have no conflict of interest. We also declare that all the experiments conducted in this work comply with the current laws of Jordan.

REFERENCES

[1] Arzhavitina A & Steckel H. Foams for pharmaceutical and cosmetic application. International Journal of Pharmaceutics. 2010;394(1–2):1–17.

[2] Diffey B. Sunscreen claims, risk management and consumer confidence. International Journal of Cosmetic Science. 2020;42(1):1–4.

[3] Battistin, Dissette, Bonetto, Durini, Manfredini, & Marcomini et al. A New Approach to UV Protection by Direct Surface Functionalization of TiO_2 with the Antioxidant Polyphenol Dihydroxyphenyl Benzimidazole Carboxylic Acid. Nanomaterials. 2020;10(2):231.

[4] Yu L. Pharmaceutical Quality by Design: Product and Process Development, Understanding, and Control. Pharmaceutical Research. 2008;25(10):2463–2463.

[5] J.D.S. Mansur, M.N.R. Breder, M.C.D.A. Mansur, & R.D. AzulayDeterminação do fator de proteção solar por espectrofotometria. An. Bras. Dermatol.1986;16:121–124.

Proceedings of the 1st International Congress on Engineering Technologies – Kiwan & Banat (Eds)
 ISBN 978-0-367-77630-5

Analysis of a zero energy building: A case study in Jordan

Suhil Kiwan
Mechanical Engineering Department, Jordan University of Science, Irbid, Jordan

Munther Kandah
Chemical Engineering Department, Jordan University of Science, Irbid, Jordan

Ghanem Kandah
Mechanical Engineering Department, Jordan University of Science, Irbid, Jordan

ABSTRACT: The concept of zero energy buildings is considered a highly demanded one because of the danger of global warming effects due to gas emissions from petroleum-based energy sources. In this work, a comprehensive study for zero energy buildings is carried out for one building in Jordan with a $240m^2$ area which consists of 10 apartments with $150m^2$each. The study focussed on developing the best design of renewable energy that achieves zero energy for each apartment using photovoltaics (PV) and thermal solar collectors. The heating and cooling loads were calculated using a commercial software package (HAP) and F-chart method for solar heating load calculations. Three different scenarios were suggested, the expected monthly saving and payback periods are 63.7 JD, 3.1 years for scenario one, 59 JD, 3.4 years for scenario two, and 55 JD, 4.1 years for scenario three, respectively. It was found that the peak power energy of photovoltaic panels for a single typical apartment is 4.62 kWp which saves 764 JD/year and the required heating and cooling peak loads are 3336 kWh/year and 7632 kWh/year respectively. Hence the conservation of energy measures and the integration of renewable energy will create energy efficiency and create a net-zero energy building.

Keywords: Zero Energy Building; F– Chart Method; PV; Thermal Solar Collector; ECMs

1 INTRODUCTION

The effects of petroleum-based energy sources on the environment due to CO & CO_2 gas emissions and the increase in petroleum prices forced researchers towards investigating and developing systems that were based on clean and renewable energy such as photovoltaics (PV) and thermal solar collectors. In recent years, low energy buildings were proposed as a solution for the mitigation of CO_2 emission [1]. In Jordan's case, as implied by the annual report submitted by the ministry of energy and mineral resources, the building sector is estimated to consume around 40% of total energy. Table 1 below shows the energy consumption of different sectors in Jordan.

Net zero energy buildings are a promising solution to maintaining energy efficiency and conservation of energy in the housing sector [3]. Low energy buildings are fully depending on renewable energy systems to achieve energy balance without the existence of conventional resources. In the current study, the integration of renewable energy within an existing building has been used to cover the energy needed for many uses such as lighting and electrical appliances as well as space cooling and heating. The kingdom of Jordan had faced difficulties from the Syrian and Iraqi refugees that came to Jordan because of war in their countries causing a very big population in Jordan. Almost 1 million refugees contribute to the high energy demand in government and private sectors [4]. Jordan depends on conventional energy sources like fossil fuels to supply the energy needed by many sectors. Energy demand is currently rising due to increasing population and industrialization. However, the limitation of fossil fuel resources and insufficient conversion capacities made the demand high. Jordan has started searching for an alternative resource to decrease the demand

 DOI 10.1201/9781003178255-23

Table 1. Distribution of energy in different sectors [2].

Sectoral Electricity Consumption (GWh)		
	Year	
Sector	2018	2019
Home & Government	7879	7964
Commercial & Hotels	2510	2508
Industrial	3910	3929
Agricultural and water pumping	2683	2719
Street lighting	402	421
Total	17504	17661

on fossil fuels [5]. However, Jordan has limited energy resources and its resources do not exceeded 3% – 4% of total energy demand which is insufficient. Then, the search for an alternative source of energy began [6]. The need for energy audits is vital in the move toward net-zero energy buildings. However, an energy audit for a building was introduced and discussed along with economic and environmental impacts by Kumar et al., [7]. The results showed good energy estimates compared to traditional energy consumption. The building has to include the utility of integrated renewable systems for hot water, heating and solar photovoltaics. Hassouneh et al., [8] discussed the influence of windows on the energy balance of apartment buildings. In this study, the energy audit was performed for an apartment building in Jordan to illustrate the impact of several modifications on the glazing, wall insulation and lighting system. The modifications saving on both electrical and fuel bills have a 3 year payback period which is a good example of how energy audit surveys can reduce energy consumption. The opportunity of zero energy buildings to building in Romanian climate conditions during heating period has been analyzed by Carutasiu et al. [9]. In this study it was found that both the high energy efficiency and smart building controllers that were applied to the house reduced the energy consumption for heating by more than 90% compared with standard buildings. Several control methods were analyzed to enable a reduction in the energy consumption associated with heating and cooling loads within the building of nearly zero energy by Fratean and Dobra [10]. This research helps net-zero energy building design collectives to assess the opportunity of implementing controls and estimate how the energy consumption reduction targets can be reached by implementing such technologies. Alajmi et al., [11] considered a passive house and studied the opportunity transforming it into a net-zero energy house. Energy analysis was used as a method to model the energy for heating, cooling, water heaters, plug loads and electric appliances using Design Builder software for model simulation. It is found that an energy consumption reduction of 814 kWh /year was recorded. In this study a system of photovoltaics and solar water heater were evaluated. It is investigated that 4053 kWh/year of PV met annual energy demand of 3936kWh/year and solar water heater reduced energy consumption by 3047 kWh/year. Jabe et al., [12] suggested the formulation of enhancing renewable energy resources in Jordan by using problem tree analysis. In this study the conducted analysis was performed by face-to-face questioner in public and private sectors. The purpose was to contribute some renewable energy scenarios in the buildings, and the target was expected to reach 10% of the total energy consumption by the year 2020.

In this current study, photovoltaics (PV) and solar thermal collector systems are designed for one building consisting of 10 apartments with 150 m^2 each in order to achieve a zero energy building. Different scenarios were used based on the energy needs and the available space area on the building.

2 METHODOLOGY

In this study the methodology is based on providing different designs according to energy efficiency and conservation of energy as stated in the Jordanian national building code. The following procedure has been applied:

1. Carrying out most recent literature survey about the project.
2. Data collection for the residential building in Jordan.
3. Estimating heating and cooling loads using HAP software.
4. Energy audit sequence approach to the concerned building.
5. Design of renewable energy system.

Collecting energy bills and carrying out energy audits are the main tasks to be performed in the facility [13]. An evaluation of energy used in the building will identify the potential of areas and energy savings. The audit process consists of three parts which are: pre-site work, site visits and post-site work that make the process more comprehensive and easier which leads to a more useful audit report. The site, rooftop area and building orientation constraints were considered in this study [14]. Energy used in the building was analyzed according to utility bills. All data collected was such that all possible measures, energy consumed and rooftop area, were used to install the renewable energy that covers the energy needed and replaced the old conventional sources of energy. In order to determine the optimum value of each parameter that influences energy consumption, a simulation optimization approach which was verified and discussed along with the integration of renewable energy to achieve a net-zero energy building. Figure 1 illustrates the process:

Figure 1 represents on-going steps to finalize net-zero energy. To decide the most appropriate design of renewable energy that will replace energy consumption,;t is important to demonstrate the energy consumed in the apartment. Energy used in the apartment represents all the sources of energy that covered by conventional energy while the energy used after auditing is the design data of renewable energy.

Optimization is the process of minimizing the total cost and energy to reach a net-zero energy building. Net-zero energy is reached by optimizing the renewable energy design based on energy audit and energy bills.

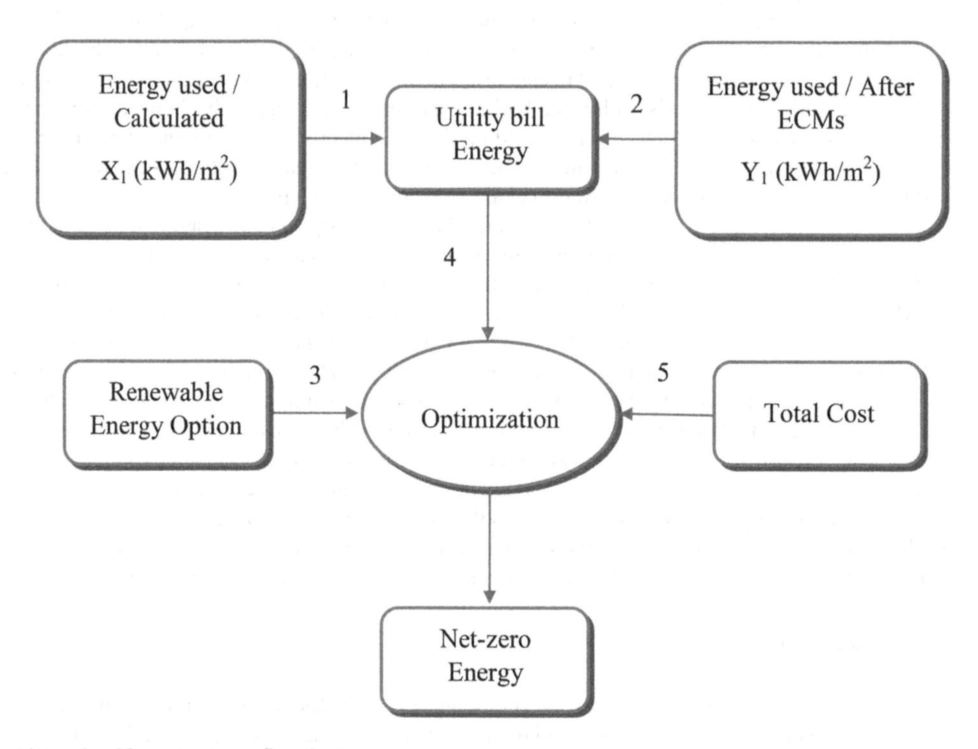

Figure 1. Net-zero energy flowchart.

3 WINDOWS, DOORS AND ROOFS

The apartments initially had old type windows with aluminium frames and single glazing. Some owners have replaced some or all of the windows with double glazing. In this study, energy conservation measures, such as using the appropriate configuration of wall construction and insulation materials that minimize energy required using optimal costs were used with various configurations of wall construction and insulation where the insulation thickness is increased and the total energy required was reduced. Figures 2 and 3 show a typical roof-wall cross section which is recommended in Jordanian buildings [15].

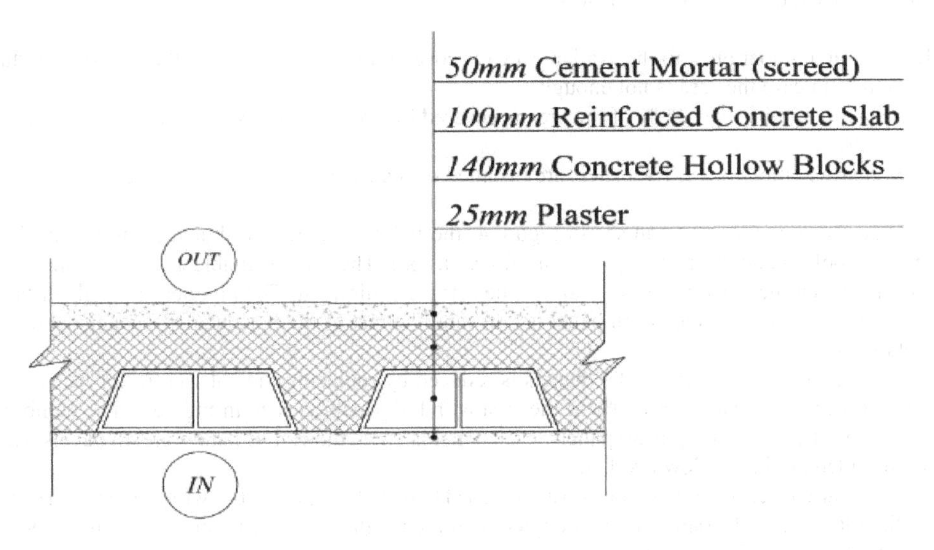

Figure 2. Typical roof construction [15].

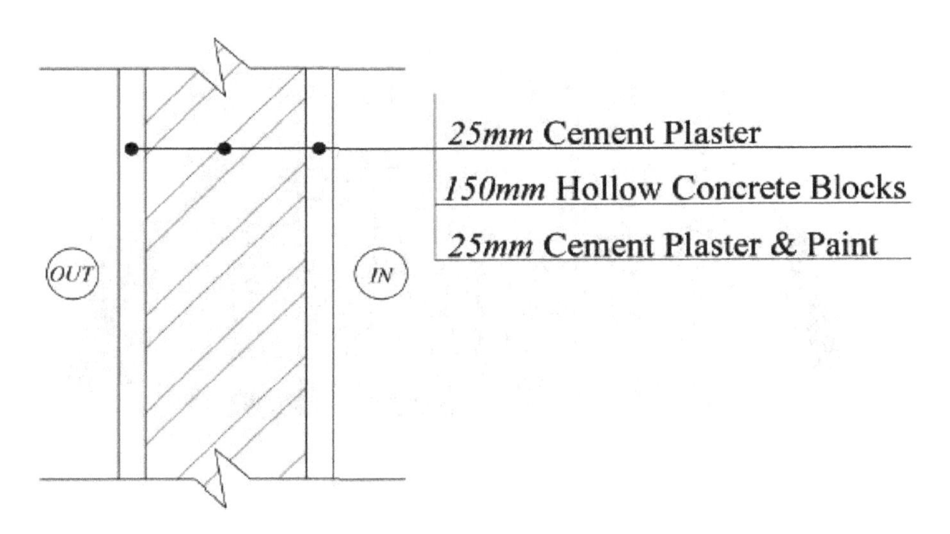

Figure 3. Interior wall – cross section [15].

4 RESULTS AND DISCUSSION

In this study, the concept of the net-zero energy building has been introduced and analyzed on a typical building. Two systems of renewable energy were suggested to cover energy consumption such as photovoltaic panels and a solar heating system. Three different scenarios of photovoltaic panels were adopted. PV_{syst} software was used to calculate energy output for each scenario.

4.1 *Photovoltaic array system*

In this study, three different scenarios are discussed to identify the most suitable one in terms of energy, area and number of PV panels.

1. Scenario number one: all the available area is covered by PV panels using the rooftop and flush on roof in case the area is not enough.
2. Scenario number two: 80% of the area is covered by PV and 20% is covered by solar collector panels.
3. Scenario number three: PV panels are installed on a structure with a specific height.

In scenario one, as you can see in Figure 4, the PV panels are distributed along the available area on rooftop and the rest of panels are flush on roof. The total available area is 150 m^2 out of 240 m^2 which includes water tanks and some service equipment. The number of panels installed on the rooftop is 75 panels: each panel is 2 m^2 and the rest of the panels are 45 panels which are installed flush on roof.

In scenario two, the area of PV panels is reduced by replacing the solar collector panels with 20% of the available area and 80% of the area with PV panels as seen in Figure 5. The number of panels installed on rooftop is 50 panels: each panel is 2 m^2, the rest of panels are 40 panels which are installed on the window overhang.

In scenario three, the whole rooftop area is used to install PV panels but with some modification on the tilt angle of the panels to be 5 degrees and the height of the structure holding the panels to be 1.5 m as shown in Figure 6.

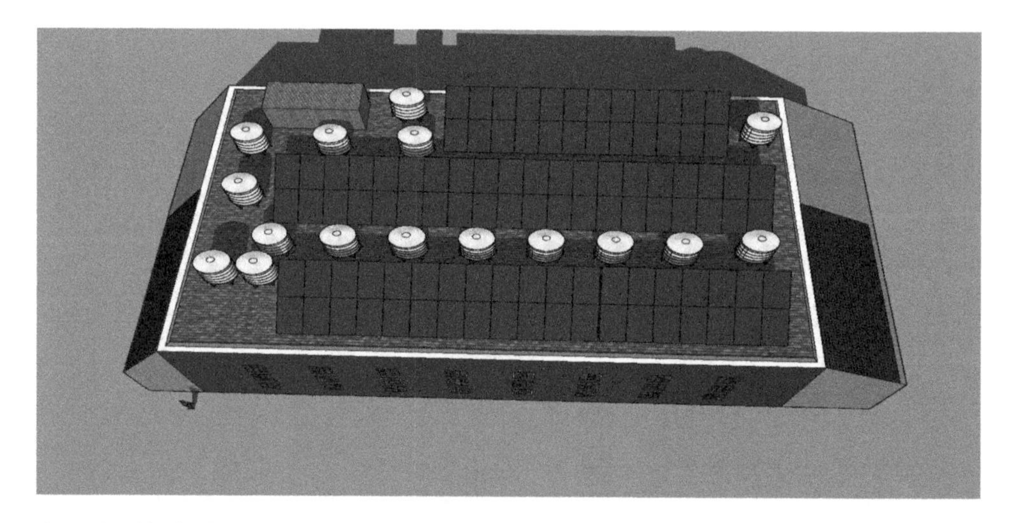

Figure 4. Distribution of PV panels on rooftop for scenario one.

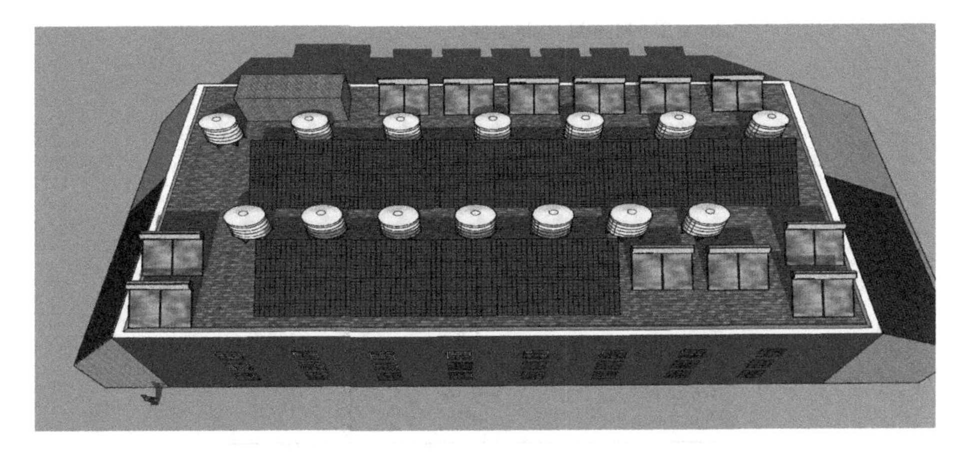

Figure 5. Distribution of PV panels and solar collector on rooftop for scenario two.

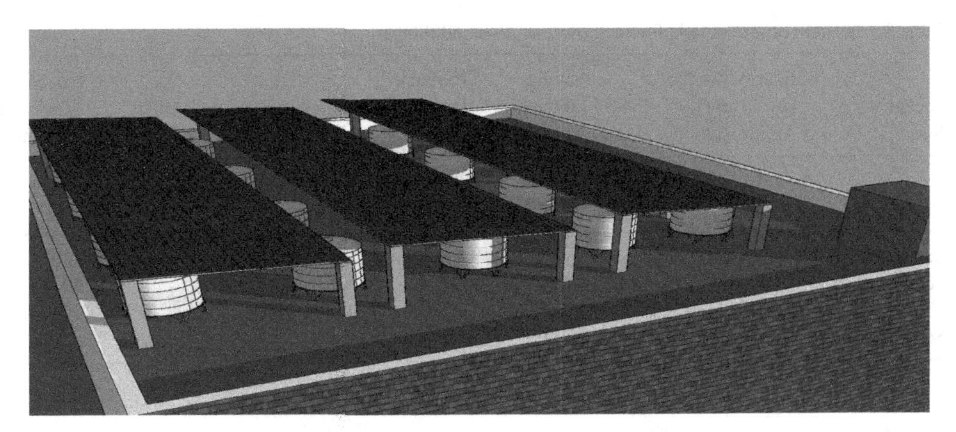

Figure 6. Distribution of PV panels on a structure on rooftop for scenario three.

4.2 *Solar thermal collector*

Hot water in the apartment will be provided by solar collectors. To estimate the required amount of hot water, it is important to calculate the area of the solar collectors. The *f*-chart method has been used to calculate the amount of hot water required through the year. The *f*-chart method provides a means for estimating the fraction of a total heating load that will be supplied by solar energy for a given solar heating system. The result is done by the calculations of the *f*-chart method. The area of the solar collector is designed based on monthly and yearly climate information and the efficiency of the solar collector. The amount of energy needed to heat the amount of water consumed is calculate by equation (1) and the results are shown in Table 2.

$$L = m.C_p.(T_H - T_C)N \tag{1}$$

L = Energy required to heat a certain amount of water (joules / month)
m = The amount of hot water consumed per day (l / day)
C_P = Specific heat (joules / kg °C) [water = 4186]
T_H = Hot water temperature (°C)
T_C = Cold water supplied to the thermostat (°C)
N = Number of days of the month
f = defines the ratio of energy supplied by solar energy

Table 2. Monthly values of f.

Month	f	fL (MJ)
January	0.24	120
February	0.4	200
March	0.6	301
April	0.9	452
May	1.0	502
June	1.0	502
July	1.0	502
August	1.0	502
September	1.0	502
October	0.7	351
November	0.34	170
December	0.28	140

Table 3. System output details for scenario one, two and three.

Scenario No.	Energy Produced (kWh/year)/ Apartment	Saving (JD/year)/ Apartment	Payback Period (year)	PV area (m^2) for building		Tilt Angle	No. of PV panels for building
				Rooftop	Windows Overhang		
Scenario one	8700	764.5	3.1	150	90	28°	120
Scenario two	**PV Panels** 7000	636	3.4	100	80	28°	90
	Solar Panels 2052	72		50	0	40°	
Scenario three	8100	669.6	4.1	240	5°	120	

Table 4. System specification for scenario one.

Inverter Capacity (kW)/ Apartment	System Capacity (kWp)/ Apartment	Number of Panels/ Apartment
4.2	4.62	12

The total energy produced by each system is 8.7 MWh/year for scenario one, 7 MWh/year for scenario two and 8.1 MWh/year for scenario three as shown in Table 3 which illustrates the result of three scenarios in term of area and number of PV panels. The available rooftop area to install PV panels is 150 m^2 which accommodates for 75 panels and the rest were installed on overhang windows. In scenario two solar collector panels were replaced so that the number of PV panels was reduced to accommodate 100 m^2 on rooftop area. In scenario three, the PV panel area needed is 240 m^2 which means that the entire rooftop area is used to install the PV panels since a structural mountain of the PV panels are used.

The energy produced in each scenario is illustrated in Tables 4–8, the simulation results were obtained by PV_{syst} software. It was found that applying some energy conservation measures to the apartment will reduce the energy required. Energy efficiency and corresponding procedures, methods and techniques are a path for reliable energy saving in the residential buildings in case of apartment buildings. However, the use of energy efficiency in the building plays a good role. It can be seen that minimizing cost using both energy conservation measures and renewable energy

Table 5. Simulated data result for scenario one.

Annual Energy Produced(kWh)/ Apartment	Saving (JD/year)/ Apartment	Payback Period	Tilt Angle	Area of system (m^2)	Monthly Energy Produced (kWh)/ Apartment
8700	764.5	3.1	28	24	725
8600	750.2	3.2	25	24	716
8400	732.6	3.3	22	24	700
8250	720.6	3.4	20	24	688
8000	701.8	3.5	15	24	666

Table 6. System specification for scenario two.

Inverter Capacity (Kw)/Apartment	System Capacity (kWp)/Apartment	Number of Panels/Apartment
3.1	3.5	9

Table 7. Simulated data result for scenario two.

Annual Energy Produced (kWh/Apartment)	Saving (JD/year)/ Apartment	Payback Period	Tilt Angle	Area of system (m^2)	Monthly Energy Produced (kWh/Apartment)
7000	636	3.4	28	18	583
6500	601.3	3.6	25	18	541
6300	590	3.7	22	18	525
6100	588	3.75	20	18	508

Table 8. Simulated data result for scenario three.

Annual Energy Produced (kWh)/Apartment	Saving (JD/year)/ Apartment	Payback Period	Tilt Angle	Area of system(m^2)/ Apartment
8100	669.6	4.1	5	24

systems reduce energy demand such that the payback period is minimal. For that, net-zero energy will be achieved meanwhile energy consumed will approximately equal the energy produced using renewable energy systems.

5 CONCLUSION

The results of this study reveal the opportunities for enhancing renewable energy, such as photovoltaic panels that provide the energy needed to cover and reduce overall energy consumption. It was found that energy saving could be achieved when some energy conservation measures have been applied. Hence, energy saving can be achieved according to energy conservation measures and the building could be converted to a net-zero energy building.

It has been concluded that the energy conservation measures through an energy audit affect the optimal design options toward net-zero energy apartment buildings. In addition, saving on both annual consumption based on ECMs has been achieved in the case of a typical building in Jordan.

It was found that mountain PV panels on a 28° tilt anglewould give the highest energy output which is 725 kWh/month/apartment. It has been concluded that the energy conservation measures through an energy audit affect the optimal design options toward net-zero energy apartment buildings. However, saving on both annual consumption based on ECMs has been achieved in case of an apartment building in Jordan.

Finally, it is found that scenario number one of renewable energy has the highest energy output which is 8700 kWh/year/apartment and has the best payback period of 3.1 years which is the best suitable scenario compared to others.

REFERENCES

[1] J. O. Jaber, M. S. Mohsen, S. D. Probert, and M. Alees, "Future electricity-demands and greenhouse-gas emissions in Jordan," *Appl. Energy*, vol. 69, no. 1, pp. 1–18, 2001.

[2] Energy and Minerals Regulatory Commission, "Energy in Figures," 2018.

[3] M. A. A. Al-busoul, "ZERO ENERGY BUILDING FOR A TYPICAL HOME," *Int. J. Recent Res. Sci. Eng. Technol.*, no. October, 2018.

[4] E. S. Hrayshat, "Analysis of renewable energy situation in Jordan," *Energy Sources, Part B Econ. Plan. Policy*, vol. 3, no. 1, pp. 89–102, 2008.

[5] M. Al-omary, M. Kaltschmitt, and C. Becker, "Electricity system in Jordan: Status & prospects," *Renew. Sustain. Energy Rev.*, vol. 81, no. August 2016, pp. 2398–2409, 2018.

[6] S. Malkawi, M. Al-Nimr, and D. Azizi, "A multi-criteria optimization analysis for Jordan's energy mix," *Energy*, vol. 127, pp. 680–696, 2017.

[7] A. Kumar, S. Ranjan, M. B. K. Singh, P. Kumari, and L. Ramesh, "Electrical Energy Audit in Residential House," *Procedia Technol.*, vol. 21, pp. 625–630, 2015.

[8] K. Hassouneh, A. Alshboul, and A. Al-Salaymeh, "Influence of windows on the energy balance of apartment buildings in Amman," *Energy Convers. Manag.*, vol. 51, no. 8, pp. 1583–1591, 2010.

[9] M. B. Carutasiu, V. Tanasiev, C. Ionescu, A. Danu, H. Necula, and A. Badea, "Reducing energy consumption in low energy buildings through implementation of a policy system used in automated heating systems," *Energy Build.*, vol. 94, pp. 227–239, 2015.

[10] A. Fratean and P. Dobra, "Control strategies for decreasing energy costs and increasing self-consumption in nearly zero-energy buildings," *Sustain. Cities Soc.*, vol. 39, pp. 459–475, 2018.

[11] A. Alajmi, S. Rodríguez, and D. Sailor, "Transforming a passive house a net-zero energy house?: a case study in the Paci fi c Northwest of the U . S .," *Energy Convers. Manag.*, vol. 172, no. June, pp. 39–49, 2018.

[12] J. O. Jaber, F. Elkarmi, E. Alasis, and A. Kostas, "Employment of renewable energy in Jordan: Current status, SWOT and problem analysis," *Renew. Sustain. Energy Rev.*, vol. 49, pp. 490–499, 2015.

[13] Y. Geng, W. Ji, B. Lin, J. Hong, and Y. Zhu, "Building energy performance diagnosis using energy bills and weather data," *Energy Build.*, vol. 172, pp. 181–191, 2018.

[14] D. K. Serghides, M. Michaelidou, M. Christofi, S. Dimitriou, and M. Katafygiotou, "Energy Refurbishment Towards Nearly Zero Energy Multi-Family Houses , for Cyprus," *Procedia Environ. Sci.*, vol. 38, pp. 11–19, 2017.

[15] A. Younis and A. Taki, "Towards Resilient Low-Middle Income Apartments in Amman , Jordan?: A Thermal Performance Investigation of Heating," *CIBSE ASHRAE Tech. Symp.*, no. April.

Proceedings of the 1st International Congress on Engineering Technologies – Kiwan & Banat (Eds)
 ISBN 978-0-367-77630-5

Circular supply chains: A comparative analysis of structure and practices

Raid Al-Aomar
Industrial Engineering, German-Jordanian University, Amman, Jordan

Matloub Hussain
College of Business, Abu Dhabi University, Abu Dhabi, UAE

ABSTRACT: This paper presents a comparative analysis of the structure and practices of Circular Supply Chains (CSC) across multiple industries. This concept has recently evolved from basic reverse logistics practices to become a comprehensive approach for reserving resources and attaining sustainability across the supply chain. It emphasizes the value of backward flow of material and information within the different elements of the supply chain. Common practices include recycling, returning, reusing, and recovery. Such practices are increasingly used in different industries to reduce waste, conserve scarce resources, and improve the overall supply chain efficiency. The paper explores and compares the CSC practices in the supply chains of three industries: furniture, aluminum, and farming. It presents the SIPOC structure of the targeted supply chains, identifies the CSC practices within the entities of each supply chain, and links these practices to specific performance measures (financial, operational, and environmental). Finally, the paper presents a comparative analysis of CSC practices across the supply chains of the targeted industries.

Keywords: Circular Supply Chains; Reverse Logistics; Comparative Analysis; Supply Chain

1 INTRODUCTION

Supply Chain Management (SCM) has been the emphasis of researchers and practitioners across many industries. This is mainly due to the increasing value of the supply chain and the growing competiveness locally and globally (Heizer *et al.*, 2016). Forward and backward flow of material and products represent the key driver of cost and value across the supply chain. The forward flow involves the upstream supplies delivered to producers from different supplier tiers and the downstream products distributed for retailers and end-users. The backward flow represents the reverse flow of materials and items amongst supply chain partners.

Traditionally, the focus of SCM studies has been on optimizing the linear supply chain (i.e., forward flow of material from suppliers to producers and from producers to customers). Circular Supply Chains (CSC) has recently emerged as a concept for integrating the backward (reverse) flow of material and information within the structure of the supply chains (Govindan & Hasanagic, 2018). The motivation is to recover the lost opportunity of material waste and to reduce the growing cost of waste handling and disposal and the environmental impact of by-products towards supply chain sustainability (Manavalan & Jayakrishna, 2019). Figure 1 depicts a generic structure for a circular supply chain.

In general, CSC supply chains are sustainable supply chains that use best practices for recycling, reuse, repair, etc. to reduce waste, save resources, and increase efficiency. As discussed in Kalmykovaa *et al.* (2018), CSC is increasingly becoming a key pillar in building an overall Circular Economy (CE) across industry sectors. However, reviewed literature has revealed a gap in analyzing

DOI 10.1201/9781003178255-24

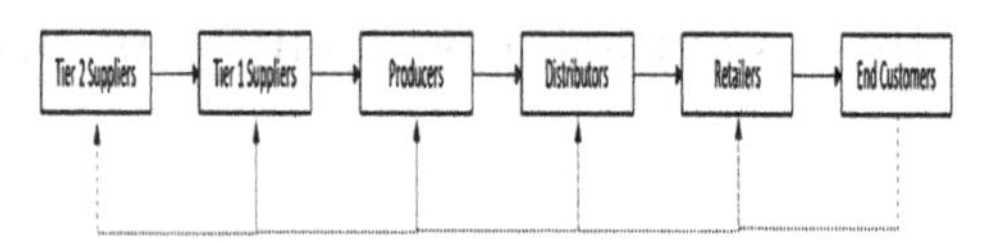

Figure 1. A generic structure of a circular supply chain.

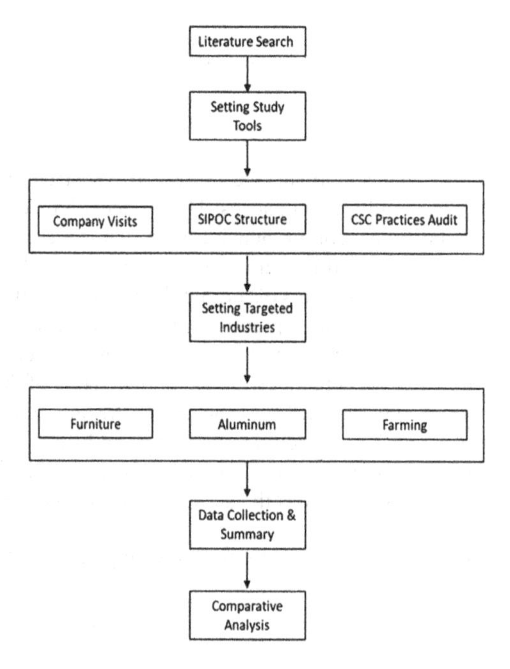

Figure 2. Study methodology.

CSC practices in particular industries. Further details of CSC structure and characteristics can be found in Batista *et al.* (2018).

This paper utilizes the existing literature on CSC to develop an audit of adopted CSC practices across multiple industries (furniture, Aluminum, and farming) in terms of structure (Farooque *et al.*, 2019), practices (Kalmykovaa *et al.*, 2018), and performance (Bracquené *et al.*, 2020). Supplier–Inputs–Process–Output–Customers (SIPOC) chart (Rasmusson, 2006) is used to present the structure of targeted supply chains and allocate CSC practices. Finally, the paper presents a comparative analysis of the three supply chains in terms of structure and adopted CSC practices. This paper is organized as follows: Section 2 present the study methodology, Section 3 summarizes the collected data and information from three case studies, Section 4 presents the results of the comparative analysis, and Section 5 presents a conclusion.

2 STUDY METHODOLOGY

This paper presents the results of an exploratory study of CSC practices. The objective is to identify the key CSC practices across the supply chains of different industries. To this end, a structured methodology is followed.

Table 1. SIPOC structure for furniture store supply chain.

Main Suppliers	Main Inputs	Main Processes	Main Outputs	Main Customers
Poland India	Storage furniture Cotton, furniture	Storing and distributing Carpenter and assembly	Kitchen products Closet	Local and international customers Business to business
China	Electronics, shades	Distribution	All electronics and lightings	Local and international customers
Brazil	Furniture	Distribution and assembly	Cabinets and living room decorations	Local and international customers
South Africa	Wood	Carpenter and distribution	All types of furniture	Local and international customers

As shown in Figure 2, a literature search was first conducted to review the latest studies on the concepts, frameworks, and applications of CSC. The review revealed a gap in analyzing CSC practices in particular industries. Thus, specific study tools were set to explore the CSC practices in the supply chains of three industries. These tools include company visits to meet supply chain specialists, understand the structure of their supply chains, and identify the adopted CSC practices. The study has targeted three industries; furniture, Aluminum, and Farming. SIPOC chart was used to develop the structure of the targeted supply chains. An audit of CSC practices, which was developed based on reviewed literature, was then applied to supply chains of the three industries to identify the used CSC practices, allocate them at SIPOC structure, and assess their impact. Finally, the collected data is summarized and a comparative analysis is conducted to identify common CSC practices and differences in their adoption across the targeted industries. Figure 2 presents the study methodology.

3 CASE STUDIES

The research methodology was applied to the supply chains of three case studies in different industries; furniture, Aluminum, and farming. The SIPOC chart and the CSC practices audit were used to collect the data and information needed for the study. SIPOC requested the supply chain manager of the company to identify its major suppliers, raw materials, processes, products, and customers. The CSC practices audit includes 18 practices collected from reviewed literature and intended to cover the diverse aspects of CSCs. The supply chain manager was asked to specify the adopted CSC practices within the supply chain, allocate the practice on the SIPOC chart, set a quantitative measure for the impact of the adopted practice, and qualitatively assess the level of that impact (Low, Medium, and High). The details of applying the methodology to the three case studies are shown in the following sections.

3.1 *Furniture store*

The first case study of CSC is in furniture industry. The targeted company owns forests to produce wood, which is typically called "foresting". Saving the environment and waste reductions are organizational values along with maintaining customer satisfaction. This represents the key motivation to adopt CSC practices. To this end, products are designed to utilize the available capacity of the materials, to use the waste of each product in the make of another product with minimum cost, and to utilize the recycled materials.

The furniture company creates smart products to be efficient and provide customized lifestyle for the customers at a competitive price. The overall process is complicated therefore it is required to maintain low price. The company does not have a storage facility, only distributions centers. The organization has 5 total warehouses around the world. Each continent has access to the closest warehouse. The SIPOC structure of the furniture store is shown in Table 1.

One major branch of the furniture company was investigated for CSC practices. The results are shown in Table 2.

Table 2. CSC practices in furniture supply chain.

General CSC Practice	Specific CSC techniques/actions	Applied at which SIPOC?	How is it measured?	Impact level (L, M, H)
Recycling	– Recycling press machine – Recovery actions – Defected products policy	Materials (Suppliers)	Finance department recovery value	H
Re-use	– Products sold for less price – Items utilized internally	Materials (Suppliers)	Number of reused items/ year	M
Reduction	– Material waste – Recovered from defective products – Democratic design	Materials (Suppliers)	Utilization of raw material to the max	M
Return	– Return policy (90 days) – Members have 120 days to return (Full refund)	Output/customer	Number of returned items per customer	H
Rejects	– Quality management department rates all the products	Input/process/output	Number of rejected raw materials per shipment	H
Repair	– 25 years warranty on selected products – Assembly team work on site free of cost	Output/Customer	Number of repairs per items returned	M
Re-distribute	– Electronics in general (extensions, lamps) – Local suppliers – Natural plants/flowers – Food (meat balls) – Seasonal products	Supplier	Items sold per customer	M
Re-sell	– Most products	Output/Customer	Returned items	L
Conservation	– Profit and loss report – Reduce utilities consumption – Awareness	Suppliers/Process	– Monthly reports – Cost of consumption	M
Restoration	– Foresting (for every plant removed new one planted)	Suppliers	The store owns farms for foresting	H

Table 3. SIPOC structure for aluminum plant supply chain.

Main Suppliers	Main Inputs	Main Processes	Main Outputs	Main Customers
Aluminum Company	Alumina	Smelting	Aluminum	External customers
Petroleum Company	Calcined Petroleum Coke	Anode Production	Baked Anode	Pot lines
Electrical Company	Gas Turbines	Power Generation	Electricity	Operations and Pot lines
Chemicals Company	Bauxite	Alumina Refinery	Alumina (raw material)	Smelters
Filters Company	CFF Filters	Aluminum casting	Purified Aluminum casting	External customers

3.2 *Aluminum plant*

The second case study is in the Aluminum industry (i.e., Aluminum plant for smelting and Alumina Refining). The company is a primary producer of Aluminum and its alloys. The production involves millions of tons of Alumina to produce Aluminum and hundreds of Alloys. This process is very costly and time consuming. The SIPOC structure of the furniture company is shown in Table 3.

The CSC in this industry is mainly focused on recycling excess material and by-products. The key CSC practices in the Aluminum plant are summarized in Table 4.

It is also worth mentioning that as the Aluminum plant aspires to be a leader in meeting its environmental and social responsibilities, it acknowledges that an effective, responsive and sustainable supply chain is essential to the continued success. The plant does not consider accepting returns. Instead, the company recycles scrap, re-melt, excess aluminum, and extracts by-products. This

Table 4. CSC practices in aluminum plant supply chain.

General CSC Practice	Specific CSC techniques/actions	Applied at which SIPOC?	How is it measured?	Impact level (L, M, H)
Recycling	– Re-melt of Aluminum scrap – Re-melt processed Dross in furnaces	Input (Pure Aluminum)	Weight	L
Re-use	– Re-use of Anode Butts	Process (Production of new Anodes)	Weight	H
Reduction	– Consumption of Natural Gas	Input (Electrical Energy)	BTU	H
Re-furbish and Remanufacture	– Used motors refurbishment and rewinding	Process (Refurbished motors)	Test certificates	M
Rejects	– Return rejected parts to suppliers	Suppliers (Unacceptable spare parts)	Specifications	L-M
Repair	– Repair of equipment/parts	Process (Used equipment/parts)	Test certificates	H
Re-sell	– Waste products/waste scrap	Output (Waste scrap material)	(weight and quantity)	L
By-products use	– Water Production from Power Generation cooling	Input (Hot Steam)	Water chemistry test	H

Table 5. SIPOC structure for farm supply chain.

Main Suppliers	Main Inputs	Main Processes	Main Outputs	Main Customers
Seeds stores	Cucumber/pepper/ tomato seeds	Farming	Vegetables	Supermarkets
Dealers	Rye	Harvesting	Rye	Grocery Stores
Packaging Suppliers	Packaging Paper	Packaging	Cartons and boxes	Schools and Institutions
Fertilizer Supplier	Fertilizers	Farming	Crops	Individuals

increases the company's economic status, and controls environmental and waste management to achieve zero landfill.

3.3 *Farm*

The third case study is for a farm supply chain (agriculture). The vegetables were planted in Green Houses and they are organic without added chemicals. The green house is glass structured house for plants which require a regulating climate conditions for plants to grow. There is a special irrigation system used in the green house where the water pipes spins all over to cover a larger area as well as there are fans that helps to regulate the temperature. The owner is looking forward to expand his business and to control of organic production, as well as products produced in compliance with national agricultural and food product quality.

All the processes used are energy efficient while maintaining requisite quality. Reduce energy intensity and emissions in all operations and the supply chain. Zero-emission manufacturing views the manufacturing system as an industrial ecosystem, utilizing minimum waste or non-reuse of by-products within the manufacturing system. The sustainability principles of the farm emphasizes a sustainable soil, reduced usage of chemical fumigants, efficient supply chain, and effective processes. The SIPOC structure of the farm is shown in Table 5.

The forward flow in the farm supply chain include procuring raw materials from upstream supplies (e.g., seeds, fertilizers, sands, minerals required for planting), reception and inspection of raw materials (quality and quantity), harvesting crops, cleaning and packaging vegetables in small packaging units that are ready for distribution to customers, and shipping: where the distributing and distributing the vegetables to downstream customers. Main backward flow includes the reuse

Table 6. CSC practices in farm supply chain.

General CSC Practice	Specific CSC techniques/actions	Applied at which SIPOC?	How is it measured?	Impact level (L, M, H)
Recycling	– Paper Packaging	Outputs	% Volume Sales	L
Re-use	– Packing Boxes	Outputs	Quantity	M
	– Average product weight	Process	Kg/bag	H
	– Water Consumption	Inputs	Liter/month	L
Reduction	– Printing paper	Inputs	Yearly Purchase	L
	– Transport	Outputs	Km/Month	H
Re-furbish and Remanufacture	– Boxes	Customers	% Returned boxes	L
Return	– Product Return	Customers	% Return/Kg	H
Rejects	– Products Reject	Customer	% Reject/Kg	H
Re-distribute	– Send reject products to Poultry	Customer	% Distributed/Kg	H

Table 7. Comparative analysis of structure and CSC practices.

Supply Chain	Furniture Store	Aluminum Plant	Farm
	Structure		
Industry	Retail	Manufacturing	Agriculture
Size and Volume	Medium	Large	Small
Suppliers	Global	Local and Global	Local
Inputs	Components and modules (subassemblies)	Raw material, Spare parts	Seeds, fertilizers, and cartons/boxes
Operations	Assembly	Production	farming
Products	Standard and customized furniture	Aluminum	Crops and vegetables
Customers	Local	Local & Global	Local
	CSC Practices		
Adopted practices	10	8	7
Specific actions	23	9	10
SIPOC focus	Inputs–Outputs	Input-Process	Inputs–Outputs
Supplier Role	Major	Major	Minor
Customer Role	Major	Minor	Major
Measures	Financial	Technical	Environmental
Impact	M-H	L-M-H	L-H

packaging materials such as cartons, the reuse of some of the returned items, and the redistribution/selling of some of the damaged or expired fruits to animal farms. The farm also strives to reduce water and electricity consumption by adopting latest technologies and recycle the used fertilizers and packaging papers. The key CSC practices in the farm are summarized in Table 6.

4 COMPARATIVE ANALYSES

Table 7 presents a comparison analysis based on the structure and the adopted CSC practices in the three supply chains. Three different industries were targeted (retail, manufacturing, and agriculture).

The furniture store represents the retail industry with a medium size, global suppliers, and local customers. The supply chain of the furniture store mainly supports assembly operations where furniture components and modules are received from suppliers and assembled and customized per customer requirements. Standard furniture is also made available to customers with short lead time. Such structure has reflected on the CSC practices within the store supply chain. As seen in Table 7, the store supply chain involves the largest number of CSC practices and actions that are mainly focused on inputs received from suppliers and products returned from customers. Both suppliers and customers are heavily involved in the adopted CSC practices. This has an overall medium to high impact especially in terms of the store financial measures.

On the other hand, the Aluminum plant represents the manufacturing industry with a large size, local/global suppliers, and local/global customers. The supply chain of the Aluminum plant is more massive as it supports diversified production operations for Aluminum and Alumina. To this end, wide range of raw materials and maintenance spare parts are received from suppliers and used in production process to meet the demand in both local and global markets. Such structure has reflected on the CSC practices within the plant supply chain. As seen in Table 7, the plant supply chain involves limited number of CSC practices and actions that are focused on inputs (excess raw material) and production by-products, with minor involvement of customers. Suppliers and production are heavily involved in the adopted CSC practices. This has a varying impact on the supply chain performance (depending on the production performance) especially in terms of the plant technical measures.

Finally, the farm represents a supply chain of agriculture industry that is small in size and dependent on local suppliers and customers. The farm receives seeds and fertilizers from local suppliers, uses them in farming, and sends produce to local markets and grocery stores. Such structure has reflected on the CSC practices within the farm supply chain. As seen in Table 7, the farm supply chain involves limited number of CSC practices and actions that are focused on inputs (minor supplier involvement in recycling seeds and fertilizers inputs) and outputs (mainly packaging and shipping), with major involvement of customers (returning packaging materials and bad produce). This can lead to either low or high impact on the supply chain performance (depending on the adopted practice and size of shipment) especially in terms of the farm environmental measures.

5 CONCLUSION

The paper has explored the structure and the practices of circular supply chains in three different industries (furniture, aluminum, and faming). To this end, a structured methodology was developed based on reviewed literature, SPOC analysis, and CSC audit. A comparative analysis was then conducted to explore the extent and impact of adopting CSC practices. Future research is encouraged to utilize quantitative methods for analyzing relationships within circular supply chains and assessing the impact of CSC practices on the overall supply chain sustainability

REFERENCES

Batista, L., Bourlakis, M., Smart, P., and Maull, R. (2018). In search of a circular supply chain archetype: a content-analysis-based literature review. Production Planning & Control, 29(6), pp. 438–451.

Bracquené, E., Dewulf, W., and Duflou, J.R. (2020). Measuring the performance of circular complex product supply chains. Resources, Conservation & Recycling, 154, 104608.

Heizer, J., Render, B., and Munson, C. (2016). *Operations Management: Sustainability and Supply Chain Management (12th Edition)*. Pearson Education.

Farooque, M., Zhang, A., Thürer, M., Qu, T., and Huisingh, D. (2019). Circular supply chain management: a definition and structured literature review. Journal of Cleaner Production, 228, pp. 882–900.

Govindan, K. and Hasanagic, M. (2018). A systematic review on drivers, barriers, and practices towards circular economy: a supply chain perspective. International Journal of Production Research, pp. 1–34.

Kalmykovaa, Y., Sadagopanb, M., and Rosado, L. (2018). Circular economy – From review of theories and practices to development of implementation tools. Resources, Conservation & Recycling, 135, pp. 190–201.

Manavalan, E. and Jayakrishna, K. (2019). An analysis on sustainable supply chain for circular economy. Procedia Manufacturing, 33, pp. 477–484.

Proceedings of the 1st International Congress on Engineering Technologies – Kiwan & Banat (Eds)

Solving fully fuzzy transportation problem with *n*-polygonal fuzzy numbers

Mahmoud Alrefaei, Noor H. Ibrahim & Marwa Tuffaha
Department of Mathematics and Statistics, Jordan University of Science and Technology (JUST), Irbid, Jordan

ABSTRACT: Recently, the interest in fuzzy transportation problems, where the variables, costs, supply and/or demands are vague and unclear, is increasing. In this paper, we propose a solution method for a general fully fuzzy transportation problem (FFTP), where fuzziness is represented by *n*-polygonal fuzzy number. The proposed method consists of two phases, Phase I is used to find a fuzzy initial basic feasible solution using a modified fuzzy version of the north-west corner method. Phase II is used to find an optimal solution starting from the initial solution found in Phase I, using an extension of the modified distribution method to deal with fuzzy numbers. Finally, a numerical example is given and solved by the proposed method, and compared with an existing method. The numerical results indicate that the proposed method gives more realistic solutions to the FFTP.

Keywords: Fuzzy Numbers; Polygonal Fuzzy Number; Fully Fuzzy Linear Programming; Fully Fuzzy Transportation Problem.

1 INTRODUCTION

The Transportation Problem (TP) is one of the most important applications of linear programming, which deals with logistics and supply-chain management for reducing costs and enhancing service. Nowadays, the increment of competition market requires more effective ways to create and deliver products to customers with the least possible cost and time while the demand is met. Transportation models are one of the good tools to model this problem. They ensure efficient movements of goods while the demand is met with least cost. In the traditional transportation model, it is assumed that a decision maker has clear values of transit costs, supply and demand. However, in real life problems, variables of transportation problems may be vague or not clear. For instance, shipping costs may vary depending on the weather, available methods of shipping or hazard of shipping. Demand and supply amounts are also vague in most cases due to the complexity in human systems. For example, demand may vary in different seasons, countries or life styles of customers. Also, supplies might get affected if new products appear by the media or many other factors. Thus, ambiguity appears in demand, supply and/or shipping costs and these vague data cause a confusion to the decision maker. To handle these ambiguities, fuzzy numbers are used and thus Fuzzy Transportation Problems (FTP) are introduced to deal with such problems. Fuzzy number are real fuzzy sets, where the concept of fuzzy sets was proposed by Zadeh [1] in 1965 to represent vagueness found in real life.

Fully Fuzzy Transportation Problem (FFTP) are transportation problems where the values of the variables, demand, availability and costs are represented as fuzzy numbers. The concept of the optimal solution for the TP with fuzzy coefficients was proposed by Chanas and Kuchta [2], who generated a crisp optimal solution by converting the FTP to a bi-criteria TP with crisp objective

 DOI 10.1201/9781003178255-25

function. Kumar et al. [3] presented an algorithm for solving FFTP with trapezoidal fuzzy numbers. They formulated it as a fuzzy linear programming problem and converted it into a crisp (unfuzzy) linear programming problem. Kaur and Kumar [4] presented a method for solving FFTP with generalized trapezoidal fuzzy numbers. They generated a fuzzy basic feasible solution (FBFS) using a generalization of three known methods for solving TP which are the north-west corner, vogel's approximation and least cost methods. Basirzadeh [5] solved the FFTP with triangular or trapezoidal fuzzy numbers using a ranking method so the problem becomes a crisp TP and solve it using the usual methods. Shanmugasundari and Ganesan [6] proposed a method for solving a FFTP without converting it into a crisp one. They used the triangular fuzzy numbers to represent the fuzziness in the problem.

Researchers vary in their choice of the fuzzy number type used to represent fuzziness in the FTP. Most of them prefer triangular or trapezoidal fuzzy numbers due to the linearity in their membership functions. In this paper, the n-Polygonal Fuzzy Number (n-PFN) is used to represent the fuzziness in a more general type of FFTP. This type of fuzzy numbers was shown by Tuffaha and Alrefaei [7] to generalize some of the mostly used types of fuzzy numbers. Moreover, the authors proposed convenient arithmetic operations on the n-PFN that satisfy the most important properties for binary operations [7, 8], in addition to preserving the ranking values. Later, Tuffaha and Alrefaei [9] applied this type of fuzzy numbers with its strong binary operations to fully fuzzy linear programming problems and proposed a general fuzzy version of the simplex method, which leads to convenient solutions. Thus, in this paper, we represent the fuzziness in FFTP by n-PFN and propose a solution method that gives more realistic solutions.

The paper is organized as follows: First, a review of the n-PFN and its needed tools is given in Section 2. Then, in Section 3, the mathematical formulation of the FFTP with n-PFN's is presented, and the solution algorithm is proposed. The algorithm has two phases. First, a fuzzy version of the North-West Corner Method is proposed to find an initial basic feasible solution for the problem. Then, for Phase 2, an algorithm to proceed starting with this solution towards reaching the optimal solution is introduced, that is a generalization of the Modified Distribution Method to a fuzzy version of it. A numerical example is then given in Section 4, and some concluding remarks are presented in Section 5.

2 THE N-POLYGONAL FUZZY NUMBER

In this section, we present the definition of the n-PFN as adopted by Tuffaha and Alrefaei [7, 8] with the ranking function and the binary operations proposed by the authors.

Definition 1. An n-Polygonal Fuzzy Number (n-PFN), $\tilde{A}$, is a fuzzy number with a membership function of the type:

$$f_{\tilde{A}}(x) = \begin{cases} \frac{1}{n}[\frac{x-p_i}{p_{i+1}-p_i}] + \frac{i}{n} & ; p_i \leq x \leq p_{i+1}, \quad i = 0, .., n-1 \\ 1 & ; p_n \leq x \leq q_0 \\ \frac{-1}{n}[\frac{x-q_i}{q_{i+1}-q_i}] + \frac{n-i}{n} & ; q_i \leq x \leq q_{i+1}, \quad i = 0, .., n-1 \\ 0 & \quad otherwise \end{cases}$$

which can be represented by its knots: $(p_0, p_1, .., p_n; q_0, q_1, .., q_n)$.

An example of a flat 4-PFN is given in Figure 1.

Remark 1. Note that the triangular and the trapezoidal fuzzy numbers, which are the most common types of fuzzy numbers in the literature, are special cases of the n-PFN with $n = 1$.

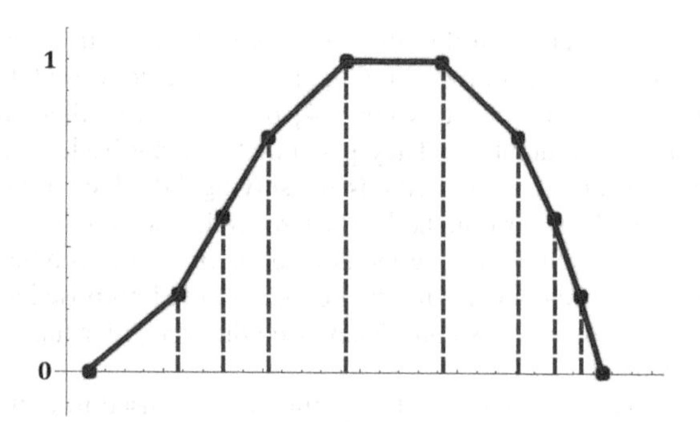

Figure 1. Example of 4-PFN.

Remark 2. Any crisp (unfuzzy) real number c can be represented in the n-PFN form as $c = (c, c, .., c; c, c, .., c)$.

Definition 2. Let $\tilde{P} = (p_0, p_1, .., p_n; q_0, q_1, .., q_n)$ be a n-PFN. Then, we will use the following ranking function:

$$\mathfrak{R}(\tilde{P}) = \frac{1}{4n}[p_0 + 2p_1 + 2p_2 + ... + 2p_{n-1} + p_n + q_0 + 2q_1 + 2q_2 + ... + 2q_{n-1} + q_n]$$

Moreover, two n-PFN's, $\tilde{P}$ and $\tilde{Q}$, are said to be equivalent if $\mathfrak{R}(\tilde{P}) = \mathfrak{R}(\tilde{Q})$. Furthermore, the minimum of two n-PFN's is the one having the minimum ranking value.

Definition 3. Let $\tilde{P} = (p_0, p_1, .., p_n; q_0, q_1, .., q_n)$ and $\tilde{Q} = (r_0, r_1, .., r_n; s_0, s_1, .., s_n)$ be two n-PFN's. Then, the binary operations are defined as follows:

1. $\tilde{P} \oplus \tilde{Q} = (p_0 + r_0, p_1 + r_1, .., p_n + r_n; q_0 + s_0, q_1 + s_1, .., q_n + s_n)$
2. $\tilde{P} \ominus \tilde{Q} = (p_0 - s_n, p_1 - s_{n-1}, .., p_n - s_0; q_0 - r_n, q_1 - r_{n-1}, .., q_n - r_0)$
3. $\tilde{P} \otimes \tilde{Q} = (t_0, t_1, .., t_n; u_0, u_1, .., u_n)$, where:

$$u_n = \frac{1}{4n}[I + \sum_{i=1}^{n}(2i-1)X_i + 2nX_{n+1} + \sum_{i=1}^{n}(2(n+i)-1)X_{n+1+i}]$$

$$u_{i-1} = u_i - X_{n+1+i}, \text{ for } i = n, n-1, .., 1$$

$$t_n = u_0 - X_{n+1}$$

$$t_{i-1} = t_i - X_i, \text{ for } i = n, n-1, .., 1$$

$$\text{and, } I = \frac{1}{4n}[(p_0 + 2p_1 + . + 2p_{n-1} + p_n + q_0 + 2q_1 + . + 2q_{n-1} + q_n)*$$
$$(r_0 + 2r_1 + . + 2r_{n-1} + r_n + s_0 + 2s_1 + . + 2s_{n-1} + s_n)]$$

$$X_i = (p_i - p_{i-1}) + (r_i - r_{i-1}), \text{ for } i = n, n-1, .., 1$$

$$X_{n+1} = (q_0 - p_n) + (s_0 - r_n)$$

$$X_{n+1+i} = (q_i - q_{i-1}) + (s_i - s_{i-1}), \text{ for } i = n, n-1, .., 1$$

3 FULLY FUZZY TRANSPORTATION PROBLEM (FFTP)

3.1 *Mathematical formulation of the FFTP*

Consider the following FFTP: Assume there are m supply sources, where the i-th source has a fuzzy number $\tilde{a}_i$ of units of some product. Suppose also that there are m destination points, where the j-th destination has a fuzzy demand of $\tilde{b}_j$ units of the product that needs to be shipped to these points from the supply sources with a minimum cost. Let c_{ij} be the unit fuzzy transportation cost from the i-th source to the j-th destination. The following table summarizes the problem:

	1	2	...	n	Supply
1	$\tilde{c}_{11}$	$\tilde{c}_{12}$	...	$\tilde{c}_{1n}$	$\tilde{a}_1$
2	$\tilde{c}_{21}$	$\tilde{c}_{22}$	...	$\tilde{c}_{2n}$	$\tilde{a}_2$
. . .	...	...	...	...	. . .
m	$\tilde{c}_{m1}$	$\tilde{c}_{m2}$	...	$\tilde{c}_{mn}$	$\tilde{a}_m$
Demand	$\tilde{b}_1$	$\tilde{b}_2$	...	$\tilde{b}_n$	

If $\tilde{\mathbf{x}}_{ij}$ is the fuzzy number of units that should be transported from the i-th source to the j-th destination, then the problem can be formulated as follows:

$$\min\ \tilde{\mathbf{z}} = \sum_{i=1}^{m}\sum_{j=1}^{n} \tilde{\mathbf{c}}_{ij} \otimes \tilde{\mathbf{x}}_{ij}$$

$$\text{subject to } \sum_{j=1}^{n} \tilde{\mathbf{x}}_{ij} = \tilde{\mathbf{a}}_i, i = 1,2,...,m$$

$$\sum_{i=1}^{m} \tilde{\mathbf{x}}_{ij} = \tilde{\mathbf{b}}_j, i = 1,2,...,n$$

$$\sum_{i=1}^{m} \tilde{\mathbf{a}}_i = \sum_{j=1}^{n} \tilde{\mathbf{b}}_j$$

$$\tilde{\mathbf{x}}_{ij} \succeq \tilde{\mathbf{0}}$$

To solve this problem, we first find an initial basic feasible solution using a modified version of the North-West Corner method, then we use a fuzzy version of the Modified Distribution method to solve the problem.

3.2 *Fuzzy Version of the North-West Corner Method*

A Fuzzy Version of North-West Corner Method (FNWCM) is used to find an initial fuzzy basic feasible solution (IFBFS) of the FFTP with n-PFN's; it proceeds as follows:

Step 1: Check if the FFTP is balanced or not, i.e., $\sum_{i=1}^{m} \tilde{\mathbf{a}}_i \approx \sum_{i=1}^{n} \tilde{\mathbf{b}}_j$ holds, If not, go to step 2, otherwise go to step 3.

Step 2: If the FFTP is unbalanced, then we will have the following two cases:

Case I: If $\sum_{i=1}^{m} \tilde{\mathbf{a}}_i \succ \sum_{j=1}^{n} \tilde{\mathbf{b}}_j$ then add a dummy column having all its costs as n-PFN's equivalent to zero. The value of the fuzzy demand at this dummy destination is $\sum_{i=1}^{m} \tilde{\mathbf{a}}_i \ominus \sum_{j=1}^{n} \tilde{\mathbf{b}}_j$. Proceed to step 3.

Case II: If $\sum_{i=1}^{m} \tilde{\mathbf{a}}_i \prec \sum_{i=1}^{n} \tilde{\mathbf{b}}_j$ then add a dummy row having all its costs as n-PFN's equivalent to zero. The value of the fuzzy supply at this dummy source is $\sum_{j=1}^{n} \tilde{\mathbf{b}}_j \ominus \sum_{i=1}^{m} \tilde{\mathbf{a}}_i$. Proceed to step 3.
Step 3: Determine the north-west corner (NWC) of the fuzzy transportation table (The upper left corner) and select the minimum of $\tilde{a}_i, \tilde{b}_j$. Then the following three cases are introduced:
Case I: If $min\{\tilde{a}_i, \tilde{b}_j\} = \tilde{a}_i$, then assign $\tilde{x}_{ij} = \tilde{a}_i$ in NWC. Delete the i-th row in the table to get a new transportation table and then find the new value of $\tilde{b}_j$ by putting $\tilde{b}_j = \tilde{b}_j \ominus \tilde{a}_i$. Proceed to step 4.
Case II: If $min\{\tilde{a}_i, \tilde{b}_j\} = \tilde{b}_j$, then assign $\tilde{x}_{ij} = \tilde{b}_j$ in NWC. Delete the j-th column in the table to get a new transportation table and then find the new value of $\tilde{a}_i$ by putting $\tilde{a}_i = \tilde{a}_i \ominus \tilde{b}_j$. Proceed to step 4.
Case III: If $\tilde{a}_i = \tilde{b}_j$, then choose case 1 or 2 and then proceed to step 4.
Step 4: Repeat step 3 until one cell is left in the fuzzy transportation table.
Step 5: Assign the value of the last cell in the table to meet the supply and demand requirements.
Step 6: The resulting fuzzy transportation table is an IFBFS and the initial fuzzy transportation cost $\tilde{z}$ is obtained by $\tilde{z} = \sum_{i=1}^{m} \sum_{j=1}^{n} \tilde{c}_{ij} \otimes \tilde{x}_{ij}$.

3.3 *Fuzzy Version of the Modified Distribution Method*

After finding an initial basic feasible solution, the Fuzzy Version of Modified Distribution Method (FMODI) method is used to find a fuzzy optimal solution to the FFTP with n-PFN's. The procedure is given as follows:
Step 0: Given an IFBFS generated by FNWCM.
Step 1: Locate the fuzzy dual variable $\tilde{u}_i$ for each row i, $i = 1, 2, 3, ..., m$ and $\tilde{v}_j$ for each column j, $j = 1, 2, 3, ..., n$.
Step 2: Find the values of $\tilde{u}_i, \tilde{v}_j$ using the formula $\tilde{c}_{ij} = \tilde{u}_i + \tilde{v}_j$ for all occupied cells (basic cells). Assign $\tilde{u}_1 = \tilde{0}$ or any one of $\tilde{u}_i, \tilde{v}_j$.
Step 3: For the rest unoccupied cells (non basic cells) find the fuzzy value of $\tilde{p}_{ij}$ using $\tilde{p}_{ij} = \tilde{u}_i \oplus \tilde{v}_j \ominus \tilde{c}_{ij}$. Note that p_{ij} represent row zero in simplex tableau, then we have two cases may arise:
Case I: If all $\tilde{p}_{ij} \preceq \tilde{0}$, then the current IFBFS is optimal.
Case II: If there exist at least one of $\tilde{p}_{ij} \succ \tilde{0}$, then the current IFBFS is not optimal and proceed to step 4.
Step 4: Determine an unoccupied cell with most positive ranked value of $\tilde{p}_{ij}$.
Step 5: Draw a closed loop contains horizontal and vertical lines for the unoccupied cell starting and ending with the cell obtained from step 4. The loop must be located in the only occupied cells except the starting point.
Step 6: Set an alternate plus and minus signs on the corner points of the drawn loop starting with a plus sign and then the opposite and so on.
Step 7: Choose the smallest value $\tilde{\theta}$ with a negative sign on the closed path. Add this value to all the corner cells of the closed loop marked with plus signs, and subtract it from those cells marked with minus signs. The basic cell whose allocation being reduced to zero, leaves the table, which means the cell become non basic, and an unoccupied cell becomes an occupied cell. Which leads to better IFBFS.
Step 8: Repeat step1 to step 7 until all $\tilde{p}_{ij} \preceq \tilde{0}$.
Step 9: The current solution is optimal and the fuzzy optimal transportation cost is $\sum_{i=1}^{m} \sum_{j=1}^{n} \tilde{c}_{ij} \otimes \tilde{x}_{ij}$.

4 NUMERICAL EXAMPLE

Example 1. Consider a company that needs to transport goods from 2 source plants to 2 distribution centers to meet their demands. The transportation cost between the plants and the distribution centers are not given precisely, so they are modeled as 2−polygonal fuzzy numbers (i.e., they are a trapezoidal fuzzy number) as shown in the table below. This problem is considered by Kumar and Kaur [3].

	D1	D2	Supply
S1	(0, 1, 3, 4)	(2, 3, 5, 6)	(10, 20, 30, 40)
S2	(1, 3, 5, 7)	(2, 6, 7, 9)	(0, 4, 8, 12)
Demand	(6, 8, 10, 20)	(10, 16, 18, 20)	

We use the proposed method to solve this problem and compare the results with that of Kumar and Kaur.

Step 1: Supply quantity $\sum_{i=1}^{2} \tilde{S}_i = (10, 24, 38, 52)$ and demand quantity $\sum_{j=1}^{2} \tilde{D}_j = (16, 24, 28, 40)$. So $\sum_{i=1}^{2} \tilde{S}_i \not\approx \sum_{j=1}^{2} \tilde{D}_j$. Which means the problem is unbalanced fuzzy transportation problem.

Step 2: Since $\sum_{i=1}^{m} \tilde{\mathbf{a}}_i \succ \sum_{j=1}^{n} \tilde{\mathbf{b}}_j$, then convert the TP into balanced one by adding a dummy column having all its costs as fuzzy zero. The assigned value at this dummy destination is $\sum_{i=1}^{m} \tilde{\mathbf{a}}_i \ominus \sum_{j=1}^{n} \tilde{\mathbf{b}}_j$.

	D1	D2	D3	Supply/RV
S1	(0, 1, 3, 4)	(2, 3, 5, 6)	(0, 0, 0, 0)	(10, 20, 30, 40)/**25**
S2	(1, 3, 5, 7)	(2, 6, 7, 9)	(0, 0, 0, 0)	(0, 4, 8, 12)/**6**
Demand/RV	(6, 8, 10, 20)/**11**	(10, 16, 18, 20)/**16**	(−6, 0, 10, 12)/**4**	

Now, we use the fuzzy north-west corner rule to obtain an initial basic feasible solution.

Step 3: Select the north-west corner cell (0, 1, 3, 4), then based on ranking values we have $\tilde{x}_{11} = min\{(10, 20, 30, 40), (6, 8, 10, 20)\} = (6, 8, 10, 20)$.

Since $(6, 8, 10, 20) \prec (10, 20, 30, 40)$, move to the second column and find

$$\tilde{x}_{12} = min\{(10, 20, 30, 40) \ominus (6, 8, 10, 20), (10, 16, 18, 20)\} = (-10, 10, 22, 34).$$

Repeat step 3 moving down towards the lower right corner cell in the table until the initial fuzzy basic feasible solution is reached. The following table represents the IFBFS:

	D1	D2	D3
S1	(6, 8, 10, 20) / (0, 1, 3, 4)	(−10, 10, 22, 34) / (2, 3, 5, 6)	0
S2	(1, 3, 5, 7)	(−24, −6, 8, 30) / (2, 6, 7, 9)	(−6, 0, 10, 12) / 0

Next, we apply the FMODI method to check if the current initial basic feasible solution is optimal or not.

	$\tilde{v}_1 = (0, 1, 3, 4)$	$\tilde{v}_2 = (2, 3, 5, 6)$	$\tilde{v}_3 = (-7, -4, -1, 4)$
$\tilde{u}_1 = (0, 0, 0, 0)$	(6, 8, 10, 20) / (0, 1, 3, 4)	(−10, 10, 22, 34) / (2, 3, 5, 6)	0
$\tilde{u}_2 = (-4, 1, 4, 7)$	(1, 3, 5, 7)	(−24, −6, 8, 30) / (2, 6, 7, 9)	(−6, 0, 10, 12) / 0

Table 1. Balanced crisp transportation problem RTP.

	D1	D2	D3	Supply/RV
S1	2	4	0	25
S2	4	6	0	6
Demand/RV	11	16	4	

Table 2. Solution of the RTP.

(a) Initial feasible solution for the RTP.

(b) MODI method for the RTP.

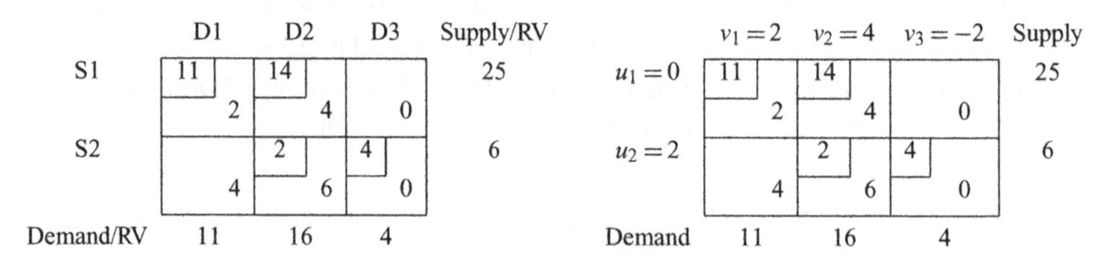

(a) Allocation shown first in each cell, followed by the cost.

	D1	D2	D3	Supply/RV
S1	11 / 2	14 / 4	0	25
S2	4	2 / 6	4 / 0	6
Demand/RV	11	16	4	

(b)

	$v_1=2$	$v_2=4$	$v_3=-2$	Supply
$u_1=0$	11 / 2	14 / 4	0	25
$u_2=2$	4	2 / 6	4 / 0	6
Demand	11	16	4	

Now, we find the values of $\tilde{u}_i, \tilde{v}_j$ for all occupied cells (i,j) using the formula $\tilde{c}_{ij} = \tilde{u}_i \oplus \tilde{v}_j$ and then compute the values of $\tilde{p}_{ij}$ for each unoccupied cell (i,j), using the relationship $\tilde{p}_{ij} = \tilde{u}_i \oplus \tilde{v}_j \ominus \tilde{c}_{ij}$. We find that $\tilde{p}_{13} = (-7,-4,-1,1), \tilde{p}_{21} = (-11,-3,4,10)$ with ranking values $\Re(\tilde{p}_{13}) = -2$, $\Re(\tilde{p}_{21}) = 0$, which means the optimality condition is satisfied.

The optimal fuzzy transportation cost is

$(6,8,10,20) \otimes (0,1,3,4) \oplus (-10,10,22,34) \otimes (2,3,5,6) \oplus (-24,-6,8,30) \otimes (2,6,7,9) \oplus (-6,0,10,12) \otimes (0,0,0,0) = (17,69,112,162)$ with ranking value equals to 90.

In real life, this means that the minimum cost is between 17 and 162. Since this is a trapezoidal fuzzy number, there is a 100% satisfaction that the minimum cost belongs to the interval [69, 112] because the membership function value is 1 for all x in this interval while this satisfaction is decreased linearly between 112 and 162 and between 69 and 17 to reach zero.

Now, we solve the Ranked Transportation Problem (RTP), which results from taking the ranking value of all the fuzzy numbers in the FFTP. The transportation cost for this problem is given in Table 1.

Table 2(a) includes the IBFS for the RTP, and Since $p_{13} = u_1 + v_3 - c_{13} = -2$ and $p_{21} u_2 + v_1 - c_{21} = 0$, then the current crisp initial basic feasible solution (Table 2(b)) is optimal. The optimal crisp transportation cost is $11 \times 2 + 14 \times 4 + 2 \times 6 + 4 \times 0 = 90$ which is the same as the ranking of the optimal solution obtained by the proposed method.

Table 3 shows a comparison between the results obtained by Kumar and Kaur's method and the proposed method. We note that, the ranking value of FTP solution equals the solution of the RTP, while this property does not hold for Kumar's solution method. Thus, the proposed method gives a more realistic optimal solution.

Table 3. Comparison between Kumar and Kaur's method and the proposed method.

Method	Solution	Ranking of Solution
Kumar	(8,38,90,166)	75.5
Proposed Method	(17,69,112,162)	90
RLP		90

5 CONCLUSION

In this paper, we have presented a method for solving the fully fuzzy transportation problem where fuzziness is represented by the n-polygonal fuzzy numbers which are a generalization of the well known fuzzy numbers such as triangular and trapezoidal fuzzy numbers. The method first finds an initial basic feasible solution using the fuzzy version of the north-west corner method, then an extension of the modified distribution method to deal with fuzzy numbers is used to obtain an optimal solution. The proposed method is implemented on a simple example of FTP and the solution is compared with the solution of onother method proposed by Kumar and Kaur's [3]. The results show that the solution obtained by the proposed method is more realistic, since it coincides with the solution of the corresponding ranking transportation problem, while the other one does not.

REFERENCES

[1] L. A. Zadeh, "Fuzzy sets," *Information and control*, vol. 8, no. 3, pp. 338–353, 1965.

[2] S. Chanas and D. Kuchta, "A concept of the optimal solution of the transportation problem with fuzzy cost coefficients," *Fuzzy sets and Systems*, vol. 82, no. 3, pp. 299–305, 1996.

[3] A. Kumar, A. Kaur, and A. Gupta, "Fuzzy linear programming approach for solving fuzzy transportation problems with transshipment," *Journal of Mathematical Modelling and Algorithms*, vol. 10, no. 2, pp. 163–180, 2011.

[4] A. Kaur and A. Kumar, "A new method for solving fuzzy transportation problems using ranking function," *Applied Mathematical Modelling*, vol. 35, no. 12, pp. 5652–5661, 2011.

[5] H. Basirzadeh, "An approach for solving fuzzy transportation problem," *Applied Mathematical Sciences*, vol. 5, no. 32, pp. 1549–1566, 2011.

[6] M. Shanmugasundari and K. Ganesan, "A novel approach for the fuzzy optimal solution of fuzzy transportation problem," *Transportation*, vol. 3, no. 1, pp. 1416–1424, 2013.

[7] M. Tuffaha and M. Alrefaei, "Arithmetic operations on piecewise linear fuzzy number," *AIP Conference Proceedings*, vol. 1991, no. 1, p. 020024, 2018.

[8] M. Tuffaha and M. Alrefaei, "Properties of binary operations of n-polygonal fuzzy numbers," *Advances in Intelligent Systems and Computing*, vol. 1111 AISC, pp. 256–265, 2020.

[9] M. Tuffaha and M. Alrefaei, "General simplex method for fully fuzzy linear programming with the piecewise linear fuzzy number," *To appear in: Nonlinear Dynamics and Systems Theory*, 2020.

Proceedings of the 1st International Congress on Engineering Technologies – Kiwan & Banat (Eds)
© 2021 Taylor & Francis Group, London, ISBN 978-0-367-77630-5

Rooftop garden in Amman residential buildings–sustainability and utilization

Naser Mughrabi, Mayyadah Fahmi Hussein & Naila Hussien Alhyari
Department of Interior Design and Department of Architecture, Faculty of Architecture and Design, University of Petra, Jordan

ABSTRACT: Adopting green roof systems in residential building is becoming necessary because of current environmental, social, and economical challenges. Although green rooftop gardens are a longstanding practice in developed countries, in Amman they are still individual trials that are far from adhering to government policies and common design practices.

The purpose of this study is to propose some indicators for designing rooftop gardens, taking into consideration the site and the age of the building to illustrate the economic, social and environmental benefits of rooftop gardens in Amman as a step toward a sustainable Amman, also to contribute to a better understanding of the potential role of green roof systems in effective planning.

This paper presents a conceptual framework to help interior designers in the Jordan to adopt green roofs in their environmental policies. To present this framework, first, researchers studied literatures that adopted green roof systems and practices in many countries, then proposed a conceptual framework for adopting green roof systems in Jordan. Second, they chose Amman, Jordan to demonstrate the applicability of this framework at city level while considering the national and local context. This demonstration provides a novel perspective for the benefits of green roof systems in energy savings and water management in Jordan.

By using descriptive analytical methodology, qualitative surveys, data collection and analysis approaches, the researchers gave a questionnaire to the residential building inhabitants that were involved with the sustainability of rooftop gardens' environmental, economic and social benefits as a step in enhancing awareness of rooftop gardens in relation to the sustainability of Amman city. The researchers collected data that includes previous studies and documented research, then conducted a systematic analysis for the collected previous studies' methodological approaches and found results. After analyzing the entire collected data, the researchers will introduce applicable indicators in Amman rooftop gardens.

The research outcomes are expected to develop indicators for the utilization of rooftop gardens in Amman, the capital city of Jordan, which is in response to principles of social, environmental and economic sustainability.

Keywords: Sustainability, Roof Gardens, Amman Utilization

1 INTRODUCTION

Rooftops are one of our cities' greatest untapped resources. They account for hundreds of acres of empty, under-utilized space, contributing to problems like the "heat island effect" and increased storm-water run-off. But rooftops could easily be turned into valuable green spaces by creating green roofs of wildflowers, trees and shrubs or vegetables on schools, apartments, homes and places of work throughout the city.

 DOI 10.1201/9781003178255-26

Expanding the green space in Amman is not a new idea; it has always been on the Amman Municipality's agenda. However, most initiatives started but were not sustained due to limited leadership, funds, awareness, and commitment, among other reasons. People were initially interested but the cost was the main limiting factor. (Whitman, 2013).

Green roofs can be divided into two types: the vegetation-covered or "inaccessible roof" where the soil and plants form another layer of the roofing system, and the rooftop garden or "accessible" roof that can become an outdoor space for wildlife. This research will introduce the ideal indicators on how to use and design the accessible type and offer practical and attainable solutions for Jordanian designers in working towards sustainable urban development. Modern green roof concepts started at the beginning of the 20th century in Germany, where vegetation was installed on roofs to mitigate the damaging physical effects of solar radiation on the roof structure. Early green roofs also functioned as fire-retardant structures. There are now several competing types of extensive green-roof systems, which provide similar functions but are composed of different materials (Oberndorfer et al., 2007).

1.1 *Benefits of green roofs (social, environmental and economic)*

A – Amenity & aesthetic social benefits include: leisure and open space, visual aesthetic value, health and therapeutic value and food production.
B – Environmental benefits include: ecological and wildlife value, water management (i.e. minimization of storm water run-off and support for rainwater collection systems), and air quality (i.e. it helps to improve air quality and absorption of carbon dioxide, sound absorption and mitigating the urban heat island effect).
C – Economic benefits include increased roof life, building insulation, energy efficiency, green building assessment and public relations. Reducing heating and cooling costs by providing a layer of insulation on buildings.

2 PROBLEM STATEMENTS

Life in Amman is becoming full of concrete, traffic and noise. The combination of these can, over time, weigh heavily on a person's physical and emotional well-being. Even without realizing it, we need open green areas filled with life and color. In this research, researchers found many problems, such as:

1. Limited technical research about green roof gardens in Amman.
2. No green roof policy exists at national or municipal levels.
3. Green roofs are not integrated into other sectors' policies (e.g., water, energy. etc.)
4. Green roofs have been integrated into the Green Building Assessment through the Jordan Green Building Council (Voluntary).
5. Green space in Amman does not exceed 2.5% of the total area, which is 1680 km2.
6. Non-exploitation of green top roof gardens for social activities in residential buildings.

3 AIMS OF THE RESEARCH

1. To conduct a quick review of the latest concepts and technology on green roofs and recommend guidelines adapted to suit local applications in Amman to promote public understanding and awareness.
2. To find indicators that help designers to create rooftop gardens in response to principles of sustainability, whether social, environmental or economic in Amman.

4 HYPOTHESES

1. Green roofs have several benefits for energy, water and pollution management.
2. Green roofs must consider specific climatic conditions.
3. Designing a green rooftop garden in Amman is an ideal response to principles of sustainability, whether social, environmental or economic.

5 METHODOLOGY

This research was based on the descriptive analytical method and follows a qualitative survey at Amman residential buildings elated to the research objective (finding indicators that help designers create a green rooftop gardens). Methods included a literary review of related research, data collected from books, articles, scientific journal web sites, and site visits to the Amman residential building,

The researchers divided the work plan in to two types: the first one involves sustainability (environmental, economic and social) as a first step in improving the sustainability of Amman city by reviewing literature and finding these indicators. The focus of the literature review was placed on qualitative analysis of the three main pillars of sustainability with respect to green roofs, as well as an analysis of case studies outlining design considerations of rooftop farming operations. A standardized list of questions was developed and data was recorded during the conversations, which each lasted approximately 30 minutes. Closed interviews with Amman inhabitants were used in the second type using a Google online questionnaire.

6 LITERATURE REVIEW

In order to better understand how rooftop gardens can benefit environmental, economic, and social sustainability, researchers performed a meta-analysis of existing literature on rooftop gardens and urban agriculture systems more generally. The results of this literature review will provide a strong argument for establishing a rooftop garden in Amman.

6.1 *Design guideline for sustainable green roof system*

In the study of Haziq Zulhabri (2011), "design guide line for sustainable green roof system," they identified performance benefits of green roof systems as well as the obstacles of green roof systems. They adopted quantitative and qualitative approaches including literature search and review and surveys among two target populations: architects and developers. Case studies were also important in their research. Methodology was designed to obtain design guide lines for green roof systems (2011, September). H. A., & Ismail, Z.

6.2 *Green roofs: A critical review on the role of components, benefits, limitations and trends*

Green roofs have been proposed as an efficient and practical tool to combat urbanization in many countries. This review paper focuses on various benefits associated with green roofs and research efforts made to promote green roofs. Through a systematic review of literature, this review also emphasizes the knowledge gaps that exist in green roof technology and highlight the need for local research to install green roofs in developing and under-developed countries. Considering that growth substrate, vegetation and drainage layers determine the success of green roofs, efforts were made to consolidate desirable characteristics for each of these components and suggests methodology to construct practical green roofs. This critical review also explores limitations associated with green roofs and recommend strategies for overcoming these limitations. Apart from stand-alone roofs, there is a huge scope for hybrid green roof systems with other established techniques which are presented and discussed. Recommendations for future study are also provided. (Vijayaraghavan, K. (2016)).

6.2.1 *Benefits of green roofs: A systematic review of the evidence for three ecosystem services*

Green roofs are often claimed to provide a range of environmental, economic and social benefits, or 'ecosystem services.' This study seeks to evaluate the documentation relating to three selected green roof ecosystem services: reduction of the urban heat island effect, reduction of urban air pollution, and reduction of building energy consumption. Analysis of the identified studies suggests that some parameters are of key importance for the effectiveness but further research is needed to clarify the complex relation between ecosystem service effectiveness and the parameters influencing it. (Francis, L.F.M., & Jensen, M.B. (2017)).

6.2.2 *Water, energy, and rooftops: Integrating green roof systems into building policies in the arab region*

New research conducted in 2017 by Alzubi & Mansour concerned with integrating green roof systems into building policies in the Arab region presented a frame work for the governments in the Arab region to adopt green roofs in their environmental policies; they studied the current international policies in t4 his field then they proposed a frame work to be adopted in the Arab region. They took Cairo, Egypt, and Amman, Jordan as case studies to illustrate their proposal. The below chart was included in their research show green roofs contribution to sustainability. See Figure 1.

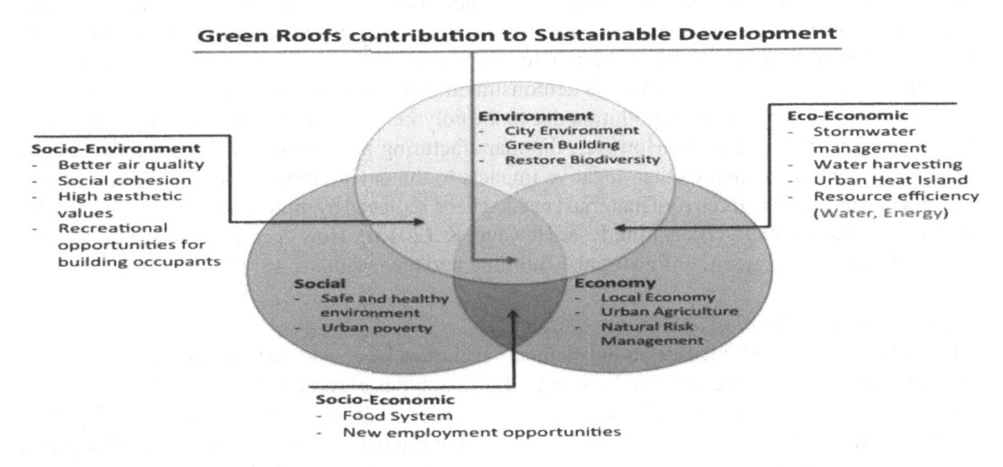

Figure 1. Green roofs contribution to sustainability Development, Al-Zu'bi, 2017.

6.2.3 *Green roof benefits, opportunities and challenges – A review*

This review paper includes the history of the green roof, green roof components and the multiple benefits (environmental, social and economic) associated with the green roof technology. This paper also emphasizes how the green roof works in different areas, their performance in reducing the storm water and energy costs, improving air and ecological performance. The benefits of green roof show that it plays an important role in making cities safe, sustainable and resilient to climate change. This paper also highlights the research challenges and research gap of the green roof. (Shafique, M., Kim, R., & Rafiq, M. 2018)).

6.3 *Green roofs contribution to more environmental sustainability*

6.3.1 *A simplified model for modular green roof hydrologic analyses and design*

Green roofs can mitigate urban rooftop storm water runoff. However, the lack of accurate, physical performance assessments and design models has hindered their wide application. Most hydrologic or hydraulic models have no direct connection to the physical properties of green roof components such as media type/depth, drainage depth, etc. In an effort to assist design engineers, a simplified yet effective physical model was developed and calibrated with pilot data in order to provide green roof hydrologic performance curves to guide design. Study results indicate that LEED criteria should be modified to require specific designer-controlled parameters of storage and media depth for the design storm to ensure desired performance. (Li, Y., & Babcock, R. (2016)).

6.3.2 *State-of-the-art analysis of the environmental benefits of green roofs*

This paper shows how green roofs may contribute to more sustainable buildings and cities. However, an efficient integration of green roofs needs to take into account both the specific climatic conditions and the characteristics of the buildings. Economic considerations related to the life-cycle cost of green roofs are presented together with policies promoting green roofs worldwide. Findings indicate the undeniable environmental benefits of green roofs and their economic feasibility. Likewise, new policies for promoting green roofs show the necessity for incentivizing programs. Future research lines are recommended and the necessity of cross-disciplinary studies is stressed. (Berardi, U., GhaffarianHoseini, A., & GhaffarianHoseini, A. 2014)).

6.3.3 *How "green" are the green roofs? Lifecycle analysis of green roof materials*

Basic layers, from bottom to top, of green roof systems usually consists of a root barrier, drainage, filter, growing medium, and vegetation layer. New technology enabled the use of low density polyethylene and polypropylene (polymers) materials with reduced weight on green roofs. This paper evaluates the environmental benefits of green roofs by comparing emissions of NO2, SO2, O3 and PM10 in green roof material manufacturing process, such as polymers, with the green roof's pollution removal capacity. The analysis demonstrated that green roofs are sustainable products on a long-term basis. In general, air pollution due to the polymer production processes can be balanced by green roofs in 13–32 years. However, the manufacturing process of low density polyethylene and polypropylene has many other negative impacts to the environment than air pollution. It was evident that the current green roof materials needed to be replaced by more environmentally friendly and sustainable products. (Bianchini, F., & Hewage, K. (2012)). How "green" are the green roofs? Lifecycle analysis of green roof materials. Building and environment, 48, 57–65.

6.3.4 *Elements of rooftop agriculture design*

This chapter of the book focuses on the elements that must be considered when designing rooftop gardens and integrating them within buildings, different types of rooftop gardens and how they can be integrated within existing and new buildings in order to enhance their environmental performance, better connect with their users and improve the urban environment are presented together with a description of necessary factors for implementation. These include: techniques and technologies for cultivation (i.e. simple planters, green roofs and hydroponics), the necessary structural loadbearing capacity of the host building and protection from wind. The chapter also gives an overview of existing innovative and experimental projects of rooftop gardens, ranging from those that require little to high investment. (Caputo, S., Iglesias, P., & Rumble, H. (2017)).

6.3.5 *Urban agriculture as a climate change and disaster risk reduction strategy*

Rooftop agriculture is the production of fresh vegetables, herbs, fruits, edible flowers and possibly some small animals on rooftops for local consumption. Productive green roofs combine food production with ecological benefits, such as reduced rainwater run-off, temperature benefits such as potential reduction of heating and cooling requirements (resulting in reduced emissions), biodiversity, improved aesthetic value and air quality. (Dubbeling, M. (2014)).

6.4 *Green roofs contribution to more economic sustainability*

6.4.1 *Study on green roof application in Hong Kong*

The study focused on the environmental, aesthetic, and economic benefits and constraints of green roofs. The latest ideas and technologies on green roofs were reviewed in order to reach for recommended guide lines to suit local applications in Hong Kong and to promote public understanding and awareness. Case studies on green roofs in Hong Kong and a plant selection matrix for green roofs in Hong Kong were presented at the end of the study (Townshend, D., & Duggie, A. (2007)).

6.4.2 *Evaluation of vegetable production on extensive green roofs*

Three growing systems—a green roof, raised green roof platforms, and in-ground—were evaluated for vegetable and herb production over three growing seasons (2009–2011). Tomatoes (Solanum lycopersicum), green beans (Phaseolus vulgaris), cucumbers (Cucumis sativus), peppers (Capsicum annuum), basil (Ocimum basilicum), and chives (Allium schoenoprasum) were studied because of their common use in home gardens. All plants, except pepper, survived and produced biomass in all growing systems and yielded crops large enough for analysis in 2009 and 2010. Overall yields and biomass of basil were higher and of better quality in-ground during 2009, the only year irrigation was applied, and similar on the roof and platforms. Variability in success was partially due to annual weather variation with the greatest impact on cucumbers. Yields of chive, a perennial crop, were not affected by the growing system after the first year. Results suggest that, with proper management, vegetable and herb production in an extensive green roof system is possible and productive. (Whittinghill, L.J., Rowe, D.B., & Cregg, B.M. (2013)).

6.4.3 *Simulation of the thermal behavior of a building retrofitted with a green roof: Optimization of energy efficiency with reference to Italian climatic zones*

Energy plus enabled the investigation of the thermal behavior variations of the building envelope, and the possible consequences, in terms of comfort, on the temperature of the internal spaces. The variation of the energy behavior of the building envelope type was assessed primarily through the analysis of the operative temperature of the elements of surface casing, the trend of the surface heat fluxes on the faces of the elements of internal and external housing, the variation of the operating temperature inside the rooms. The energy savings achieved with a green roof varies considerably in relation to the reference performance obtained without this kind of insulating structure. The main parameters useful to defining the contribution of the green roof to the reduction of the loads of cooling plants consist of the specific climate and the thermal isolation level of the initial coverage. (Gargari, C., Bibbiani, C., Fantozzi, F., & Campiotti, C.A. 2016)).

6.4.4 *A review of energy aspects of green roofs*

This paper intends to run a review on the application of green roof strategy. The review scans a time frame from 2002 through early 2012 with a focus on the energy related topics of green roofs. The review discussed various types of green roof, components of a green roof, economic revenues, and technical attributes. Also discussed were many advantages and a few general disadvantages of green roofs on one hand and pros and cons of green roofs with respect to energy utilization on the other. Some recommendations for future study are also proposed (Saadatian, 2013).

6.4.5 *Probabilistic social cost-benefit analysis for green roofs: A lifecycle approach*

This paper is based on an extensive literature review in multiple fields and reasonable assumptions for unavailable data. The Net Present Value (NPV) per unit of area of a green roof was assessed by considering the social-cost benefits that green roofs generate over their lifecycle. Two main types of green roofs – i.e. extensive and intensive – were analyzed. Additionally, an experimental extensive green roof, which replaced roof layers with construction and demolition waste (C&D), was assessed.

A probabilistic analysis was performed to estimate the personal and social NPV and payback period of green roofs. Additionally, a sensitivity analysis was also conducted. The analysis demonstrated that green roofs are short-term investments in terms of net returns. In general, installing green roofs is a low risk investment. Furthermore, the probability of profits out of this technology is much higher than the potential financial losses. It is evident that the inclusion of social costs and benefits of green roofs improves their value. (Bianchini, F., & Hewage, K. (2012)).

6.5 *Green roofs contribution to more social sustainability*

6.5.1 *Amenity & aesthetic benefits of green roofs*

Open space flat roofs present enormous potential in providing urban dwellers with the amenity and recreational space essential for healthy living. The sights, fragrances and sounds of a garden add immeasurably to the richness of experience and quality of life. Communal gardens also offer opportunities for social interaction between neighbors that might not otherwise be available, in both residential and commercial developments. These roof open spaces may be private, for the sole use of occupants of the development, or public spaces for the use of the general populace. (Townshend, D., & Duggie, A. (2007)).

6.5.2 *Visual aesthetic value*

An obvious and significant benefit of a green roof (subject to good maintenance) is the potentially attractive view offered to overlooking buildings. This is of great importance in a dense urban environment such as Amman, where views of roofs are often associated with grey concrete slabs and various pipes, electrical and mechanical equipment and maintenance tools that usually clutters roof spaces. (Townshend, D., & Duggie, A. (2007), p10).

Visual aesthetic benefits are offered by both intensive and extensive green roofs. However, for the same plan area, intensive green roofs offer potentially greater visual benefit than extensive green roofs, because the former may include large trees and shrubs which offer a three-dimensional greening effect, whereas the latter comprise only a thin two-dimensional 'skin' of green which may not be visible unless viewed from above.

6.5.3 *Health and therapeutic value*

Visual contact with vegetation has proven direct health benefits. Psychological studies have demonstrated that the restorative effect of natural scenery holds the viewer's attention, diverts their awareness away from themselves and worrisome thoughts and elicits a meditation-like state. People living in high density developments are known to be less susceptible to illness if they have a balcony or terrace garden. This is partly due to additional oxygen, air filtration and humidity control supplied by plants. The variety of sounds, smells, colors and movement provided by plants, although not quantifiable, can add significantly to human health and wellness. This in turn can lead to some potential savings in community expenditure on healthcare. (Townshend, D., & Duggie, A. (2007). p11).

6.5.4 *Attitudes and aesthetic reactions toward green roofs in the Northeastern United States*

This study conducted visitor surveys at seven green roofs in the Northeastern US to assess visitors' aesthetic reactions to different types of green roofs. Attitudes toward green roofs were positive with higher importance being placed on green roof benefits than costs. Aesthetic reactions were, in general, positive. Principle component analysis showed that negative aesthetic reactions were associated primarily with a perception of messiness. Furthermore, respondents felt that the grass-dominated roofs blended less well with the building and surrounding landscape. Aesthetic reactions were positively correlated with attitudes and importance placed on the benefits of green roofs. (Jungels, J., Rakow, D.A., Allred, S.B. & Skelly, S.M. (2013)).

7 RESULTS

Related to the aims of research, the research finding three main indicators involved with sustainability (environment, economy and social) and helps the potential designer create a green roof in Amman.

7.1 *Environmental indicators:*

1. Rooftop gardens, being established in a soil substrate, provide a porous material that absorbs rainwater landing on buildings.
2. Lined green roof systems can reduce storm water runoff by 50–70%
3. Rooftop gardens can reduce the urban heat island effect of cities.
4. Rooftop plants maintain a lower overall roof temperature and reduce cooling costs during the summer months.
5. Insulation effects of the soil substrate can also lower building heating costs during the winter.
6. The degree of energy savings in green roof garden depends on the amount of vegetative coverage, local climate, and type of building heating/cooling system.
7. Jordanian residential buildings having widespread rooftop gardens and green roofs could have a significant impact in reducing the buildings' energy consumption and environmental impact.

7.2 *Economic indicators:*

1. Savings can be accrued by growing food locally and by energy savings provided through a green roof system.
2. Green roofs can help provide roof cooling in the summer and insulation in the winter.
3. Green roofs increase the lifespan of roofs and reduce maintenance or replacement costs by protecting the structure from severe weather and wind.
4. Tomatoes, green beans, cucumbers, peppers, basil, and chives were studied because of their common use in home gardens. All plants, except pepper, survived and produced biomass in all growing systems.
5. The yields and biomass of basil were higher and of better quality in-ground and results suggest that, with proper management, vegetable and herb production in an extensive green roof system is possible and productive. (Whittinghill, L.J., Rowe, D.B., & Cregg, B.M. (2013)).
6. The analysis demonstrated that green roofs are short-term investments in terms of net returns and green roofs is a low risk investment.
7. The longevity of green roofs contributes to cost savings for the building operator.

7.3 *Social indicators:*

1. Rooftop gardens contribute to social sustainability and local food systems by providing increased food security and access to better nutrition.
2. The objects of sustainable Jordanian cultural heritage, like handmade products and reused and recycled materials in labor market can be used in rooftop gardens in both residential buildings and villas.
3. Cultural value was assessed by breaking it down into five components: aesthetic, social, symbolic, spiritual and educational value.
4. Aesthetic value is straight forward, being related to beauty, harmony, visual appeal, etc. Guyer, P. (2005). While the social value is linked to cultural identity and an understanding of the role of culture in society. (Baumeister, R.F., & Muraven, M. (1996)).
5. Intensive green roofs offer a potentially greater visual benefit than extensive green roofs, and they have the potential to aid visual green space continuity.

6. Visual contact with vegetation has proven direct health benefits. People living in high density developments are known to be less susceptible to illness if they have a balcony or terrace garden. This is partly due to additional oxygen, air filtration and humidity control supplied by plants.
7. Aesthetic reactions were positively correlated with attitudes and importance was placed on the benefits of green roofs. (Jungels, J., Rakow, D.A., Allred, S.B., & Skelly, S.M. (2013)).

8 CONCLUSIONS

1. Cities can have an impact on the surrounding fragile ecosystem and increase pressure on already scarce natural resources (water, energy, and food).
2. Implementing sustainable and environmental friendly techniques is becoming more essential and their introduction into the Arab region urban environment are much more recent.
3. Cities with prolonged summers and limited water resources like Jordan require custom-made designs, innovative irrigation techniques, plant species that are particularly heat and drought tolerant, and longer periods of initial maintenance with irrigation.
4. Overall attitudes and aesthetic reactions to green roofs are positive, while more negative aesthetic reactions are associated with a sense of messiness.
5. Recycled roof tiles, green materials, select insulation materials and waterproofing membranes still have a large market in Jordan as in other Mediterranean countries.

REFERENCES

[1] Allhoff, F., & Cooper, D. E. (2011). *Gardening-Philosophy for Everyone: Cultivating Wisdom* (Vol. 38). John Wiley & Sons.

[2] Al-Zu'bi, M., & Mansour, O. (2017). Water, energy, and rooftops: integrating green roof systems into building policies in the Arab region. *Environment and Natural Resources Research*, 7(2), 11–36.

[3] Aziz, H. A., & Ismail, Z. (2011, September). Design guideline for sustainable green roof system. In *2011 IEEE Symposium on Business, Engineering and Industrial Applications (ISBEIA)* (pp. 198–203). IEEE

[4] Baumeister, R. F., & Muraven, M. (1996). Identity as adaptation to social, cultural, and historical context. *Journal of adolescence*, 19(5), 405–416.

[5] Berardi, U., GhaffarianHoseini, A., & GhaffarianHoseini, A. (2014). State-of-the-art analysis of the environmental benefits of green roofs. *Applied energy*, 115, 411–428.

[6] Bianchini, F., & Hewage, K. (2012). How "green" are the green roofs? Lifecycle analysis of green roof materials. *Building and environment*, 48, 57–65.

[7] Bianchini, F., & Hewage, K. (2012). Probabilistic social cost-benefit analysis for green roofs: a lifecycle approach. *Building and Environment*, 58, 152–162.

[8] Caputo, S., Iglesias, P., & Rumble, H. (2017). Elements of Rooftop Agriculture Design. In *Rooftop Urban Agriculture* (pp. 39–59). Springer, Cham.

[9] Dubbeling, M. (2014). Urban agriculture as a climate change and disaster risk reduction strategy. *UA Magazine*, 27, 3–7.

[10] Dubbeling, M., & Massonneau, E. (2014). Rooftop agriculture in a climate change perspective. *Urban Agriculture*, 28.

[11] Francis, L. F. M., & Jensen, M. B. (2017). Benefits of green roofs: A systematic review of the evidence for three ecosystem services. *Urban forestry & urban greening*, 28, 167–176.

[12] Gargari, C., Bibbiani, C., Fantozzi, F., & Campiotti, C. A. (2016). Simulation of the thermal behaviour of a building retrofitted with a green roof: optimization of energy efficiency with reference to italian climatic zones. *Agriculture and agricultural science procedia*, 8, 628–636.

[13] Guyer, P. (2005). *Values of beauty: Historical essays in aesthetics*. Cambridge University Press.

[14] Jungels, J., Rakow, D. A., Allred, S. B., & Skelly, S. M. (2013). Attitudes and aesthetic reactions toward green roofs in the Northeastern United States. *Landscape and Urban Planning*, 117, 13-21.

[15] Li, Y., & Babcock, R. (2016). A simplified model for modular green roof hydrologic analyses and design. *Water*, 8(8), 343.

[16] Oberndorfer, E., Lundholm, J., Bass, B., Coffman, R. R., Doshi, H., Dunnett, N., & Rowe, B. (2007). Green roofs as urban ecosystems: ecological structures, functions, and services. *BioScience*, 57(10), 823–833.

[17] Saadatian, O., Sopian, K., Salleh, E., Lim, C. H., Riffat, S., Saadatian, E., & Sulaiman, M. Y. (2013). A review of energy aspects of green roofs. *Renewable and sustainable energy reviews*, 23, 155–168.

[18] Shafique, M., Kim, R., & Rafiq, M. (2018). Green roof benefits, opportunities and challenges–A review. *Renewable and Sustainable Energy Reviews*, 90, 757–773.

[19] Townshend, D., & Duggie, A. (2007). Study on green roof application in Hong Kong. *Architectural services department*.

[20] Vijayaraghavan, K. (2016). Green roofs: A critical review on the role of components, benefits, limitations and trends. *Renewable and sustainable energy reviews*, 57, 740–752.

[21] Whitman, M. E., Mattord, H. J., & Green, A. (2013). *Principles of incident response and disaster recovery*. Cengage Learning.

[22] Whittinghill, L. J., & Rowe, D. B. (2012). The role of green roof technology in urban agriculture. *Renewable Agriculture and Food Systems*, 27(4), 314–322.

[23] Whittinghill, L. J., Rowe, D. B., & Cregg, B. M. (2013). Evaluation of vegetable production on extensive green roofs. *Agroecology and sustainable food systems*, 37(4), 465–484.

Proceedings of the 1st International Congress on Engineering Technologies – Kiwan & Banat (Eds)

Analysis of speed variance on multilane highways in Jordan

Ahmad H. Alomari
Yarmouk University, Irbid, Jordan

Bashar H. Al-Omari & Mohammad E. Al-Adwan
Jordan University of Science & Technology (JUST), Irbid, Jordan

ABSTRACT: This research aimed to study the relationship between speed limits and speed variance in the traffic stream and to explore the significant factors that might affect speed variance on multilane highways in Jordan. This study also developed models to get the most accurate prediction with a higher coefficient of determination and to help agencies better-set speed limits to reduce speed variance and their associated crashes. Field measurements and observations were performed at twenty-five multilane highway segments during off-peak hours. Data include spot speeds, speed limits, lane width, number of lanes, and presence of roadside barriers, which were used to determine if its effect was statistically significant on speed variance.

Multiple Linear Regression and One-Way ANOVA were used to analyze the data. It was found that the speed variance was mainly dependent on design speed minus speed limit (DS-SL), the presence of roadside barriers, and the number of lanes with $R^2 = 0.835$ at $\alpha = 0.05$. Also, results indicated that as the number of lanes increases, the speed variance decreases. Moreover, the presence of roadside barriers reduces speed variance. This paper provides beneficial information to authorities and policymakers who set speed limits on multilane highways.

Keywords: Speed Variance, Speed Limit, Design Speed, Multilane Highway, Regression.

1 INTRODUCTION

Highways are public streets, especially the major ones, that connect two or more destinations. The roads that have at least two or more lanes in each direction for the exclusive use of traffic, with no or partial control of access, and may have periodic interruptions of flow at signalized intersections not closer than 3.0 km are called multilane highway (Mathew, 2014).

Roadway speed limits are used in most countries to set the maximum speed, or minimum in some cases, at which vehicles may legally travel on a particular section of roadway. Speed limits may be variable, and it is typically indicated on a traffic sign. The legislative bodies, decision-making specialists, and empowered authorities usually set speed limits and enforce it by police (patrol vehicles) or using automated roadside systems such as speed cameras (visible or hidden). Moreover, several authorities use variable speed limits (VSL) to control posted speed limits according to the prevailing traffic and weather conditions to improve the level of service and enhance safety (Khondaker and Kattan, 2015).

Designing a roadway with a desired speed for any facility can be reached by considering the relationship between three essential types of speed; posted speed limit, design speed, and operating speed. While the relationship between operating speed and posted speed limit can be well defined, the relationship between the design speed and either operating speed or posted speed limit cannot be determined with the same level of confidence (Fitzpatrick et al., 2003). Himes et al. (2013) suggested that the posted speed limit should be considered as an exogenous variable in operating

 DOI 10.1201/9781003178255-27

speed models, speed magnitude, and speed dispersion. They concluded that the posted speed limit had a "positive association with both mean speed and speed deviation" (Himes et al., 2013).

Several studies have shown that higher speed variance is associated with higher accident rates and severities (Garber and Gadiraju, 1989; Pisarski, 1986; Graves et al. 1993; Quddus, 2013; Wang et al., 2018). When all vehicles are moving at a similar speed in the same direction, the distance between vehicles stays identical so that no accidents may occur. However, the chance of accident's occurrence increases when vehicles are traveling at different speeds due to the interaction between vehicles, which increases opportunities for crashes. As a summary, accidents risk increases with the increase in speed variance. Besides, speed variance grows up when the difference between design speed and the posted speed limit is high, and undoubtedly higher speeds lead to higher severity levels of accidents. Furthermore, Graves et al. (1993) determined that a higher speed limit could reduce the speed variance, which may lead to a reduction in traffic accidents. Yet, many other researchers concluded that increasing the speed limit will rise the speed variance, accident rate, and severity (Rock, 1995; Malyshkina and Mannering, 2008; Hu, 2017). Additional studies showed that geometric characteristics do have a major effect on the operating speed. Elmberg (1960) explained that drivers chose the speed, which they considered the most appropriate for the prevailing geometric conditions. Also, Ma et al. (2010) concluded that lane width affects the variance of vehicle speed, especially near intersection exits, while the variance of speed increased as the lane width increased at mid-block points.

Previous research showed that many studies have found that increased speed variance amplified the accident rate. Additionally, higher speed limits tend to result in higher speed variance and more accidents. This research aimed to study the factors that affect speed variance in traffic streams on multilane highways in Jordan. This research also intended to investigate the relationships between speed variance, design speed, and speed limits.

One of the primary factors considered was the difference between design speed and speed limit to see if roadways with higher differences have higher speed variance. Finally, this research was designed to develop statistical models that can represent the relationships between speed variance and influencing factors. Understanding the factors that influence speed variance and the relationship between speed variance, design speed, and the posted speed limit are essential to set speed limits effectively and make roadway modifications to reduce speed variance.

2 METHODOLOGY

Data collected in this study were obtained through field measurements during off-peak periods. In addition to the speeds, further field data were recorded to cover geometric and control characteristics such as lane width (ranged from 3.5m to 3.65m), number of lanes (2 or 3 lanes), speed limit (varied from 60 km/hour to 100 km/hour), design speed (ranged from 80 km/hour to 120 km/hour), and presence of roadside barriers. Twenty-five divided multilane highways were selected from five major governorates in Jordan: Amman (Capital), Irbid, Jerash, Mafraq, and Zarqa. All selected highways were examined to locate a mix of highways with different geometric and control characteristics but with good pavement conditions, level grade, clear sight distance, free of interruptions, and free of vertical and horizontal curves.

The speeds of vehicles (passenger cars only) were detected using speed traps and radar throughout sunny days with dry pavement conditions. At each selected site, 150 passenger car speeds were collected. Average speed, standard deviation, and speed variance were then computed from the measured field data. The true mean of measured speeds in a section was determined with its true standard deviation for determining spot speed sample size. The degree of confidence was specified to determine the number of standard variations. The following equation was usually used for determining the speed sample sizes (Garber and Hoel, 2014):

$$\mathrm{N} = \left(\frac{Z * S}{d}\right)^2 \tag{1}$$

Where:

- N = Minimum sample size.
- Z= Number of standard deviations corresponding to the required confidence.
- S = Sample standard deviation (km/h).
- d = acceptable error in the speed estimate (km/h).

It was observed that the sample size of the collected data (150 vehicles at each location) was larger than the required minimum sample size. Table 1 shows the datasheet for the collected data from all 25 highway segments.

Since one of the main factors that impact speed variance is speed itself (as shown in previous studies), the design speed minus speed limit (DS–SL) was considered as a variable to investigate speed variance. All geometric and traffic data were considered carefully to explore the main possible factors that affect speed variance and to develop statistical models that can represent the relationships between speed variance and influencing factors. Data were coded, characterized, and statistically analyzed using SPSS statistical software (IBM, 2020).

The sample variance of a set of observations is defined as the "average squared deviation of the individual observations from the mean and varies across samples" (Washington et al., 2011). It is given by:

$$S^2 = \frac{\sum (x_i - \overline{X})^2}{n - 1} \tag{2}$$

Where:

- S^2 = sample variance.
- $\overline{X}$ = Sample mean.
- n = sample size.

Data were checked by various tests to explore the main possible factors that affect speed variance, as shown in Figure 1. Linear regression models were developed to describe and represent the relationship between geometric characteristics, traffic conditions, and speed variance. Additionally, One-Way Analysis of Variance (ANOVA) was used to determine if the number of lanes, lane width, and presence of roadside barriers have statistically significant effects on the speed variance.

3 MULTIPLE LINEAR REGRESSION

Speed variance is the dependent variable (DV), and the independent variables (IV) considered were lane width, number of lanes, presence of roadside barriers, design speed, speed limit, and design speed minus speed limit. Using linear regression, stepwise analysis eliminated all variables except design speed minus speed limit, barrier presence, and number of lanes. Three different Multiple Linear Regression (MLR) models (Equations 3, 4, and 5) were obtained as follows:

$$\mathbf{SV = 90.645 + 2.956(DS - SL)} \tag{3}$$

$$\begin{aligned}\mathbf{SV} &= \mathbf{93.799 + 2.956(DS - SL)}\\ &\quad \mathbf{-39.426(B)}\end{aligned} \tag{4}$$

$$\begin{aligned}\mathbf{SV} &= \mathbf{139.774 + 3.176(DS - SL)}\\ &\quad \mathbf{-41.525(B) - 24.132(No.L)}\end{aligned} \tag{5}$$

Where:

- SV = Speed Variance $(km/hr)^2$

Table 1. Data collection sheet.

No.	Coordinates	No. of Lanes	Lane Width LW (m)	Design Speed DS (km/hr)	Speed Limit SL (km/hr)	DS – SL (km/hr)	Average Speed (km/hr)	Standard Deviation (km/hr)	Speed Variance SV $(km/hr)^2$	Barrier Presence (B)
1	32.452675, 35.929641	2	3.60	100	90	10	89.7	11.1	123.2	No
2	32.500926, 35.876882	2	3.60	100	80	20	86.3	12.9	166.4	No
3	32.509600, 35.925287	2	3.60	120	100	20	93.6	12.6	158.8	No
4	32.517704, 35.996715	2	3.50	100	70	30	86.0	13.6	184.9	No
5	32.528106, 36.000200	2	3.55	100	70	30	85.1	13.9	193.2	No
6	32.501136, 35.862093	2	3.60	100	60	40	79.6	14.9	222.0	No
7	32.322035, 36.062645	2	3.60	100	80	20	85.1	13.0	169.0	No
8	32.286300, 36.092800	2	3.60	100	80	20	86.7	12.5	156.3	No
9	32.097527, 36.115817	2	3.55	100	80	20	78.8	12.5	156.3	No
10	32.084542, 36.121449	2	3.60	100	70	30	87.0	14.2	201.6	No
11	32.065715, 36.129021	3	3.55	100	70	30	89.0	13.0	169.0	No
12	32.020372, 36.079593	2	3.60	100	90	10	88.6	10.7	114.5	No
13	32.016548, 36.069628	2	3.65	100	90	10	90.0	10.5	110.3	No
14	32.017116, 35.902616	3	3.60	110	80	30	89.1	12.5	156.3	No
15	32.034284, 35.863886	2	3.60	90	70	20	76.0	12.8	163.8	No
16	32.007897, 35.836824	2	3.65	100	80	20	85.2	13.2	174.2	No
17	32.042487, 35.836568	2	3.60	80	70	10	69.7	9.6	92.2	No
18	32.077126, 35.838749	2	3.60	100	80	20	86.5	10.9	118.8	Yes
19	32.077235, 35.838637	2	3.60	100	80	20	86.0	10.4	108.2	Yes
20	32.114595, 35.853749	2	3.60	100	60	40	77.3	13.5	182.3	No
21	32.189130, 35.854548	2	3.60	100	90	10	84.3	10.8	116.6	No
22	32.248875, 35.892714	2	3.60	100	90	10	85.0	11.3	127.7	No
23	32.269253, 35.910655	2	3.60	110	100	10	89.9	11.0	121.0	No
24	32.297741, 35.914110	2	3.60	110	100	10	86.5	11.5	132.3	No
25	32.369081, 35.928065	2	3.60	100	90	10	86.8	11.2	125.4	No

Table 2. ANOVA[a] for developed MLR models.

Equation		Sum of Squares	df	Mean Square	F	Sig.
3	Regression	19228.189	1	19228.189	57.461	.000[b]
	Residual	7696.561	23	334.633		
	Total	26924.750	24			
4	Regression	22088.315	2	11044.158	50.238	.000[c]
	Residual	4836.435	22	219.838		
	Total	26924.750	24			
5	Regression	23045.858	3	7681.953	41.589	.000[d]
	Residual	3878.892	21	184.709		
	Total	26924.750	24			

a. Dependent Variable: Speed Variance
b. Predictors: (Constant), Design Speed – Speed Limit
c. Predictors: (Constant), Design Speed – Speed Limit, Barriers
d. Predictors: (Constant), Design Speed – Speed Limit, Barriers, Number of lanes.

- DS-SL = Design Speed – Speed Limit (km/hr).
- B = Roadside Barriers (without barriers = 0; with barriers = 1)
- No. L = Number of Lanes (2 or 3).

The regression models, intercept and variables were found to be significant at 95% confidence, with R^2 = 0.702 and standard error of estimate (SEE) = 18.29 for Equation 3, R^2 = 0.804 and standard error of estimate (SEE) = 14.83 for Equation 4, and R^2 = 0.835 and SEE = 13.59 for Equation 5. The R^2 values indicate that all models fit the data well. The ANOVA results for these models (Table 2) showed that their F values are 57.461, 50.238, and 41.589, respectively, with p-values of 0.000. Thus, they are all significant at 95 % confidence. Table 3 shows the coefficients, significance, and multicollinearity parameters for these models. This table shows that tolerance is greater than 0.1, and the Variation Inflation Factor (VIF) is less than 10 for all three models, indicating there is no collinearity in either of the three models.

The histogram of standard residuals is presented in Figure 2A, where it can be noticed that the observed and expected cumulative probabilities are almost normally distributed. Figure 2B shows a scatter plot between the normal score versus the residuals.

The plot in Figure 2C is approximately a straight line, indicating that the normality assumption is reasonable. Figure 2B also showed that the residuals are falling with a horizontal band centered around '0', so the relationship between the dependent and independent variables is linear. Therefore, the developed models are appropriate and valid conclusions can be drawn from them. The models' coefficients show that speed variance increases as the difference between design speed and speed limit increases. In contrast, speed variance decreases for roads with roadside barriers present or more lanes. This information is useful for transportation planners and operating agencies so they can modify existing highways and better design future highways to reduce speed variance proactively.

Table 4 shows speed variance predictions and the relative error of the developed models. It can be seen that, overall, Equation 5 had the best protection results with the lowest relative error compared with the actual speed variance.

4 ONE WAY ANOVA

In addition to the modeling, One-Way ANOVA was used to find out if the number of lanes, lane width, and presence of roadside barriers were statistically significant factors on the speed variance. The results of this analysis are shown in Table 5.

Table 3. Coefficients[a] for developed MLR models.

Equation		Unstandardized Coefficients		Standardized Coefficients	t	Sig.	95.0% Confidence Interval for B		Correlations			Statistics Statistics	
		B	Std. Error	Beta			Lower Bound	Upper Bound	Zero-order	Partial	Part	Tolerance	VIF
3	(Constant)	90.645	8.616		10.521	.000	72.822	108.467					
	DS-SL	2.956	.390	.845	7.580	.000	2.150	3.763	.845	.845	.845	1.000	1.000
4	(Constant)	93.799	7.038		13.328	.000	79.204	108.394					
	DS-SL	2.956	.316	.845	9.352	.000	2.301	3.612	.845	.894	.845	1.000	1.000
	Barriers	−39.426	10.931	−.326	−3.607	.002	−62.095	−16.757	−.326	−.610	−.326	1.000	1.000
5	(Constant)	139.774	21.198		6.594	.000	95.691	183.856					
	DS-SL	3.176	.305	.908	10.400	.000	2.541	3.811	.845	.915	.861	.900	1.111
	Barriers	−41.525	10.062	−.343	−4.127	.000	−62.449	−20.600	−.326	−.669	−.342	.992	1.008
	No. L	−24.132	10.599	−.199	−2.277	.033	−46.173	−2.091	.116	−.445	−.189	.894	1.119

a. Dependent Variable: Speed Variance.

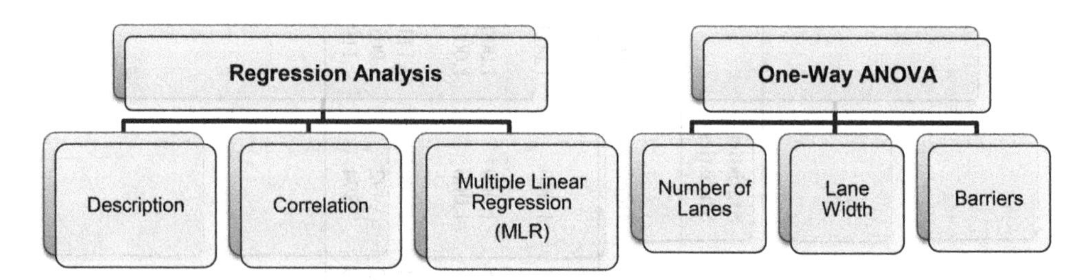

Figure 1. Regression analysis and one-way ANOVA flow chart.

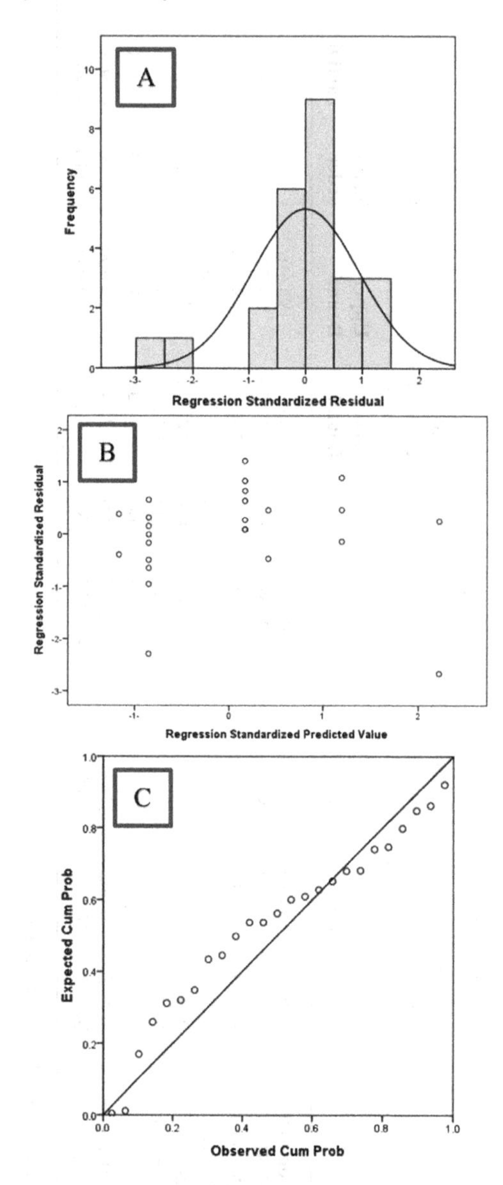

Figure 2. Regression standardized residual histogram (A), Scatterplot (B), and normal P-P plot of regression standardized residual (C) for MLR model.

Table 4. Speed variance predictions and relative error of developed models.

No.	Actual Speed Variance SV $(km/hr)^2$	Equation 3 Prediction $(km/hr)^2$	Equation 3 Error (%)	Equation 4 Prediction $(km/hr)^2$	Equation 4 Error (%)	Equation 5 Prediction $(km/hr)^2$	Equation 5 Error (%)
1	**123.2**	120.205	2%	123.359	0%	123.27	0%
2	**166.4**	149.765	10%	152.919	8%	155.03	7%
3	**158.8**	149.765	6%	152.919	4%	155.03	2%
4	**184.9**	179.325	3%	182.479	1%	186.79	1%
5	**193.2**	179.325	7%	182.479	6%	186.79	3%
6	**222**	208.885	6%	212.039	4%	218.55	2%
7	**169**	149.765	11%	152.919	10%	155.03	8%
8	**156.3**	149.765	4%	152.919	2%	155.03	1%
9	**156.3**	149.765	4%	152.919	2%	155.03	1%
10	**201.6**	179.325	11%	182.479	9%	186.79	7%
11	**169**	179.325	6%	182.479	8%	162.658	4%
12	**114.5**	120.205	5%	123.359	8%	123.27	8%
13	**110.3**	120.205	9%	123.359	12%	123.27	12%
14	**156.3**	179.325	15%	182.479	17%	162.658	4%
15	**163.8**	149.765	9%	152.919	7%	155.03	5%
16	**174.2**	149.765	14%	152.919	12%	155.03	11%
17	**92.2**	120.205	30%	123.359	34%	123.27	34%
18	**118.8**	149.765	26%	113.493	4%	113.505	4%
19	**108.2**	149.765	38%	113.493	5%	113.505	5%
20	**182.3**	208.885	15%	212.039	16%	218.55	20%
21	**116.6**	120.205	3%	123.359	6%	123.27	6%
22	**127.7**	120.205	6%	123.359	3%	123.27	3%
23	**121**	120.205	1%	123.359	2%	123.27	2%
24	**132.3**	120.205	9%	123.359	7%	123.27	7%
25	**125.4**	120.205	4%	123.359	2%	123.27	2%

The number of lanes was found to have a significant effect on speed variance at $\alpha = 0.05$. For Leven's test, the null hypothesis that the error variance of the dependent variable is equal, F = 0.047 and significance = 0.843 (more than 0.05), which is not significant, and this means equal variance is meet. The One-Way ANOVA had F = 15.299 and significance = 0.03 (less than 0.05), which is significant. The multiple comparison test showed that the difference between 2 and 3 lanes with significance = 0.03 (less than 0.05) is significant.

There was insufficient evidence to conclude that lane width has a significant effect on speed variance at $\alpha = 0.05$. The One-Way ANOVA had F = 0.18 and significance = 0.982 (more than 0.05), which is not significant. The multiple comparison test showed that the difference between all categories of lane width is not significant (p-value > 0.05).

Multilane highways can be separated from surrounding areas by roadside barriers to increase safety, especially when a multilane highway passes through crowded zones. Two locations with roadside barriers were selected during data collection to investigate the effect of roadside barriers on speed variance. Each of these selected sites had a design speed of 100 km/hr, speed limit of 80 km/hr, two lanes, and lane widths of 3.6 meters. One-Way ANOVA showed that the presence of roadside barriers had a significant effect on speed variance at $\alpha = 0.05$. For Leven's test, the null hypothesis that the error variance of the dependent variable is equal, F = 0.26 and significance = 0.632 (more than 0.05), which is not significant, and this means equal variance is meet. The One-Way ANOVA had F = 60.063 and significance = 0.001 (less than 0.05), which is significant.

Table 5. One-Way ANOVA results for lane width, number of lanes, and barrier presence.

		Homogeneity of Variances		ANOVA Table		Multiple Comparisons	
DV	Factor	Levene Statistic (F)	Sig.[a]	F	Sig.[a]	i~j	Sig.[a]
Speed Variance	Number of Lanes	0.047	0.843	15.299	0.03	2~3	0.03
						3~2	0.03
	Lane Width	–	–	0.18	0.982	3.5~3.55	0.915
						3.5~3.6	0.867
						3.55~3.5	0.915
						3.55~3.6	0.941
						3.6~3.5	0.867
						3.6~3.55	0.941
	With or Without Barriers	0.26	0.632	60.063	0.001	With~ without	0.001
						With~ without	0.001

a. The difference is significant at the 0.05 level; <0.05 means the difference is significant and > 0.05 means the difference is not significant.

The multiple comparison test showed that the difference between with and without barriers with significance = 0.001 (less than 0.05) is significant.

5 CONCLUSIONS

This research aimed to study the relationship between speed limits and speed variance in the traffic stream and to explore the main factors that might affect speed variance on multilane highways in Jordan. This study also designed to develop models for predicting speed variance as functions of geometric, operating, and traffic characteristics. Field measurements and observations were performed at 25 multilane highway segments in Jordan. Data include speeds, speed limits, design speeds, lane width, number of lanes, and presence of roadside barriers, which was used to know if its effect was statically significant on speed variance. Different variables were considered to get the most accurate prediction with a higher coefficient of determination.

Multiple Linear Regression (MLR) was used to predict speed variance as a function of geometric, operation, and traffic characteristics with R^2 = 0.702, 0.804, and 0.835 for Equations 3, 4, and 5, respectively. The R^2 values indicate that all models fit the data well. It was found that the speed variance was mainly dependent on design speed minus speed limit (DS-SL), the presence of roadside barriers, and the number of lanes. Moreover, One-Way ANOVA was used to explore if lane width, number of lanes, and presence of roadside barriers or not are statically significant or not on speed variance. Based on the developed models and statistical analyses conducted in this paper, it was found that as the number of lanes increases, the speed variance decreases. Likewise, the presence of roadside barriers reduces speed variance.

Results of this research will provide valuable information to authorities that can be used to estimate changes in speed variance due to differences in geometric and traffic characteristics, and therefore provide traffic engineers a means for controlling speed variance to minimize accidents. Also, this study targets to illustrate the importance of different factors that impact speed variance.

Future work for this research includes data collection for additional speed samples on different highways in Jordan with different conditions to estimate the studied variables better and any other possible factors on speed variance. However, the research discussed in this paper provides a significant understanding of the major factors that impact speed variance.

REFERENCES

Elmberg, C. M. (1960). Effects of Speed Zoning in Urban Areas. *Doctoral dissertation, Purdue University*, Lafayette, Indiana, United States.

Fitzpatrick, K., Carlson, P., Brewer, M. A., Wooldridge, M. D., & Miaou, S. P. (2003). Design speed, operating speed, and posted speed practices. *NCHRP Report 504*. Transportation Research Board of the National Academies, TRB (No. Project 15-18 FY'98).

Garber, N. J., & Gadirau, R. (1989). Factors Affecting Speed Variance and Its Influence on Accidents. *Transportation Research Record*, Vol. 1213, No. 5, pp. 64–71.

Garber, N. J., & Hoel, L. A. (2014). Traffic and highway engineering. Fourth Edition. Cengage Learning, Canada.

Graves, P. E., Lee, D., & Sexton, R. L. (1993). Speed variance, enforcement, and the optimal speed limit. *Economics Letters*, Vol. 42, pp. 237–243.

Himes, S. C., Donnell, E. T., & Porter, R. J. (2013). Posted speed limit: To include or not to include in operating speed models. *Transportation research part A: policy and practice*, 52, 23–33.

Hu, W. (2017). Raising the speed limit from 75 to 80 mph on Utah rural interstates: Effects on vehicle speeds and speed variance. *Journal of Safety Research*, Vol. 61, pp. 83–92.

IBM SPSS Software. (2020). Version 25. https://www.ibm.com/analytics-/us/en/spss/spss-statistics-version/ (Accessed on May, 2020).

Khondaker, B. & Kattan, L. (2015). Variable speed limit: an overview. *Transportation Letters: The International Journal of Transportation Research*, 7(5), 264–278.

Ma, Y., Zeng, Y., & Yang, X. (2010). Impact of lane width on vehicle speed of urban arterials. *In ICCTP 2010: Integrated Transportation Systems: Green, Intelligent, Reliable* (pp. 1844–1852).

Malyshkina, N. V., & Mannering, F. L. (2008). Analysis of the Effect of Speed Limit Increases on Accident-Injury Severities. *Transportation Research Board*, Paper 08-0056.

Mathew, Tom V. (2014). Transportation Systems Engineering; Chapter 23: Multilane Highways. pp. 23.1–23.19.

Pisarski, A. E. Deep-Six 55. (1986). *Reason Foundation*, Vol. 17' No. 6, Nov.1986, pp. 32–35.

Quddus, M. (2013). Exploring the relationship between average speed, speed variation, and accident rates using spatial statistical models and GIS. *Journal of Transportation Safety & Security*, 5(1), 27–45.

Rock, S. M. (1995). Impact of the 65-mph speed limit on accidents, deaths, and injuries in Illinois. *Accident Analysis & Prevention*, Vol. 27(2), pp. 207–214.

Washington, S., Karlaftis, M. G., and Mannering, F. (2011). Statistical and econometric methods for transportation data analysis. Second Edition. CRC press.

Wang, X., Zhou, Q., Quddus, M., & Fan, T. (2018). Speed, speed variation and crash relationships for urban arterials. *Accident Analysis & Prevention*, Vol. 113, pp. 236–243.

LIST OF ABBREVIATIONS

- ANOVA: Analysis of Variance.
- DS: Design Speed.
- DS-SL: Design Speed minus Speed Limit.
- DV: Dependent Variable.
- IV: Independent Variables.
- LW: Lane Width.
- MLR: Multiple Linear Regression.
- No. L: Number of Lanes.
- S: Standard Deviation.
- SEE: Standard Error of Estimate.
- SL: Speed Limit.
- SV: Speed Variance.
- VIF: Variation Inflation Factor.
- VSL: Variable Speed Limits.

Proceedings of the 1st International Congress on Engineering Technologies – Kiwan & Banat (Eds)
© 2021 Taylor & Francis Group, London, ISBN 978-0-367-77630-5

Author index

Mosharaka for Research and Studies International Conference Proceedings

1. Proceedings of the 1st International Congress on Engineering Technologies (2021)
   Edited by Suhil Kiwan & Mohammad M. Banat
   ISBN 978-0-367-77630-5

9780367776305